Life in Neolithic Farming Communities

Social Organization, Identity, and Differentiation

FUNDAMENTAL ISSUES IN ARCHAEOLOGY

EMERGENCE AND CHANGE IN EARLY URBAN SOCIETIES
Edited by Linda Manzanilla

FOUNDATIONS OF SOCIAL INEQUALITY
Edited by T. Douglas Price and Gary M. Feinman

LEADERSHIP STRATEGIES, ECONOMIC ACTIVITY, AND INTERREGIONAL INTERACTION
Social Complexity in Northeast China
Gideon Shelach

LIFE IN NEOLITHIC FARMING COMMUNITIES
Social Organization, Identity, and Differentiation
Edited by Ian Kuijt

Life in Neolithic Farming Communities
Social Organization, Identity, and Differentiation

Edited by

IAN KUIJT
University of Notre Dame
Notre Dame, Indiana

KLUWER ACADEMIC / PLENUM PUBLISHERS
New York Boston Dordrecht London Moscow

Library of Congress Cataloging-in-Publication Data

Life in noelithic farming communities: social organization, identity, and differentiation/ edited by Ian Kuijt

p. cm. — (Fundamental issues in archaeology)

Includes bibliographical references and index.

ISBN 0-306-46122-6

1. Neolithic period—Middle East. 2. Agriculture, Prehistoric—Middle East 3. Land settlement patterns, Prehistoric—Middle East. 4. Social archaeology—Middle East 5. Middle East—Antiquities. I. Kuijt, Ian II. Series.

GN776.32M628 L54 1999
306.3'0939'4—dc21 99-052091

ISBN: 0-306-46122-6

233 Spring Street, New York, N.Y. 10013

http://www.wkap.nl/

10 9 8 7 6 5 4 3 2 1

A C.I.P. record for this book is available from the Library of Congress

Printed in the United States of America

Contributors

Ofer Bar-Yosef • Department of Anthropology, Harvard University, Cambridge, Massachusetts 02138

Anna Belfer-Cohen • Department of Prehistory, Institute of Archaeology, Hebrew University, Jerusalem, Israel 91905

Brian F. Byrd • ASM Affiliates, 543 Encinitas Boulevard, Suite 114, Encinitas, California 92024 and Department of Anthropology, University of California, San Diego, La Jolla, California 92093-0532

Jacques Cauvin • Centre National de la Recherche Scientifique, Institut de Prehistoire Orientale, Jalès, France

Avi Gopher • Institute of Archaeology, University of Tel Aviv, Ramat Aviv, Israel 69978

Nigel Goring-Morris • Department of Prehistory, Institute of Archaeology, Hebrew University, Jerusalem, Israel 91905

Frank Hole • Department of Anthropology, Yale University, New Haven, Connecticut 06520-8277

Ian Kuijt • Department of Anthropology, University of Notre Dame, Notre Dame, Indiana 46556

Estelle Orrelle • Institute of Archaeology, University of Tel Aviv, Ramat Aviv, Israel 69978

Richard W. Redding • Museum of Anthropology, University of Michigan, Ann Arbor, Michigan 48109

Gary O. Rollefson • 'Ain Ghazal Research Institute, 64372 Ober-Ramstadt, Germany

Michael Rosenberg • Department of Anthropology, University Parallel Program, University of Delaware, Newark, Delaware 19716

Alan H. Simmons • Department of Anthropology and Ethnic Studies, University of Nevada, Las Vegas, Nevada 89154-5012

Mary M. Voigt • Department of Anthropology, College of William and Mary, Williamsburg, Virginia 23187-8795

Preface

This volume emerged from an increasing awareness among archaeologists that while researchers have explored some of the technological, subsistence, and economic dimensions of the Near Eastern Neolithic, far less attention has been paid to understanding the nature of social organization for this important period. In relation to other topics, it has only been in the last 20 years or so that researchers have started to study the nature of Neolithic social organization in any detailed fashion. Given that the Neolithic provides our earliest case studies for how food production, social differentiation, and population aggregation and growth are interrelated, it is all that much more surprising to recognize that as archaeologists we do not have a comprehensive understanding of some of the social foundations within Neolithic communities. Archaeologists, for example, have only a limited understanding of how the household served as a social and economic unit, how kinship might have been organized, or the degree to which leadership was identified, shared, and allocated within communities. The breadth of research in this volume furthers our understanding of the Neolithic as an economic event, opening up what is unquestionably the Pandora's box of the Neolithic: studying the dynamic nature of social arrangements, how these behaviors were linked to material culture, and how they help us understand the trajectory of life within Neolithic communities. Ultimately, addressing these issues is not only challenging, but it requires focusing new attention on issues of social agency and understanding how different social practices may have been employed to define, shape, and manipulate identities at the household, kin-group, and community level. This tentative exploration of human agency, while still in its infancy, represents an important departure from previous studies, requiring an interpretive framework based on Neolithic data sets.

Six of the chapters in this volume were originally written for a symposium, titled "Social Configurations of Near Eastern Early Neolithic: Community Identity, Heterarchical Organization, and Ritual," held in 1995 at the Sixtieth Annual Meeting for the Society of American Archaeology in Minneapolis, Minnesota. Building on this foundation, these original studies have been expanded and five other chapters have been incorporated into the volume to address complementary dimensions of Neolithic social organization. In this collection, researchers synthesized recent anthropological and archaeological thought concerning the variation within—and the nature of—Neolithic social arrangements. Drawing on both the results of recent archaeological research as well as anthropological theory, the authors reconstruct key aspects of ritual practices, labor organization, and collective social identity at the scale of the household, community, and region. The chapters encompass a range of methodological and theoretical perspectives and utilize innovative analytic approaches in the study of mortuary, settlement pattern, and architectural data to better understand the processes of (and reasons for) specific social arrangements and ritual and mortuary practices. As such, each of the contributing authors struggles with the highly complex, and often avoided, interpretive interface between archaeological data sets and social interpretation of the Neolithic of the Near East. The goals of this volume, therefore, are not to reject traditional or other important research agendas nor to enforce a specific theoretical or methodological approach. Rather, this collection is envisioned as a vehicle by which discussion of other social dimensions of the Neolithic can be brought to the attention of archaeologists, anthropologists, and prehistorians to enhance the existing reconstruction of this fascinating period of time.

My thanks go to the participants in the original symposium, the participation by the audience during the symposium, and the other researchers who agreed to contribute to this book. The addition of these papers has greatly expanded the scope and nature of discussion, debate, and reflection on a wide range of theoretical and methodological issues related to the nature of social organization of the Near Eastern Neolithic. The preparation of this volume, as well as the task of organizing the original symposium, has been facilitated by the enthusiasm, interest, and energy of all of the participants. Beyond this list, I would like to thank Ofer Bar-Yosef, Herman Makler, C. C. Lamberg-Karlovsky, T. Douglas Price, and Gary Feinman for their active support with the publication of this volume. It was Ofer who first pushed me to organize the original symposium and T. Douglas who pushed this work at Plenum and introduced me to Eliot Werner. Thanks are also give to Eliot Werner, Archaeology Czar at Plenum, who has made the processes of negotiation and publication a direct, honest, and enjoyable task. Publication of this work was facilitated by critical financial support from the American School of Prehistoric Research, support that I am most grateful for. The cover artwork kindly provided by Nigel Goring-Morris and Michael Rosenberg is from their excavations at Kfar HaHoresh and Hallan Çemi. Finally, I want to express my thanks to Meredith S. Chesson, my wife, friend, and mate, who serves as a continuing source of advice, help, patience, and support. It is to her that this book is dedicated.

Contents

PART IV. SOCIAL RELATIONS AND MATERIAL CULTURE: SYMBOLISM AND MEANING

Part I

Introduction

Chapter 1

Life in Neolithic Farming Communities

An Introduction

Ian Kuijt

INTRODUCTION

While previous research on the Near Eastern Neolithic has addressed many dimensions of food production and environmental change, relatively few studies have explored the social context of these processes, the nature of political, economic, and religious practices in Neolithic communities, and how these specific patterns aid archaeologists and anthropologists in understanding some of the material and symbolic ways in which people were identified, how these are reflected through archaeological data sets, and the process by which social differentiation emerged in our earliest prehistoric agricultural contexts. Given that these communities existed some 10,000 to 8,000 years ago, it should come as no surprise that the exploration of the social context of Neolithic life is complex, somewhat contentious, and very complicated. To expand upon the metaphor employed by Bruce Winterhalder (1981): exploring the topic of Neolithic social organization is like juggling a porcupine—whatever way you throw it up, it comes down prickly. In light

IAN KUIJT • Department of Anthropology, University of Notre Dame, Notre Dame, Indiana 46556.

Life in Neolithic Farming Communities: Social Organization, Identity, and Differentiation, edited by Ian Kuijt. Kluwer Academic/Plenum Publishers, New York, 2000.

of the inherent difficulties in juggling this intellectual beast, it is understandable that this topic has been bypassed, if not avoided, by archaeologists working in the Neolithic of the Near East. While there are clear exceptions to this trend, for the most part research on the Near Eastern Neolithic has traditionally focused on the origins of agriculture, descriptive accounts of material culture, and shifts in economic practices, and archaeologists have adopted a position where our discussions of social organization are usually appended to detailed material studies rather than directly addressed. This prioritizing of research is understandable, especially considering the need to document changes in material culture through time. At the same time, however, this has resulted in the current condition under which researchers are only now beginning to explore the complexity, contradictions, and richness within Neolithic communities in early agricultural and horticultural contexts.

Viewing this gap optimistically, issues of social organization are now among the best-kept secrets of the Neolithic of the Near East and, at the same time, a topic in which there is a phenomenal research potential for those willing to take their turn at juggling this problematic beast. The researchers contributing to this volume have accepted this challenge and, by engaging in such debate, further our understanding of the multiple pathways toward social inequality and changing social organization in early agricultural contexts. Ultimately, such a reconceptualization of the Neolithic as a social and an economic process is important to a broader readership, for it helps us to understand some of the possible links between population aggregation, sedentism, and social change, how social differentiation may have emerged from within an egalitarian ideology, the mechanisms by which authority was ceded by individuals to groups in precomplex chiefdoms, and how changing social relations were negotiated through the built environment, and mortuary and ritual practices. This challenging discussion augments and enriches our understanding of the ecological, subsistence, and economic dimensions of the transition from foraging to farming and, perhaps more importantly, presents other researchers with the opportunity to reflect on data sets from other geographical and temporal contexts and the theoretical models currently employed in interpreting the emergence of social differentiation.

The primary objective of the research in this volume, then, focuses on the examination of the nature of social systems during different periods of the Neolithic by archaeologists working in the Near East. The authors address fundamental questions concerning social relations during a period of great social, economic, and technological change. They explore dimensions of Neolithic social organization on the basis of current archaeological data rather than labeling or interpreting Neolithic social systems through the lens of previously defined ethnographic categories such as chiefdoms. This does not imply a rejection of important ethnographic categories, nor the use of a comparative approach. I believe it reflects the recognition that any interpretation of Neolithic social organization must be founded on archaeological evidence and require integration of broader ethnographic and anthropological understandings of the human context in agricultural communities. Such intellectual balancing of archaeological data of the Neolithic with that of the anthropological understanding of social systems presents the contributors with a very challenging task and, by its very nature, illustrates a range of methodological approaches archaeologists adopt to study the scale, nature, and relative complexity of relationships within and between Neolithic communities.

THE NEAR EASTERN NEOLITHIC: REGIONAL CHRONOLOGY FRAMEWORKS

It can be argued that behavioral or social interpretations in archaeology stand or fall on the basis of the material data, that is, their careful placement in time and space. For the Neolithic Near East there is no clear consensus as to which data sets should be used to create cultural–historical sequences or the duration of individual periods or subperiods of the Pre-Pottery Neolithic. The reasons underlying these debates are complex and linked to the geographic scale and perceived cultural differences between Neolithic occupations from different areas of the Near East. In general, archaeologists disagree over what types of material evidence, such as architecture or lithic technology, can and should be employed to subdivide the Neolithic into different cultural–historical phases (see Figures 1 and 2).

A number of regional syntheses of the Neolithic of the Near East have illustrated that there are subtle, yet important, differences within regional cultural–historical sequences in different areas of the Near East that are only now becoming clear to researchers (Bar-Yosef 1981; Cauvin 1987; Moore 1985; Rollefson 1989). These nuances of timing and regionalism are particularly important when discussing cultural–historical sequences of the northern Levant and Anatolia, those areas defined by the borders of modern Turkey, Iraq, western Iran, and northern Syria and Lebanon. In comparison to the database of the south-central Levant, that area generally defined as including southern Syria and Lebanon, Israel, Jordan, and the Sinai peninsula of Egypt, less is known about the transition into the Neolithic in the northern Levant and Anatolia. Only in the last twenty years has the crucial field research at Mureybet, Nemrik 9, Qermez Dere, Jerf el-Ahmar, and Hallan Çemi (Cauvin 1987; Kozlowski 1989; Rosenberg and Davis 1992;

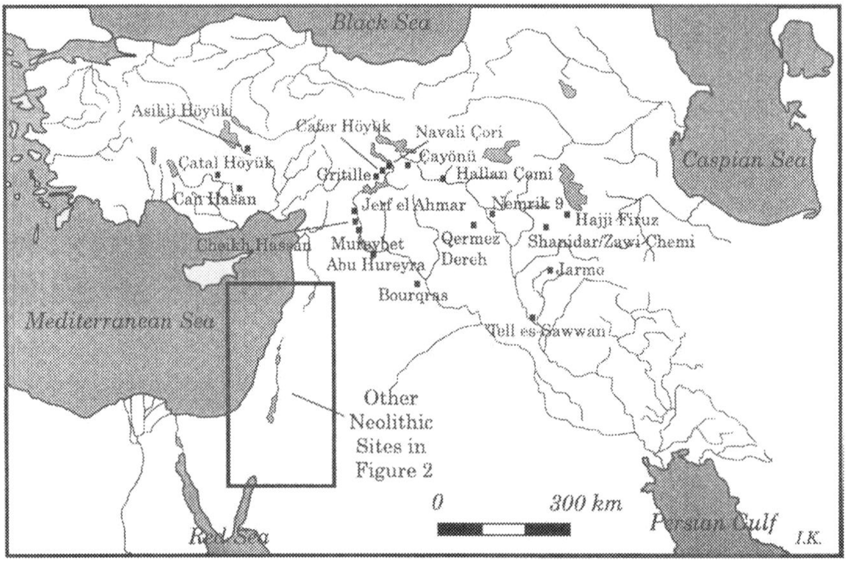

Figure 1. Location of Pre-Pottery and Pottery Neolithic period archaeology sites in the Near East.

Stordeur et al. 1996; Watkins et al. 1989) provided us with an understanding of the development of a distinctive culture along the Taurus-Zagros flanks during the eleventh millennium and how these may have been culturally articulated with later communities in areas of Iraq, Turkey, and Syria. Combined with the exciting field work at other settlements with later occupations, such as Nevali Çori, Göbekli Tepe, and Gürcütepe (Hauptmann 1993; Schmidt 1996), and the very important regional Neolithic research project centered on Çatal Höyük (Hodder 1996), archaeologists are now beginning to explore variations in material and social adaptations within and between different geographical areas of the northern Levant and Anatolia.

Although not without disagreement, there is a general consensus among researchers working in the Near East that the Pre-Pottery Neolithic period, characterized by diagnostic stone tool technology, the development of plant and animal domestication, and the first aggregate villages, emerged around 10,500 to 10,300 years ago and continued until approximately 7,750 bp. In the case of the south-central Levant, researchers generally accept that the Pre-Pottery Neolithic A period (PPNA), lasting from approximately 10,500–10,300 bp to 9,500–9,200 bp, followed by the Pre-Pottery Neolithic B period (PPNB), lasting from approximately 9,500–9,200 bp to 8,000 bp, and the Pre-Pottery Neolithic C period (PPNC), also referred to as the Final Pre-Pottery Neolithic B period, which lasted from approximately 8,000 to 7,750

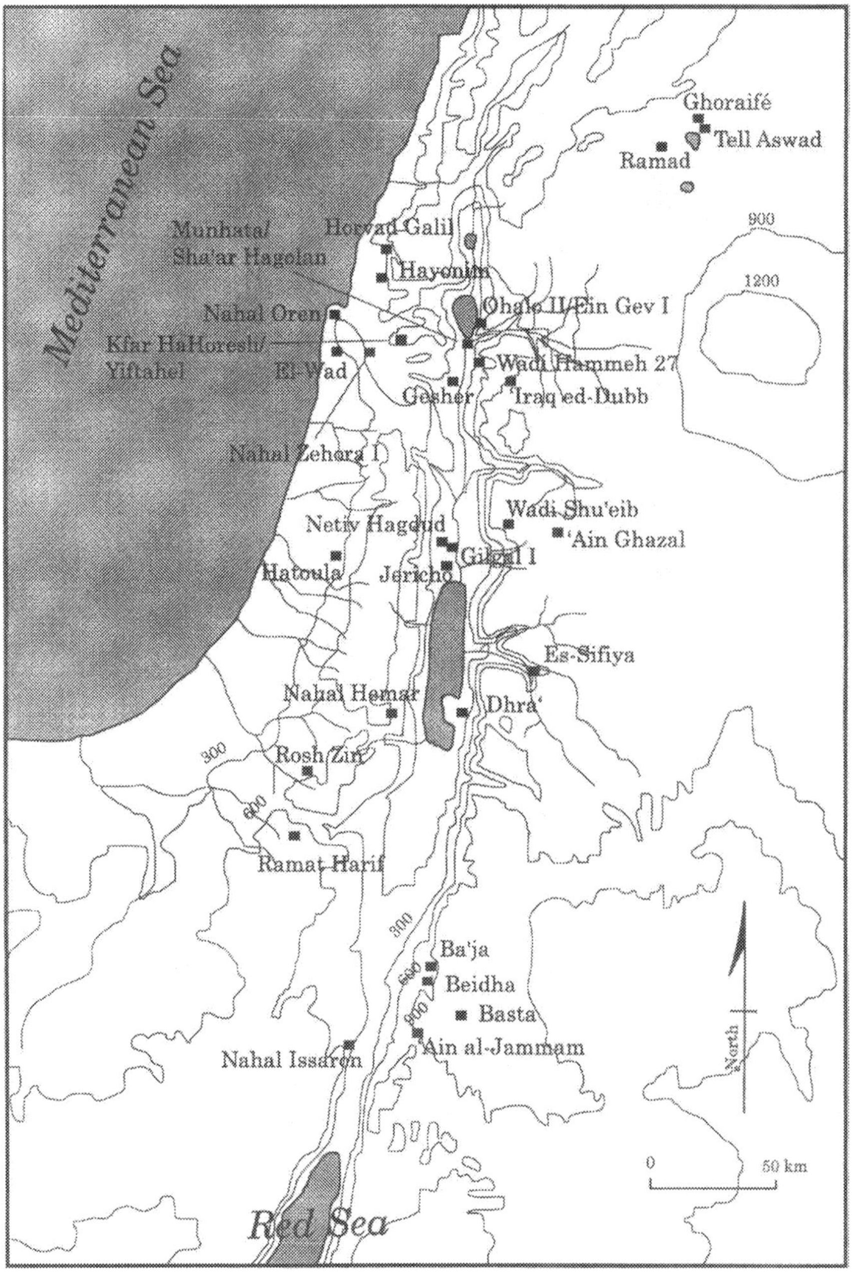

Figure 2. Location of Pre-Pottery and Pottery Neolithic period archaeology sites in the south-central Levant.

bp. Prioritizing different lines of archaeological evidence, researchers continue to debate how, or if, these periods should be further subdivided or labeled. For the PPNA, for example, some disagreement exists whether this period should be subdivided into two different phases or treated as a single cultural–historical unit. Similarly, researchers disagree if the broader PPNB should be subdivided into two or three phases. (For detailed discussion of cultural–historical sequences of the PPNA readers are referred to Bar-Yosef 1991; Crowfoot-Payne 1976, 1983; Garfinkel and Nadel 1989; Kuijt 1997a; Nadel 1990. Description and discussion of cultural–historical sequences for the general PPNB are covered in Bar-Yosef 1981; Gopher 1994; Kuijt 1997b; Rollefson, 1989. Discussion of the cultural–historical period of the PPNC, which is alternatively termed the Final PPNB, is presented in Nissen 1993; Perrot 1993; Rollefson 1989; Rollefson and Köhler-Rollefson 1993a, 1993b.) In the northern Levant and Anatolia archaeologists have uncovered evidence for similar, yet different, material patterning for the development of the Pre-Pottery Neolithic period. According to Cauvin (1987), a similar expression of the PPNA appears to have lasted from 10,300 to approximately 9,600 bp, as illustrated by the excavations at Mureybet, Qermez Dereh, and Nemerik 9. In the northern Levant and Anatolia the Pre-Pottery Neolithic B period can be tentatively divided into (1) the Early Pre-Pottery Neolithic B (EPPNB) (9,600–8,500 bp), (2) the Middle Pre-Pottery Neolithic B (MPPNB) (9,300–8,500 bp), (3) the Late Pre-Pottery Neolithic B (LPPNB) (8,500–8,000 bp), and, (4) the Final Pre-Pottery Neolithic B period (8,000–*c.* 7,500 bp), although also see Moore (1985) for an alternative cultural-historical system (Figure 3).

In placing archaeological materials in time and space, researchers working in the prehistory of the Near East are faced with the unenviable task of searching for consensus among their colleagues in constructing cultural-historical sequences, often drawing on data from different geographical areas before moving on to the more rewarding task of reconstructing past social phenomena. In many cases, such as the south-central Levant, disagreements do not focus on material patterning from individual archaeological sites; rather they rest on what is perceived to be appropriate data sets for creating cultural–historical sequences. In recognition of the ongoing nature of this discussion, no attempt is made in this volume to present a revised all-encompassing chronological sequence for the Near East. Moreover, these chronological debates, although important to researchers for understanding subtle changes through time, by no means detract from the general consensus as to the overall reconstruction of the transition from various Epipaleolithic through Neolithic cultures in different areas of the Near East.

Finally, it is necessary to briefly comment on the continued employ-

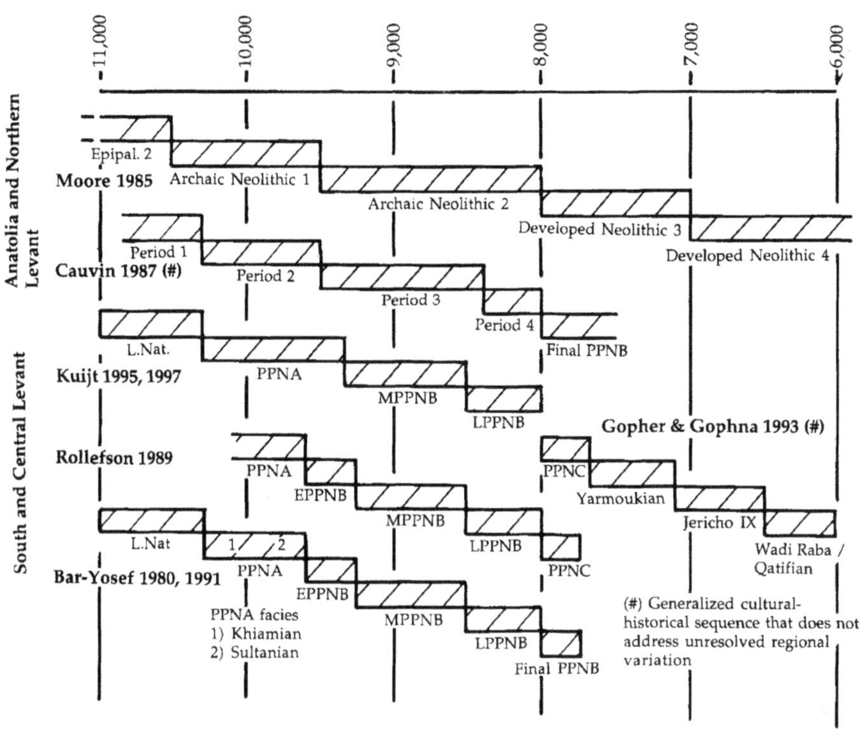

Figure 3. Different cultural–historical organizational schemes for the Near Eastern Neolithic.

ment of the chronological–historical term *Pre-Pottery Neolithic* and on how radiocarbon data are reported in these essays. In the next few years it is likely that our understanding and reconstruction of cultural-historical sequences for the Neolithic of the Near East will be revised and refined through the calibration of all radiocarbon dates for the Aceramic Neolithic. This development will unquestionably render some of the debate on chronological sequences obsolete, and it should help us define individual periods of time and their time length and, perhaps most importantly, resolve existing data gaps. It will also allow us to determine how existing chronological sequences for the Near East are linked to "plateaus" in radiocarbon curves that hinder detailed chronological resolution (Bar-Yosef and Meadow 1995). It is, however, not possible to realistically anticipate such future changes; hence, it is necessary to present all cited dates in this volume in uncalibrated radiocarbon years before 1950, employing the 5,568-year half-life.

For readers not familiar with the cultural–historical framework of the Near East, it will probably come as somewhat of a surprise to learn that, despite Dame Kenyon's original intention, the term *Pre-Pottery Neolithic* does not suggest that ceramics were absent. Subsequent to the proposal and adoption of this cultural–historical sequence, archaeological field research has illustrated that in the PPNC–Final Pre-Pottery Neolithic B period fired clay was employed, albeit in a very limited way. While this labeling system is therefore somewhat ironic, the authors in this volume continue to employ the term *Pre-Pottery Neolithic* for reasons of historical precedent and continuity in the literature. Ultimately, the documentation of limited ceramic materials for this period does not diminish the utility of this organizational framework as long as one is willing to recognize that the term is employed in a way that is more flexible than originally envisioned. (See Bar-Yosef 1991; Cauvin 1987; Moore 1985; Rollefson 1989 for more detailed discussions.)

SUMMARY

Collectively, the chapters in this volume illustrate how current archaeological research on the Near Eastern Neolithic helps us understand the economic and political contexts from which food production emerged, and the pathways and archaeological evidence for social differentiation in early agricultural contexts. Ultimately it is at the intersection of these different, yet clearly interrelated, questions of how and why social differentiation was minimized, how power and authority were ceded to groups, and how social differentiation in precomplex chiefdoms emerged from within an egalitarian ethic, that the importance of the archaeological study of the Near Eastern Neolithic becomes clear. In many ways the examination of the Neolithic of the Near East provides us with a unique opportunity to explore these questions, for it combines many of the important elements widely perceived to be important in later complex chiefdoms: increased population aggregation, food production, abundant surplus subsistence, and economic resources coupled with the social need for maintaining existing social frameworks in rapidly growing communities (e.g., Bar-Yosef and Belfer-Cohen 1989; Bar-Yosef and Kislev 1989; Cauvin 1994; Hayden 1995; Price 1995). Modeling the origins of agriculture in the Near East, changing Late Pleistocene environmental conditions, and the geographical and temporal location of the first domestication of plant and animal species continues to be widely perceived by the professional academy as being central to our understanding of the past (e.g., Braidwood and Braidwood 1953; Hillman and Davis 1990; McCorriston and Hole 1991). In conjunction with this recognition is a growing awareness that the Neolithic was also an important social process, one in which we see a remarkable increase in global population levels, the first appearance of large villages occupied on a year-round basis, and some of our earliest evidence for social differentiation.

As to be expected, collections such as this one are bound to raise more questions than they answer, for that is both the beauty and horror of such debate: with more detail we are drawn even deeper into the search for greater understanding. This volume, for example, leads us toward other important aspects to the changing nature of social relations in the human past. Does the process of social differentiation occur through the actions of charismatic individuals, as is commonly held, or did this process originally focus on a group of people? What were the social, economic, and political loci for such changing social arrangements? How was the process of changing social arrangements linked to that of ritual beliefs? Needless to say, the archaeological and anthropological exploration of these topics is in its infancy and, beyond pointing out that none of these processes need be mutually exclusive to others, it is important to recognize that the preliminary examinations of some of these questions by the researchers in this volume represent an important line of study. To return to Winterhalder's image of the prickly porcupine, this volume presents a juggling act that complements our rich and still developing awareness of the economic side to the emergence of social inequality in precomplex chiefdoms.

REFERENCES

Bar-Yosef, O., 1981, The "Pre-Pottery Neolithic" period in the Southern Levant. In *Préhistoire du Levant*, edited by J. Cauvin and P. Sanlaville, pp. 551–570. CNRS, Paris.

Bar-Yosef, O., 1991, The Early Neolithic of the Levant: Recent Advances. *The Review of Archaeology* 12(2): 1–18.

Bar-Yosef, O., and Belfer-Cohen A., 1989, The Levantine "PPNB" Interaction Sphere. In *People and Culture in Change: Proceedings of the Second Symposium of Upper Paleolithic, Mesolithic, and Neolithic Populations of Europe and the Mediterranean Basin,* edited by I. Hershkovitz, pp. 59–72. British Archaeology Review, International Series 508. BAR, Oxford.

Bar-Yosef, O., and Kislev, M., 1989, Early Farming Communities in the Jordan Valley. In *Foraging and Farming: The Evolution of Plant Exploitation,* edited by D. R. Harris and G. C. Hillman, pp. 632–640. Hyman and Unwin, London.

Bar-Yosef, O., and Meadow, R., 1995, The Origins of Agriculture in the Near East. In *Last Hunters-First Farmers: New Perspectives on the Prehistoric Transition to Agriculture*, edited by T. D. Price and A. G. Gebauer, pp. 39–94. School for American Research, Santa Fe, New Mexico.

Braidwood, R., J., and Braidwood L., S., 1953, The Earliest Village Communities of Southwest Asia. *Journal of World History* 1(2): 278–310.

Cauvin, J., 1987, Chronologie Relative et Absolue dans le Néolithique du Levant Nord et d'Anatolie Entre 10,000–8,000 B.P. In *Chronologies in the Near East,* edited by O. Aurence, F. Evin, and F. Hours, pp. 325–342. British Archaeology Review, International Series 508. BAR, Oxford.

Cauvin, J., 1994, *Naissance des Divinités, Naissance de l'agriculture*. CNRS, Paris.
Crowfoot-Payne, J., 1976, The Terminology of the Aceramic Neolithic Period in the Levant. In *Terminology of Prehistory of the Near East*, edited by F. Wendorf, pp. 131–137. UISPP Congress IX, Nice.
Crowfoot-Payne, J., 1983, The Flint Industries of Jericho. In *Excavations at Jericho, Vol. V*, edited by K. M. Kenyon and T. A. Holland, pp. 622–759. The British School of Archaeology in Jerusalem, London.
Garfinkel, Y., and Nadel, D., 1989, The Sultanian Flint Assemblage from Gesher and Its Implications for Recognizing Early Neolithic Entities in the Levant. *Paléorient* 15(2): 139–152.
Gopher, A., 1994, *Arrowheads of the Neolithic Levant*. American Schools of Oriental Research: Dissertation Series, Volume 10. Eisenbrauns, Winona Lake, Indiana.
Hauptmann, H., 1993, Ein Kultgebäude in Nevali Çori. In *Between the Rivers and over the Mountains*, edited by M. Frangipane, H. Hauptmann, M. Liverani, P. Matthiae, and M. Mellink, pp. 37-69. Dipartimento di Scienze Storiche Archeologiche e Anthropologiche dell'Antichità, Università di Roma "La Sapienza," Rome.
Hayden, B., 1995, A New Overview of Domestication. In *Last Hunters–First Farmers: New Perspectives on the Transition to Agriculture*, edited by T. Douglas Price and A. B. Gebauer, pp. 273–300. Schools of American Research Press, Santa Fe, New Mexico.
Hillman, G. C., and Davis M., S., 1990, Measured Domestication Rates in Wild Wheats and Barley under Primative Cultivation, and Their Archaeological Implications. *Journal of World Prehistory* 4(2): 157–222.
Hodder, I., (ed.), 1996, *On the Surface: Çatalhöyük 1993–1995*. McDonald Institute of Archaeological Research and the British Institute of Archaeology at Ankara, Cambridge.
Kozlowski, S. K., 1989, Nemrik 9, A PPN Neolithic Site in Northern Iraq. *Paléorient* 15(1): 25–31.
Kuijt, I., 1997a, Interpretation, Data and the Khiamian of the South-central Levant. *Neo-Lithics* 3:3–6.
Kuijt, I., 1997b, Trying to Fit Round Houses into Square Holes: Re-examining the Timing of the South-central Levantine Pre-Pottery Neolithic A and Pre-Pottery Neolithic B Cultural Transition. In *The Prehistory of Jordan II: Perspectives from 1997*, edited by H. G. Gebel, Z. Kafafi, and G. O. Rollefson, pp. 193-202. Ex Oriente, Berlin.
McCorriston, J., and Hole F., 1991, The Ecology of Seasonal Stress and the Origins of Agriculture in the Near East. *American Anthropologist* 93(1): 46–94.
Moore, A., 1985, The Development in Neolithic Societies in the Near East. In *Advances in World Archaeology*, edited by F. Wendorf and A. E. Close, pp. 1–69. Academic Press, New York.
Nadel, D., 1990, The Khiamian as a Case of Sultanian Intersite Variability. *Mitekufat Haeven* 23: 86–99.
Nissen, H. J., 1993, The PPNC, the Sheep and the "Hiatus Palestinien." *Paléorient* 19(1): 177–183.
Perrot, J., 1993, Commentaires. *Paléorient* 19(1): 175–177.
Price, T. D., 1995, Social Inequality at the Origins of Agriculture. In *Foundations of Social Inequality*, edited by T. D. Price and G. M. Feinman, pp. 129–151. Plenum, New York.
Rollefson, G. O., 1989, The Late Aceramic Neolithic of the Levant: A Synthesis. *Paléorient* 15(1): 168–173.
Rollefson, G. O., and Köhler-Rollefson, I., 1993a, PPNC Adaptations in the First Half of the 6th Millennium B. C. *Paléorient* 19(1): 33–42.
Rollefson, G. O., and Köhler-Rollefson, I., 1993b, Reply to Comments by Hans Nissen and Jean Perrot. *Paléorient* 19(1): 183–185.
Rosenberg, M., and Davis M., 1992, Hallan Çemi Tepesi, an Early Aceramic Neolithic Site in

Eastern Anatolia: Some Preliminary Observations Concerning Material Culture. *Anatolica* 18: 1–18.
Schmidt, K., 1996, The Urfa-Project 1996. *Neo-Lithics* 2: 2-3.
Stordeur, D., Jammous, B., Helmer, D., and Wilcox, G., 1996, Jerf el-Ahmar: A New Mureybetian Site (PPNA) on the Middle Euphrates. *Neo-Lithics* 2: 1–2.
Watkins, T., Baird, D., and Betts, A., 1989, Qermez Dere and the Early Aceramic Neolithic in Northern Iraq. *Paléorient* 15(1): 19–24.
Winterhalder, B., 1981, Optimal Foraging Strategies and Hunter-Gatherer Research in Anthropology: Theory and Models. In *Hunter-Gatherer Foraging Stategies*, edited by B. Winterhalder and E. A. Smith, pp. 13–35. University of Chicago Press, Chicago, Illinois.

Part II

Regional Issues, Settlement Practices, and Sedentism

Traditionally arguments of the nature of Neolithic social organization have been founded, in part, on considerations of regional settlement practices and an understanding of the process by which communities become increasingly sedentary in the Epipaleolithic and Neolithic periods. Although limited by the relatively small number of excavated early Neolithic settlements in some of the larger countries in the Near East, such as Syria, Iraq, and Turkey, exploration of regional and interregional settlement practices provides researchers with important broader considerations of how community-level practices might be interlinked with those from other neighboring communities. Similarly, consideration of a broader trajectory in which increasing sedentism occurs in the transition from Epipaleolithic to Neolithic household and communities levels improves our understanding of the overall process to sedentism and how this shift might be linked to broader social and economic changes. Expanding this intellectual foundation, several researchers, in the next three chapters, address a series of interrelated regional-scale issues, including the social dimensions of the move toward increased sedentism leading into the Neolithic and how the nature of the household changed through time with increased sedentism.

In an essay that focuses on the periods immediately before, and leading into the earliest stages of the Pre-Pottery Neolithic, Anna Belfer-Cohen and Ofer Bar-Yosef focus on the questions of when, how, and why Near Eastern communities became more sedentary. They note that sedentism occurred independently from the development of plant domestication with the appearance of large residential structures occupied year-round in the Early Natufian period. Archaeological remains from this period illustrate a cultural bridge linking two ways of life, one that focused on mobile hunter–gathering and one that relied primarily on sedentary agricultural cultivation, domestication of plants and animals, and herding. They argue that Natufian communities were specialized hunter–gatherer/collectors and, therefore, not based on stable social units. Specifically, sedentism and foraging are primarily opposed to each other, as limited mobility triggers progressive population growth and, by extension, resource and social stresses. The authors link these social stresses with simultaneous changes in the use of zoomorphic art as a means of alleviating scalar stress and changing mortuary practices. Moving to a consideration of how these practices change with the onset of the Pre-Pottery Neolithic at around 10,300 bp, Belfer-Cohen and Bar-Yosef discuss how, for the first time, we see the emergence of large agricultural communities of several hundred people, large communal buildings, and a new emphasis on anthropomorphic art.

Focusing on the excavations at Hallen Çemi, Michael Rosenberg and Richard Redding explore the nature of community organization leading into the Neolithic, in specific, and the evolution of social inequality, sedentism, and ownership among sedentary hunter–gatherers, in general. As with other Neolithic settlements, the authors note that at the site of Hallan Çemi there is clear evidence for deliberate organization of space at the structural and settlement level, in this case in the construction and maintenance of a large courtyard area with the entrances of surrounding residential structures facing away from the courtyard. Drawing on the presence of public buildings, as well as the presence of highly stylized carved stone bowls, pestles, and batons, the authors argue that this relatively small community provides insights into the existence of a social system within which elements of social inequality were balanced with the symbolic and physical employment of aspects of material culture (such as formalized food preparation and consumption) for conflict resolution and promotion of group cohesion. From the standpoint of these chapters, then, researchers have begun to explore some of the relationships between social organization, sedentism, and settlement practices in ways that address the complexities and subtleties of archaeological data and social systems, such as the spatial organizations of individual settlements and/or the existence of communal structures such as those identified at Çayönü, Hallan Çemi, Nevali Çori, and 'Ain Ghazal.

Concentrating on the Epipaleolithic through Pre-Pottery Neolithic periods in the south-central Levant, in the fourth chapter Brian Byrd explores the degree to which, and timing of, changes in the size and organization of households in the transition to settled, food-producing villages. In brief, he notes that previous studies of Neolithic social organization have largely focused on social complexity and ritual behavior with minimal attention given to the understanding of the nature of household organization. Furthermore, he argues that early agricultural villages in the Near East were characterized by a restriction in the social network for sharing production and consumption activities and by the development of more institutionalized mechanisms for integrating the community as a whole. Drawing on ethnographic sources, Byrd suggests that communities were composed of nuclear or extended households and discusses how such inferences can be linked to specific archaeological correlates. One aspect of this change appears to be the continued reinforcement of community social order by leaders while inequality was expanded within and between households. In discussing exceptions to this pattern, he noted that several other factors may have influenced the temporal and spatial patterning for Pre-Pottery Neolithic period households, including site setting, degree of settlement permanence, extent of reliance on domesticates, degree of household autonomy, and total settlement population.

Chapter **2**

Early Sedentism in the Near East

A Bumpy Ride to Village Life

ANNA BELFER-COHEN AND OFER BAR-YOSEF

The emergence of sedentary communities in the Near East, as reflected in bioarchaeological data sets, occurred between 13,000 and 10,000 years ago. The process involved two major archaeological entities, the Natufian and the earliest Neolithic period, termed Pre-Pottery Neolithic A (PPNA). These were followed by the cultural entities of the Pre-Pottery Neolithic B period (PPNB) with extensive evidence for sedentary village communities as well as domesticates (both plants and animals). Sedentism can be envisioned as a response to particular environmental and/or social conditions (discussed in detail elsewhere, see Bar-Yosef and Belfer-Cohen 1989a, 1989b, 1991, 1992; Bar-Yosef and Meadow 1995; Bar-Yosef 1998). This chapter does not revolve around the issue of why sedentism occurred in the southern Levant, nor with the particular changes and adaptations occurring at the later part of the Pre-Pottery Neolithic (i.e., the PPNB). Rather, we focus on what can be inferred from the archaeological data regarding the course and tempo of

---

ANNA BELFER-COHEN • Department of Prehistory, Institute of Archaeology, Hebrew University, Jerusalem, Israel 91905. **OFER BAR-YOSEF** • Department of Antrhropology, Harvard University, Cambridge, Massachusetts 02138.

Life in Neolithic Farming Communities: Social Organization, Identity, and Differentiation, edited by Ian Kuijt. Kluwer Academic/Plenum Publishers, New York, 2000.

the critical transformation from mobile to sedentary ways of life and, by extension, changes in social organization in southern Levant.

Interestingly, there is clear evidence to indicate that sedentism occurred independently of the development of an agricultural way of living in the Near East. For example, it has been argued that evidence for sedentism can already be observed in certain Middle Paleolithic Mousterian sites in Israel (Hietala and Stevens 1977; Lieberman 1993; Lieberman and Shea 1994). Moreover, the Early Natufian, in which evidence of sedentism was clearly observed (see below), was culturally a complex hunter–gatherer society that predated the agricultural communities by almost 3,000 radiocarbon years. The transition to sedentism was not a smooth process but a bumpy ride along a course that obviously was not planned, and its consequences were unforeseen. A number of crossroads can be identified along this route, where turns have been taken by the human group involved, according to the specific socioeconomic circumstances prevailing at some particular time.

Sedentism can be very difficult to identify by strictly archaeological evidence, (i.e., the cultural material remains of the people involved, such as architecture, lithics, bone and groundstone implements, burials, etc.). As an alternative approach, during the last decade or so, researchers (e.g., Bar-Yosef 1983; Henry 1989) have shown preference for bioarchaeological evidence for sedentism such as high frequencies of human commensals—the house mouse, the house sparrow, and the rat (Auffray et al. 1988; Tchernov 1991); indications of year-round hunting of gazelle, based on cementum increment analyses (Lieberman 1993); or the particular age profiles of hunted specimens—a steep rise in the young specimens (Davis 1983). In contrast, archaeological remains provide secondary evidence for the existence of sedentism. Stone and mud brick structures, heavy groundstone utensils, silos, and storage pits (mainly from the PPNA)—all these indicate a mode of existence combining year-round occupation of sites, interspersed with short, seasonal spells of anticipatory mobility, both of task groups and even of entire communities. Other indications of sedentism are certain demographic features deduced from the archaeological record, such as scalar stress indicated by the rise in the number of children, signs of primary social stratification, inter- and intragroup conflicts, and diseases (Agelarakis 1993; Bar-Yosef and Belfer-Cohen 1991; Belfer-Cohen 1991a, 1991b; Cohen 1985). Most of these phenomena have been observed in the archaeological record of the Natufian culture, although no indications of intrasite or intergroup violence were found, at least in the core-area sites where hundreds of human burials were uncovered (Belfer-Cohen, Schepartz and Arensburg 1991) .

NATUFIAN (12,800–10,300 BP)

The Natufian, the most extensively dated Late Quaternary complex in the Levant (Byrd 1994a), is the archaeological entity bridging two different ways of life: the first (hunting-gathering) prevailed more or less since the evolution of humans and is now replaced by the second (agriculture), which has basically sustained most of humanity to this day. The Natufian archaeological entity, therefore, is identified with the transformation from mostly mobile foraging to mostly sedentary agricultural cultivation, domestication of plants and animals, and herding. Initially identified and described by D. A. E. Garrod (1932), following her excavations at the Shukba Cave in the Judean hills, Natufian occurrences have subsequently been uncovered within the boundaries of what is now known as the Natufian "homeland" in the central Levant as well as other parts of the Near East (for a detailed description see Bar-Yosef 1983; Bar-Yosef and Belfer-Cohen 1989a; Belfer-Cohen 1989b, 1991b; Byrd 1989; Cauvin 1991, 1994; Henry 1989 and references therein) (Figure 1). The Natufian differs strikingly from preceding Epi-Paleolithic Levantine archaeological entities in its material culture, namely in the profusion of groundstone utensils, bone tools, art and decoration objects, and jewelry. It is also the first archaeological entity in the Levant featuring relatively substantial architectural remains as well as large burial grounds from which more than 400 individuals have so far been recovered (Byrd and Monahan 1995). Interestingly, the lithic industry remains basically similar to the preceding Epi-Palaeolithic ones, in that the microlithic component reaches 40% or more in all the assemblages. Of note are the picks (the forerunners of the future Neolithic ax-adzes group) and the sickle blades, which were most probably used for harvesting cereals (Unger-Hamilton 1991).

Natufian sites are clearly dichotomized into base camps—large sites, demonstrating most of the features listed above, and smaller sites that contain little architecture, no burials, and poor bone and groundstone assemblages (Bar-Yosef 1983). The material recovered from sites of the latter type consists mainly of lithics, and the sites are commonly interpreted as ephemeral camps of various functions since they are usually located in marginal environments. In recent years, there has been a growing awareness among researchers that the Natufian culture as a whole should be subdivided according to different modes of adaptations and different environmental settings (Belfer-Cohen 1989b, 1991a, and see further discussion below). By extension, these subdivisions inform us as to the links between the emergence of sedentism and Natufian social structure.

At first Early Natufian (12,800–11,000 bp) communities were most probably composed of hunter–gatherers practicing anticipatory mobility, whose

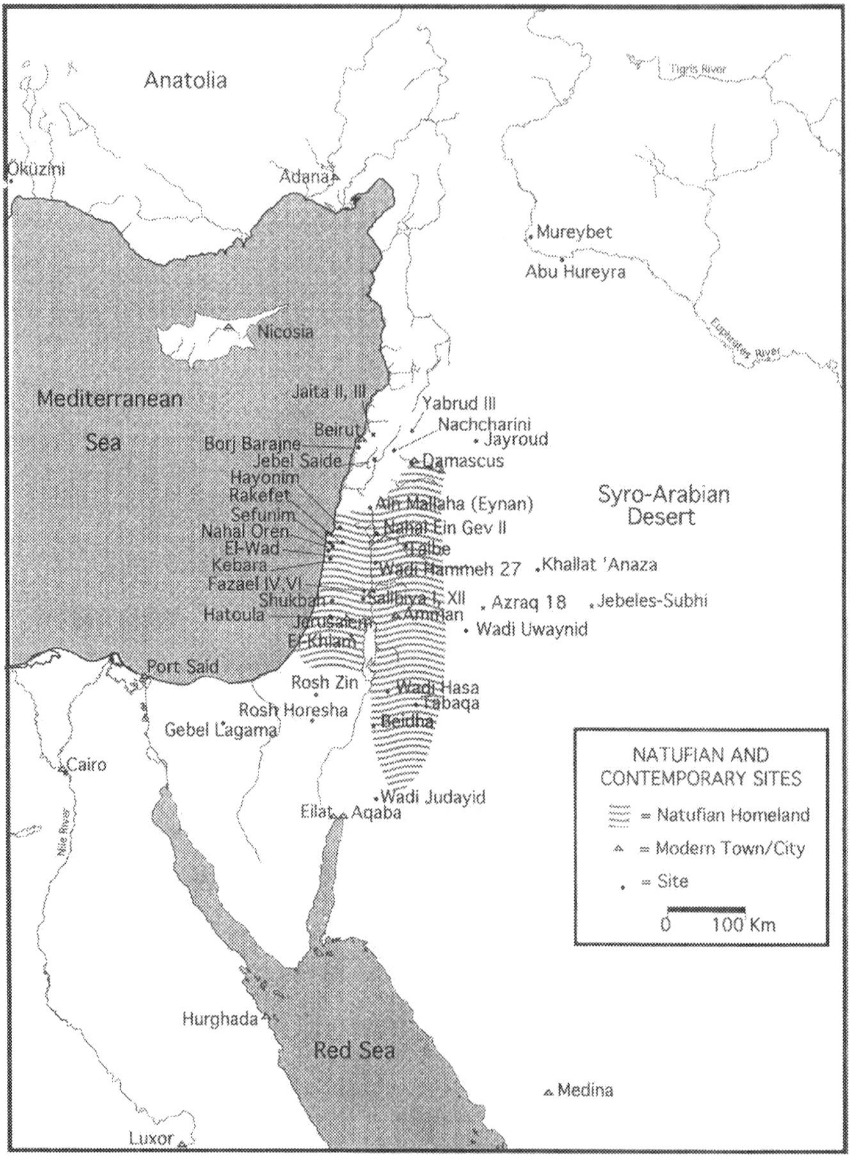

Figure 1. Map indicates location of Natufian sites in the Levant. The stippled area delineates the proposed Natufian homeland.

aggregations gradually became permanent settlements, changing their way of life. By the onset of the Late Natufian (11,000–10,300 bp), there were already, at least in the core area, sedentary hunter–gatherers specializing in particular food procurement strategies, namely intensive collection of cereals and/or acorns and lentils. The faunal remains, recovered from the various Natufian sites (see, for example Bar-Yosef et al. 1974; Bouchud 1987; Edwards 1987), reflect a 'broad spectrum' exploitation of food resources (Flannery 1973). Evidence of severe depletion of the genetic pool of the local gazelle has been suggested to indicate intensive hunting of this species by the Natufians, which had purportedly led to 'inbreeding' (Cope 1991). Although rejected by Dayan and Simberloff (1995), this interpretation implies that the Natufians exerted some degree of cultural control on herds in their immediate vicinity, resulting in "protodomestication". We should bear in mind that there are other explanations for the size reduction in the gazelles, as for example the dry and cold conditions of the Younger Dryas.

Most researchers agree that the Natufian, from its inception, can be defined as a complex society of specialized hunter–gatherers/collectors (Henry 1985). This particular social system influenced and modified much of the Natufian behavioral makeup: social structure, social behaviors and subsistence-acquiring practices. Note, however, that complex hunter–gatherer societies are not considered stable social units, and the Natufian was no exception (Henry 1991; Byrd 1994b). Henry (1991), for example, argues that complex societies of foragers ultimately provided their members with greater economic security. Yet, like most of the researchers cited herein, he agrees that even when such socioeconomic systems were fully established, they did not come near to attaining real stability but instead were "destined to fail over the not-so-long run" (Henry 1991:355). Sedentism and foraging are primarily opposed to each other with respect to the long term success of a population, since limited mobility triggers progressive population growth that cannot be sustained by fixed natural-resource ceilings. As noted by Keeley (1988), complexity creates resource stress—resource stress does not create complexity. Shnirelman (1992) supports this contention, maintaining that social evolution is more closely related to the degree of efficiency of a subsistence strategy than to the particular form of the economic activity. Thus, while the complexity of societies of specialized hunter–gatherers promoted sedentism and population growth, and intensified resource exploitation and the formation of social hierarchies, it also ultimately brought about the failure of "sedentary foraging."

The fragility of complex groups is also discussed by Kosse (1994). She maintains that, while advantageous under certain conditions, socioeconomic complexity is not always essential for small-group survival. At the same time, however, groups had to grow in size in order to best adapt to a changing and heterogeneous environment and to compete effectively. Inevitably, beyond a certain population size threshold these groups became complex. Since large groups were most probably more advantageous in competition than small ones, and the "large size of a local community was not only a consequence, but also a prerequisite for a highly efficient food-gathering economy" (Shnirelman 1992:188), it would seem that, contrary to what Henry (1991) proposes, complex hunter–gatherer societies were a relatively common phenomenon. Indeed, according to Schnirelman, there were many nonagricultural societies in the past, rather than only a few exceptions, which underwent social differentiation processes that progressed far beyond the early stages of complex hunter–gatherers (see Chapter 3, this volume). It is important to note that while competition favors larger groups, cost considerations favor smaller ones. Large groups, with their additional decision-making and coordinating institutions, are more expensive to maintain than simpler, smaller groups, and they become economically disadvantageous when marginal returns diminish. One option under such circumstances is to curtail growth and attempt to maintain some sort of equilibrium between population levels and resources. An economically more expensive option is to solve the problem of diminishing returns through the expansion of social networks. Fish and Fish (1991:403) suggest that "in view of its distributional contiguity and shared stylistic and technological attributes, the unifying structure of the Natufian could also be compared to an interaction sphere or an environmentally based commonality in economic orientation." They support their argument with the example of the Piman from southern Arizona and adjacent portions of Sonora who share a closely related family of dialects and maintain extensive social interactions, including frequent intermarriage, yet at the same time these different Piman groups still demonstrate a diversity of subsistence modes. This supports the possibility that the Natufian culture may have indeed developed a number of simultaneous, yet different, answers to the problems of daily existence.

All these observations have to be taken into consideration while looking for the various ways and means Natufian groups employed to alleviate conditions of stress. Indeed, the instability characterizing the Natufian is also reflected in the differences between the Early and Late Natufian as well as between Natufian groups from different geographical regions. Different responses were dictated by the different circumstances and ultimately were linked to the social ordering of the Natufians, their mode of subsistence, and the development of sedentism.

Archaeological data, scanty as they may be, reveal some of the measures taken by the Natufians at various stages of their existence in order to withstand stress, stemming from external causes, such as changes in environmental conditions, and internal ones, such as continuing intergroup and intragroup rivalries. The intensification of artistic activities during the Natufian, compared to its predecessors, is a good example. Intensive artistic activity can be regarded as a means for alleviating scalar stress caused by intensive social interactions, which a sedentary mode of living entails (Bar-Yosef 1998; Belfer-Cohen 1988 and references therein). Since sedentary people are forced into sharing territories or even living quarters with other, not immediately related, people with no option of splitting into smaller, nuclear groups, the need gradually arises for a mechanism that would create a sense of group identity and group loyalty (Belfer-Cohen 1991b). Artistic activity in the Natufian provided such a mechanism and is illustrated by significant differences in the distribution of bead types among various Natufian sites (Figure 2). For example, beads and pendants common in certain other assemblages were rarely found at Hayonim Cave, while the most typical of Hayonim Cave beads are rarely, if ever, found at other settlements (Bar-Yosef and Tchernov 1970; Belfer-Cohen 1991a; Stordeur 1981). Such differences can be interpreted as implying the existence of distinct social microunits, such as small bands or extended families, within the Natufian macrostructure

The territoriality observed among Natufian sites may also explain why the Natufians made use of sickles rather than beaters and baskets in harvesting cereals. It has been demonstrated (Hillman and Davies 1990) that the most convenient method of harvesting wild cereals is beating, yet it is efficient only in terms of the amount harvested per unit time, while in terms of amount harvested per unit area beating is no better than uprooting and sickling. If the limitations on the local groups were the small size of fields of wild stands, the methods used for harvesting would have necessarily been adapted so as to maximize energy returns per unit area instead of maximizing returns per unit time. The climatic episode of the "Younger Dryas" (10,800–10,300(?) bp), whose effects are well recorded from the Gahb and the Hula Valley in the northern and central Levant (Baruch and Bottema 1991), was probably a period of stress caused by the cold and dry conditions that also affected the Mediterranean coastal ranges of the Levant (Bar-Yosef and Belfer-Cohen 1989a; Moore and Hillman 1992). It seems that during this period the yield of wild cereals, which are C3 plants, may have decreased (Bar-Yosef and Meadows 1995), a development that could have motivated the Late Natufians to cultivate wild cereals. Similar conclusions have been reached by Henry (1989, 1991) on the basis of site distributions, which he interpreted as reflecting major shifts in the resource base of wild cereals, and are supported by the study of the Natufian botanical material from Abu Hureyra (Hillman et al. 1989).

The Late and Final Natufian periods have been shown to differ from the Early Natufian in various aspects of material culture and social practices. This includes a decrease in the manufacture of groundstone utensils and

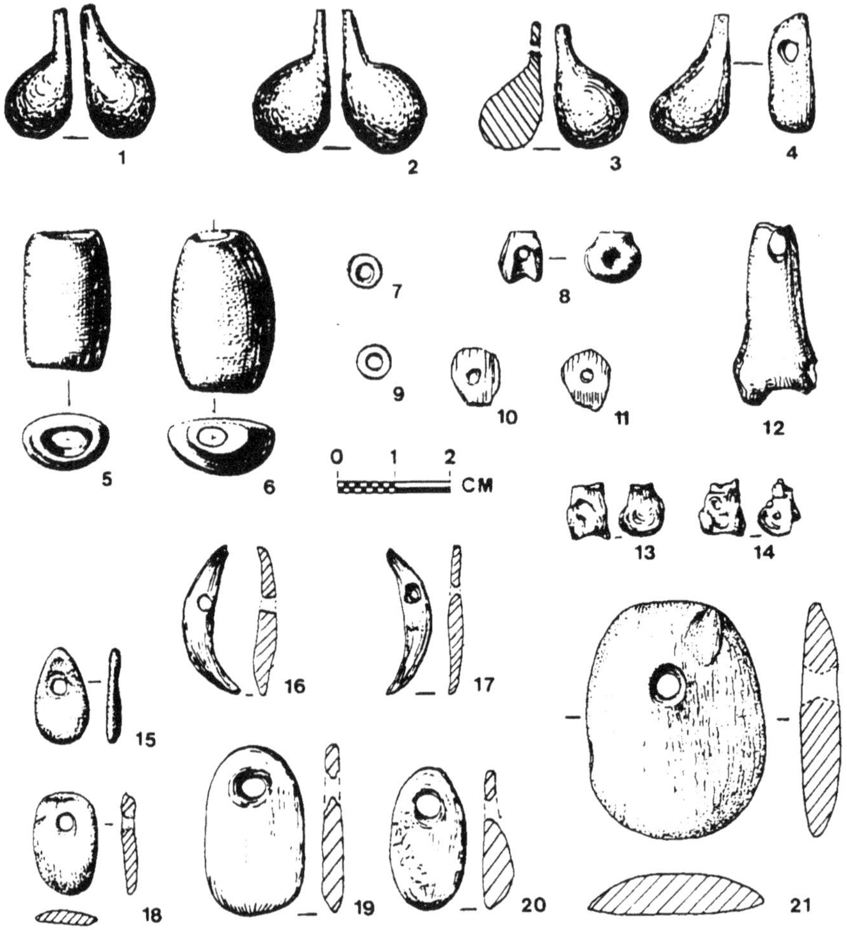

Figure 2. Different types of jewelry made of bone from three Natufian sites: 1–4-El Wad (redrawn after Garrod and Bate 1937); 5–12-Ain Mallaha (after Perrot 1966); 13–21-Hayonim cave (after Bar-Yosef and Tchernov 1972; Belfer-Cohen 1991).

ornaments and number and type of artistic objects and architecture, as well as an increase in the number of secondary and group burials (Belfer-Cohen 1989a; Belfer-Cohen et al. 1991; Garrod 1957). While most of the Early Natufian burials were primary, in the Late Natufian an increasing number of secondary burials are found along with primary burials of axial skeletons. At Hayonim Cave for instance, most of the Late Natufian group graves contain a single primary burial alongside several secondary ones. Indeed, each such primary burial may have been merely the "last one in," which was to become a secondary burial when the grave was reopened. On the other hand, the increase in the frequency of secondary burials may reflect a change in the function of certain sites during the Late Natufian. At that time some sites may have been used more extensively as human bone repositories, as for example the site of Eynan (Mallaha), which at the end of the Natufian occupation might have functioned solely as a graveyard (Valla 1981, 1987; Perrot et al. 1988). Individuals may have been initially buried elsewhere and later transferred to a site, possibly in sacks (Perrot et al. 1988), in order to be reburied in their traditional home range (Valla 1991). This practice could then be regarded as solid evidence of greater mobility than that which is attributed to the Early Natufian (Belfer-Cohen et al. 1991).

As observed by Garrod (1957), Natufian decorated burials are exclusively associated with the Early Natufian, and indeed, all of the decorated burials uncovered since are dated to the Early phase (Belfer-Cohen 1989a, 1995; Byrd and Monahan 1995). Perhaps the greater mobility of the later Natufian may be partly responsible for the disappearance of the decorated burials. Most of the Late Natufian burials were recovered from Eynan (Mallaha), Hayonim Cave, and el-Wad, where the numbers of burials dwindle and most of them are secondary. The same also holds for the burials at the Late/Final Natufian site of Nahal Oren (Grognier and Dupouy-Madre 1974). The decorative items recovered from graves in the Early Natufian did not constitute burial offerings but were personal jewelry or, more probably, garment ornaments. Therefore it is quite possible that the increasing tendency toward secondary burials contributed to the disappearance of this custom which was quite rarely practiced to begin with. It should be emphasized that production of beads and other decorative items did not stop during the Late Natufian, yet they were made for the use of the living rather than for the dead (Belfer-Cohen 1991a, b). The 'cemetery evidence' from some sites suggests not only increased mobility but also more overt expressions of territorial boundaries in the Late Natufian. Specifically, the larger number of multi-individual graves from this phase can be interpreted as an expression of group cohesion, since they indicate more concentrated efforts toward reburial in specified locations serving as territorial markers or means of limiting social inequality (Perrot et al. 1988; Kuijt 1996).

Consideration of Natufian skeletal data raises the possibility that the stature of the Late Natufians may reflect circumstances of stress (Belfer-Cohen et al. 1991). Inter-group differences in mean male/female stature can indicate environmental stress conditions. Under such conditions, taller individuals achieve less of their potential stature than shorter individuals, which in turn tends to reduce sexual dimorphism within the population under study by the decrease in male stature. Thus, reduction in the difference in male/female stature between the Early and the Late Natufian (11.5–12.9 cm in the former and 8.0–8.6 cm in the latter, Belfer-Cohen et al. 1991), is indicative of possible stress conditions during the later stages of the Natufian. It should be noted, however, that this varies between the Natufian populations, and no portion of these populations shows any evidence of serious stress (such as signs of malnutrition, violence, dramatic reduction in height, etc.).

There is a clear-cut dichotomy in the Early Natufian between the sedentary–permanent, semipermanent, seasonal sites in the Natufian Levantine homeland and the non-Natufian hunter–gatherers living in surrounding regions of the Saharo-Arabian vegetation belt. In contrast, by the Late Natufian it is difficult to differentiate between permanent and seasonal sites. A major shift can be observed at the end of the Late Natufian in the size and location of settlements and in their distribution, as illustrated by the Late Natufian sites in the Negev (Goring-Morris 1987) and southern Jordan (Henry 1995). These settlements are ephemeral and contain only scanty architectural remains. It seems as if the Natufians were forced into a more mobile settlement pattern than that observed in the core area, adopting a mobile way of life similar to that of earlier hunter–gatherers in this region, as if groups inhabiting the region reverted to a life of greater mobility while trying to adapt to particular local/territorial conditions. This is well illustrated by the case of the Harifian, located mainly in the Negev, which represents a short lived attempt at an arid adaptation that lasted for only about 500 years (Goring-Morris 1987, 1991; Bar-Yosef 1987).

In their aforementioned comparison of the Piman and the Natufian, Fish and Fish (1991) note that among the former there is evidence for continuous subsistence reversals (from hunter–gatherers to cultivators, to agriculturist, and back to mobile foragers), that occurred within very short intervals. They remark (1991:407) that

> The potential for subsistence shifts away from sedentary and intensified tendencies would likely be greatest among groups whose social and demographic patterns had been least shaped by long traditions of such experience. During initial portions of the transition to domestication economies in both regions, commitments to emerging food producer stances are likely to have been more ephemeral, and multidirectional shifts more commonplace.

PRE POTTERY NEOLITHIC A PERIOD (10,300–9,300 BP)

Though of shorter duration than the preceding Natufian, the Levantine Pre-Pottery Neolithic A period (PPNA), dated to between *ca.*10,300 and 9,300 years bp, represents a period of dramatic change, during which crucial phenomena appeared for the first time in human history (for more detailed descriptions see Bar-Yosef 1991, 1992; Bar-Yosef and Belfer-Cohen 1989a; Cauvin 1994; Kuijt 1995; Moore 1985). The PPNA comprises two archaeological entities; the Khiamian and the Sultanian (Figure 3), of which the

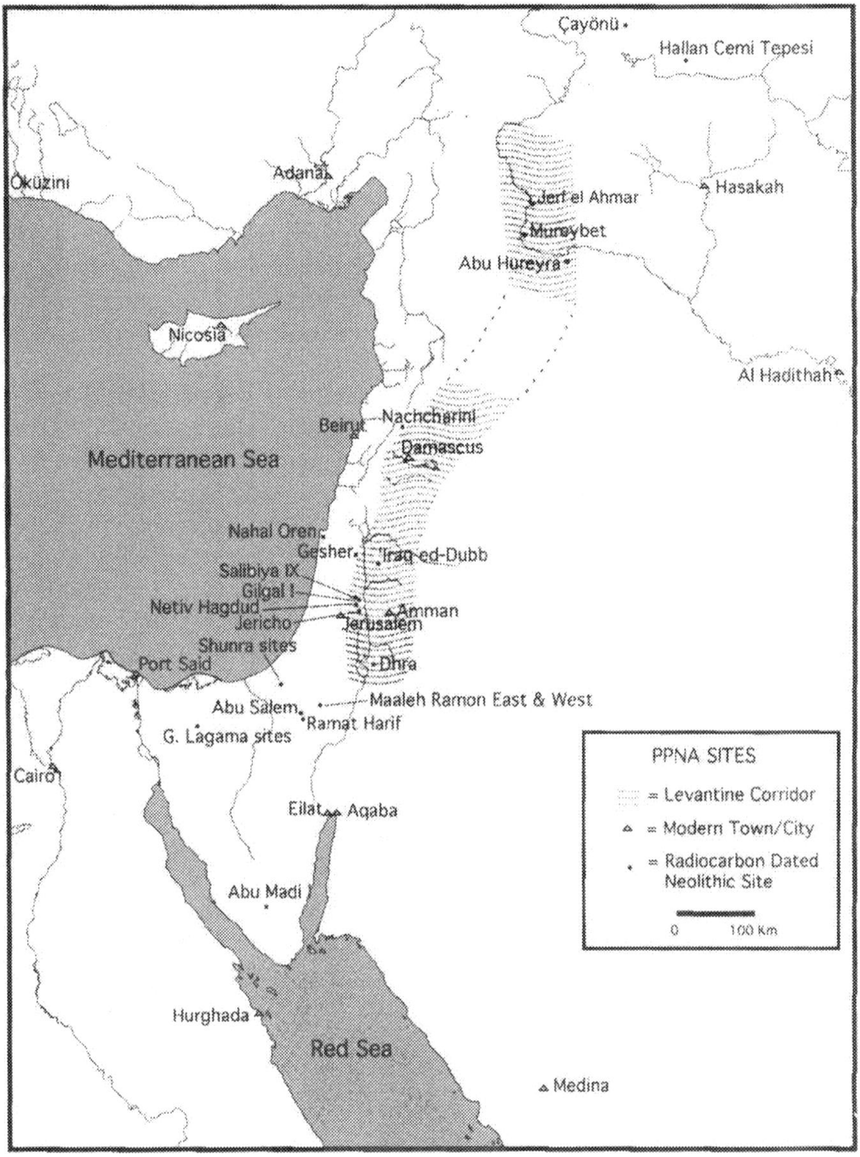

Figure 3. Map of PPNA sites in the Levant. The stippled area marks the Levantine Corridor.

former seems to have been a transitional short-term stage between the Natufian communities and the fully farming Sultanian ones. Most of the PPNA sites known are located in the Jordan Valley, and none are known in the Negev (Fig. 10.3). Regardless of possible sampling problems, it is important to note the change in settlement pattern from the Late Natufian to the PPNA. Whereas the larger Natufian sites were all located along the western ecotones, the larger PPNA sites were situated in the Jordan Valley and the smaller ones (such as Nahal Oren) on its western flanks (Bar-Yosef 1991; Kuijt 1995; Chapter 9, this volume). The abandonment of some of these areas, or at least the increased focus on the Jordan Valley, suggests that not all of the Natufians managed to transform to the next stage of sedentism and food production.

Subsistence practices of the PPNA communities were based on intensive collection or cultivation of cereals and legumes, although gathering of wild fruits and seeds as well as hunting of wild animals, particularly gazelle, continued. PPNA sites vary in size, but they are clearly larger, on average, than the Natufian settlements. Structures in the PPNA are rounded or oval and have stone foundations and superstructures of unbaked mud bricks. Evidence of communal building is illustrated by the walls and tower at Jericho, and the presence of silos—small ones (built of stone) or larger ones (built mainly of mud bricks)—found in most sites. The emergence of long-distance exchange of Anatolian obsidian in the PPNA is seen from excavations at some Jordan Valley sites, such as Jericho and Netiv Hagdud. Art also changes, with the human figure—mainly female—gaining predominance (Bar-Yosef 1991).

The subsistence strategy of food-producing communities in the PPNA still had more in common with complex foraging groups than with simple foraging ones (Woodburn 1982 in Henry 1991), as it aimed at attaining greater security through intensified resource exploitation and development of surpluses. Population growth, coupled with scarcity or depletion of resources, increases the importance of social organizations and mechanisms that principally function to "close down" and formalize access to resources and intensify their exploitation. Thus, in many ways the structure and dynamics of social organization in complex foraging societies mirror those of agricultural societies. Much of this can be observed in the archaeological record of the earliest Neolithic period in the Levant.

Studying the mechanism of cultural transmission and the viability of nuclear families helps us to understand how individuals survived the collapse of complex societies (Kosse 1994) or how they successfully moved from one social framework to another during the PPNA. It also explains how certain cultural traits and cultural units survived the collapse of the Natufian interaction sphere: unlike genetic traits, cultural elements can be adopted, and small groups can be incorporated into larger ones while preserving much of their cultural distinctiveness (Kosse 1994). Natufian innovations were only sporadically embraced by their descendants in the PPNA, although Kuijt (1996) has recently presented arguments for considerable continuity in mortuary practices between the Late Natufian and PPNA. This is as can be expected, since the mere availability of a novelty or invention does not guarantee its widespread adoption, which does not seem to follow " . . . automatically upon the inception of the new process or form but depends in a complicated way upon individual choice governed by social and other factors" (Renfrew 1978:396). It seems that whenever individuals are faced with the necessity of making a choice, they prefer to stick to what is known rather than "moving at once to a new global optimum" (Renfrew 1978:396). For one thing, opting for a novel course of action is risky, as one cannot be entirely sure of the consequences of such a choice. In addition, there is the sheer difficulty of abandoning old habits for new ones, which likewise tends to encourage "conservatism." All this is in agreement with the view that most innovations involve, at least initially, little more than new ways of doing old things (Renfrew 1978). Only later do technical developments amplify the existing divergence to such an extent that it comes to be recognized as an innovation, as for example the technology of groundstone and bone tool production (for detailed discussion see Bar-Yosef and Belfer-Cohen 1992; Bar-Yosef and Meadow 1995; Cauvin 1994; Moore 1985).

Toward the end of the Natufian and throughout the PPNA, various loci of incipient sedentism and cultivation were forming elsewhere in the Near East, wherever local conditions and the climate were appropriate (Bar-Yosef and Meadow 1995, Chapter 3, this volume). This is well illustrated in southeast Anatolia and the northen Levant by the transition from mobile hunter–gatherers to sedentary foragers in Hallan Çemi (Rosenberg and Davis 1992) or the establishment of agricultural communities in Qermez Dere (Watkins et al. 1989, 1991), and the base of Çayönü (Özdogan and Özdogan 1989). This seems to confirm the view that "residential corporate groups probably emerge under conditions where access is restricted to important resources and/or where there is frequent need for many people to exploit these resources most effectively" (Hayden 1990:38). It seems that some of the developments in the Levant were influencing the course of events in neighboring areas of the Near East through some sort of domino effect. In light of the great geographical distance between such sites as the Late Natufian Mount Carmel sites and those in northwest Iraq or southeast Turkey, it appears that these occurrences developed independently of one another. Apparently, human groups throughout this region were susceptible to these changes, the main difference being the matter of timing (determined by such factors as the degree of complexity of social structure within the particular group, the length of time that it sustained scalar stress, etc.) and local conditions (i.e., the reliability, accessibility, and distribution of specific resources, such as wild stands of cereals).

Indeed, the dimensions of time and scale are of crucial importance here. While similarities observed between the Natufian assemblages can be interpreted as indicative of some central, coherent social force, the cultural diversity observed in the southern Levant by the end of this period seems to indicate a disintegration of this social system. Similar tendencies are observed within the much later PPNB societies in this region (Bar-Yosef and Belfer-Cohen 1989b), but, contrary to the situation during the late PPNB (Rollefson 1987, 1989; Rollefson and Kohler-Rollefson 1989), in the Late/Final Natufian there is a strong feeling of decentralization in the core area and adjacent regions, as well as in regions further away. As the various vegetation zones in the Levant differ markedly in environmental conditions, there is hardly any room for speculation regarding the Natufians' alleged role as the instigators of Near Eastern sedentism (Watkins 1992). Thus, although the earliest evidence for sedentism comes from the Early Natufian core-area sites, later instances of sedentism in the Levant are sporadic and unconnected, seemingly reflecting local adaptations. Indeed, it seems that we are presented with instances of parallel cultural evolution, and the similarity observed stems from a common denominator shared by all these communities.

The fact that all these groups are basically similar (all of them being complex hunter–gatherer communities) explains why eventual similarity in environmental circumstances can be expected to have led them to similar responses, including sedentism or semisedentism. Hayden (1990) argues that complex societies are primed for agriculture, and its actual appearance is dependent merely on the availability of suitable plants for cultivation or animals to domesticate. In much the same way, it can be argued that the Northwest Coast Indian societies were ripe for agriculture at the time of the European conquest, as demonstrated by the extremely rapid spread of potatoes to all coastal groups within a brief fifteen-year period. This contrasts markedly with the frustration experienced by missionaries and government administrators who tried to introduce agriculture to more generalized hunter–gatherers in Australia, Africa and North America, to no avail. It is of interest to note that the basic inclination of human groups in the face of resource shortage has been found to be almost anything but food production. Only after all the other options were explored and found unsatisfactory, will a group resort to producing its own food. Most explanations for this phenomenon center on the relation between costs and returns in the matter of food production (Hayden 1990:35).

Viewed collectively, studying settlement practices and variation in material culture within communities helps us understand the transition from foragers to agriculturists of the Neolithic period. Economic and social stresses during the late Younger Dryas were relieved through the adoption of strategies such as aggregation, territoriality, intensified resource exploitation, and probably incipient agriculture. It is worth noting that these reactions of the Final Natufian/Early PPNA communities were different from those of the Early Natufians some 2000 years earlier. The transition to full sedentism was not smooth, and sometimes progress was hindered and reversals occurred. Thus, the Harifian culture from the Negev (dated to 10,700–10,100 bp) and the Abu Madi I entity (10,100–9,700 bp) from southern Sinai (Bar-Yosef 1985), which are partially contemporaneous with the PPNA of the Jordan Valley, preserved more elements of the core-area Natufian tradition in lithic technology and settlement pattern than have contemporaneous assemblages recovered from the core-area itself and its neighboring regions along the Levantine Corridor. These sites, located in the more marginal areas of the Levant, though having well-built dwellings and permanent installations, were occupied seasonally as part of residential moves, similar to those of historical Bedouin groups (Bar-Yosef 1984). The same is true also for the succeeding PPNB cultures, which show a clear dichotomy and time lag between full-fledged agricultural sites in the fertile lands and hunter–gatherer sites in the arid zones. While the process began with the Early Natufian, about 13,000 years ago, it is only some 6,000 radiocarbon years later that we can unequivocally state that the entire Near East is populated by agricultural societies, whether farmers, herders, or both.

REFERENCES

Agelarakis, A., 1993, The Shanidar Cave proto Neolithic human population: aspects of demography and paleopathology. *Human Evolution* 8: 235–253.

Auffray, J.-C., Tchernov, E. and Nevo, E., 1988, Origine du commensalisme de la souris domestique (*Mus musculus domesticus*) vis-à-vis de l'homme. *Comptes Rendus de l'Academie des Sciences Paris* 307: 517–522.

Baruch, U., and Bottema, S., 1991, Palynological evidence for climatic changes in the Levant ca. 17,000–9,000 B.P. In *The Natufian Culture in the Levant*, edited by O. Bar-Yosef and F. R. Valla, pp. 11–20. International Monographs in Prehistory, Ann Arbor, Michigan.

Bar-Yosef, O., 1983, The Natufian of the southern Levant. In *The Hilly Flanks and Beyond,* edited by C. T. Young, P. E. L. Smith, and P. Mortensen, Studies in Ancient Oriental Civilization, pp. 11–42. University of Chicago Press, Chicago, Illinois.

Bar-Yosef, O., 1984, Seasonality among Neolithic Hunter-Gatherers in Southern Sinai. In *Animals and Archaeology: Herders and Their Flocks,* edited by J. Clutton-Brock and C. Grigson, pp. 145–160. British Archaeological Reports International Series 202, Oxford.

Bar-Yosef, O., 1985, The Stone Age of the Sinai Peninsula. In *Studi di Paleontologia in Onore i Salvatore M. Puglisi,* edited by M. Liverani, A. Palmieri, and P. Peroni, pp. 107–122. Universita di Rome "La Sapienza," Rome.

Bar-Yosef, O., 1987, The Late Pleistocene in the Levant. In *The Pleistocene Old World: Regional Perspectives,* edited by O. Soffer, pp. 219–236. Plenum Press, New York.

Bar-Yosef, O., 1998, The Natufian Culture in the Levant, threshold to the Origins of Agriculture. *Evolutionary Anthropology* 6(5): 159–177.

Bar-Yosef, O., 1991, The Early Neolithic of the Levant: Recent Advances. *The Review of Archaeology* 12:1–18.

Bar-Yosef, O., 1992, The Neolithic Period. In *The Archaeology of Ancient Israel,* edited by A. Ben-Tor, pp. 10–39. Yale University Press, New Haven, Connecticut.

Bar-Yosef, O., 1997, Symbolic Expressions in Later Prehistory of the Levant—Why Are They So few? In *Beyond Art: Pleistocene Image and Symbol,* edited by M. D. Conkey, O. Soffer, and D. Stratman, pp. 161–187. California Academy of Sciences, San Francisco, California.

Bar-Yosef, O., and Belfer-Cohen, A., 1989a, The Origins of Sedentism and Farming Communities in the Levant. *Journal of World Prehistory* 3: 447–498.

Bar-Yosef, O., and Belfer-Cohen, A., 1989b, The Levantine "PPNB" Interaction Sphere. In *People and Culture in Change,* edited by I. Hershkovitz, pp. 59–72. British Archaeological Reports International Series 508, Oxford.

Bar-Yosef, O., and Belfer-Cohen, A., 1991, From Sedentary Hunter-Gatherers to Territorial Farmers in the Levant. In *Between Bands and States,* edited by S. A. Gregg, pp. 181–202. Southern Illinois University, Carbondale, Illinois.

Bar-Yosef, O., and Belfer-Cohen, A., 1992, From Foraging to Farming in the Mediterranean Levant. In *Transitions to Agriculture in Prehistory,* edited by A. B. Gebauer and T. D. Price, pp., 21–48. Prehistory Press, Madison, Wisconsin.

Bar-Yosef, O., and Meadow, R. H., 1995, The Origins of Agriculture in the Near East. In *Last Hunters–First Farmers: New Perspectives on the Prehistoric Transition to Agriculture,* edited by T. D. Price and A. B. Gebauer, pp. 39–94. School of American Research Press, Santa Fe, New Mexico

Bar-Yosef, O., and Tchernov, E., 1970, The Natufian Bone Industry of Hayonim Cave. *Israel Exploration Journal* 20: 141–50.

Bar-Yosef, O., Arensburg, B., and Tchernov, E., 1974, Hayonim Cave: Natufian Cemetery and Settlement Remains. In *Bemaaravo Shell Hagalil,* edited by M. Yeda'aya. Regional Councils of Sulam Zor and Ga'aton (in Hebrew).

Belfer-Cohen, A., 1988, The Appearance of Symbolic Expression in the Upper Pleistocene of the Levant as Compared to Western Europe. In *L'Homme de Neandertal,* Volume 5, La Pense, edited by M. Otte, pp. 25–29. ERAUL 32. Etudes et recherches aarchéologiques de l'Université de Liège, Liège.

Belfer-Cohen, A., 1989a, The Natufian Graveyard in Hayonim Cave. *Paléorient* 14: 297–308.

Belfer-Cohen, A., 1989b, The Natufian issue: A Suggestion. In *Investigations in South Levantine Prehistory,* edited by O. Bar-Yosef and B. Vandermeersch, pp. 297–307. British Archaeological Reports International Series 497, Oxford.

Belfer-Cohen, A., 1991a, Art in Hayonim Cave, a Case Study of Art in a Natufian Context. In *The Natufian Culture in the Levant,* edited by O. Bar-Yosef and F. R. Valla, pp. 569–588. International Monographs in Prehistory, Ann Arbor, Michigan.

Belfer-Cohen, A., 1991b, The Natufian in the Levant. *Annual Review of Anthropology* 20: 167–186.

Belfer-Cohen, A., 1995, Rethinking Social Stratification in the Natufian Culture: the Evidence from Burials. In *The Archaeology of Death in the Ancient Near East,* edited by S. Campbell and A. Green, pp. 9–16. Oxbow Monographs 51. Oxbow Books, Oxford..

Belfer-Cohen, A., Schepartz, L. A., and Arensburg, B. 1991, New Biological Data for the Natufian Populations in Israel. In *The Natufian Culture in the Levant,* edited by O. Bar-Yosef and F. R. Valla, pp. 411–424. International Monographs in Prehistory, Ann Arbor, Michigan.

Bouchud, J. (ed.) 1987, *La Fauna du Gisement Natoufien de Mallaha (Eynan) Israel*. Memoires et Travaux du centre de Recherche Francaise de Jerusalem 4. Association Paléorient, Paris.

Byrd, B. F., 1989, The Natufian: Settlement Variability and Economic Adaptations in the Levant at the End of the Pleistocene. *Journal of World Prehistory* 3: 159–197.

Byrd, B. F., 1994a, Late Quaternary hunter-gatherer complexes in the Levant between 20,000 and 10,000 BP. In *Late Quaternary Chronology and Paleoclimates of the Eastern Mediterranean,* edited by O. Bar-Yosef and R. S. Kra, pp. 205–226. Tucson, Arizona.

Byrd, B. F., 1994b, From Early Humans to Farmers and Herders—Recent Progress on Key Transitions in Southwest Asia. *Journal of Archaeological Research* 2: 221–253.

Byrd, B. F., and C. M. Monahan, 1995, Death, Mortuary Ritual, and Natufian Social Structure. *Journal of Anthropological Archaeology* 14: 251–287.

Cauvin, J. 1994, *Naissance des Divinités, Naissance de L'Agriculture.* CNRS, Paris.

Cauvin, M. C., 1991, Du Natoufien au Levant nord? Jayroud et Mureybet (Syrie). In *The Natufian Culture in the Levant,* edited by O. Bar-Yosef and F. R. Valla , pp. 295–314. International Monographs in Prehistory, Ann Arbor, Michigan.

Cohen, M. N., 1985, Prehistoric Hunter-Gatherers: The Meaning of Social Complexity. In *Prehistoric Hunter-Gatherers: The Emergence of Cultural Complexity,* edited by T. D. Price and J. A. Brown, pp. 99–22. Academic Press, New York.

Cope, C., 1991, Gazelle Hunting Strategies in the Southern Levant. In *The Natufian Culture in the Levant,* edited by O. Bar-Yosef and F. R. Valla, pp. 341–458. International Monographs in Prehistory, Ann Arbor, Michigan.

Davis, S., 1983, The Age Profiles of Gazelles Predated by Ancient Man in Israel. *Paléorient* 9: 57–64.

Dayan, T., and Simberloff, D., 1995, Natufian Gazelles: Proto-Domestication Reconsidered. *Journal of Archaeological Science* 22: 671–676.

Edwards, P. C., 1987, *Late Pleistocene Occupation in Wadi al-Hammeh, Jordan Valley.* Unpublished Ph.D. Thesis, University of Sydney, Sydney.

Fish, S. K., and Fish, P. R., 1991, Comparative Aspects of Paradigms for the Neolithic Transition in the Levant and the American Southwest. In *Perspectives on the Past. Theoretical Biases in Mediterranean Hunter-Gatherers Research,* edited by G. A. Clark, pp. 396–410. University of Pennsylvania Press, Philadelphia, Pennsylvania.

Flannery, K. V., 1973, The origins of agriculture. *Annual Review of Anthropology* 2: 271–310.

Garrod, D. A. E., 1932, A New Mesolithic Industry: The Natufian of Palestine. *Journal of Royal Anthropological Institute* 62: 257–269.

Garrod, D. A. E., and Bate, D. M. A., 1937, *The Stone Age of Mount Carmel* (Vol. 1). Clarendon Press, Oxford.

Garrod, D. A. E., 1957, The Natufian Culture: The Life and Economy of a Mesolithic People in the Near East. *Proceedings of the British Academy* 43: 211–227.

Goring-Morris, A. N., 1987, *At the Edge: Terminal Pleistocene Hunter-Gatherers in the Negev and Sinai.* British Archaeological Reports International Series 361, Oxford.

Goring-Morris, A. N., 1991, The Harifian of the Southern Levant. In *The Natufian Culture in the Levant,* edited by O. Bar-Yosef and F. R. Valla, 173–216. International Monographs in Prehistory, Ann Arbor, Michigan.

Grognier, E. and Dupouy-Madre, M., 1974, Les Natoufiens du Nahal Oren, Israel. Etude anthropologique. *Paléorient* 2: 103–121.

Hayden, B., 1990, Nimrods, Piscators, Pluckers, and Planters: the Emergence of Food Production. *Journal of Anthropological Archaeology* 9: 31–69.

Henry, D. O., 1985, Preagricultural Sedentism: The Natufian Example. In *Prehistoric Hunter-Gatherers: The Emergence of Cultural Complexity,* edited by T. D. Price and J. A. Brown, pp. 365–384. Academic Press, New York.

Henry, D., 1989, *From Foraging to Agriculture. The Levant at the End of the Ice Age.* University of Pennsylvania Press, Philadelphia, Pennsylvania.

Henry, D., 1991, Foraging, Sedentism, and Adaptive Vigor in the Natufian: Rethinking the Linkages. In *Perspectives on the Past. Theoretical Biases in Mediterranean Hunter-Gatherers Research,* edited by G. A. Clark, pp. 353–370. University of Pennsylvania Press, Philadelphia, Pennsylvania.

Henry, D., 1995, *Prehistoric Cultural Ecology and Evolution.* Plenum Press, New York.

Hietala, H. J., and Stevens, D. E., 1977, Spatial Analysis: Multiple Procedures in Pattern Recognition Studies. *American Antiquity* 42: 539–559.

Hillman, G. C., and Davies, M. S., 1990, Measured Domestication Rates in Wild Wheats and Barley under Primitive Cultivation, and Their Archaeological Implications. *Journal of World Prehistory* 4: 157–222.

Hillman, G. C., Colledge, S., and Harris, D. R., 1989, Plant Food Economy During the Epi-Palaeolithic Period at Tell Abu Hureyra, Syria: Dietary Diversity, Seasonality and Modes of Exploitation. In *Foraging and Farming: The Evolution of Plant Exploitation,* edited by G. C. Hillman and D. R. Harris, pp. 240–266. Unwin Hyman, London.

Keeley, L. H., 1988, Hunter-Gatherer Economic Complexity and Population Pressure: A Cross-Cultural Analysis. *Journal of Anthropological Archaeology* 7: 373–411.

Kosse, K., 1994, The Evolution of Large, Complex Groups: A Hypothesis. *Journal of Anthropological Archaeology* 13: 35–50.

Kuijt, I., 1995, Pre-Pottery Neolithic A Settlement Variability: Evidence for Sociopolitical Developments in the Southern Levant. *Journal of Mediterranean Archaeology* 7: 165–192.

Kuijt, I., 1996, Negotiating Equality through Titual: A Consideration of Late Natufian and Prepottery Neolithic A Period Mortuary Practices. *Journal of Anthropological Archaeology* 15: 313–336.

Lieberman, D. E., 1993, The Rise and Fall of Seasonal Mobility among Hunter-Gatherers: The Case of the Southern Levant. *Current Anthropology* 34: 599–632.

Lieberman, D. E., and Shea, J. J., 1994, Behavioural Differences between Archaic and Modern Humans in the Levantine Mousterian. *American Anthropologist* 96: 300–332.

Moore, A. M. T., 1985, The Development of Neolithic Societies in the Near East. In *Advances in World Prehistory, Volume 4,* edited by F. Wendorf and A. Close, pp. 1–69. Academic Press, New York.

Moore, A. M. T., and Hillman, G. C., 1992, The Pleistocene to Holocene Transition and Human Economy in Southwest Asia: The Impact of the Younger Dryas. *American Antiquity* 57: 482–494.

Özdogan, M., and Özdogan, A., 1989, Çayönü: A Conspectus of Recent Work. *Paléorient* 15: 65–74.

Perrot, J., Ladiray, D., and Soliveres-Massei, O., 1988, *Les Hommes de Mallaha (Eynan) Israel.* Memoires et Travaux du Centre de Recherche Francais de Jerusalem No. 71. Association Paleorient, Paris.

Renfrew, C., 1978, The Anatomy of Innovation. In *Approaches to Social Archaeology,* edited by C. Renfrew, pp. 390–419. Harvard University Press, Cambridge, Massachusetts.

Rollefson, G., 1987, Local and External Relations in the Levantine PPN Period: 'Ain Ghazal as a Regional Center. In *Studies in the History and Archaeology of Jordan III,* edited by A. Hadidi, pp. 29–32. Department of Antiquities, Amman.

Rollefson, G., 1989, The Aceramic Neolithic of the Southern Levant: The View from 'Ain Ghazal. *Paléorient* 15: 135–140.

Rollefson, G., and Köhler-Rollefson, I., 1989, The collapse of Early Neolithic Settlements in the Southern Levant. In *People and Culture in Change: Proceedings of the Second Symposium on Upper Paleolithic, Mesolithic and Neolithic Populations of Europe and the Medi-*

terranean Basin, edited by I. Hershkovitz, pp. 73–89. British Archaeological Reports International Series 508, Oxford.

Rosenberg, M., and Davis, M., 1992, Hallan Çemi Tepesi, an Early Aceramic Neolithic Site in Eastern Anatolia: Some Preliminary Observations Concerning Material Culture. *Anatolica* 18: 1–18.

Shnirelman, V. A., 1992, Complex Hunter-Gatherers: Exception or Common Phenomenon? *Dialectical Anthropology* 17: 183–196.

Stordeur, D. 1981, La Contribution de l'Industrie de l'os à la Delimitation des Aires Culturelles: l'Exemple du Natoufien. In *Préhistoire du Levant,* edited by J. Cauvin and P. Sanlaville, pp. 433–437. CNRS, Paris.

Tchernov, E., 1991, Biological Evidence for Human Sedentism in Southwest Asia during the Natufian. In *The Natufian Culture in the Levant,* edited by O. Bar-Yosef and F. R. Valla, pp. 315–340. International Monographs in Prehistory, Ann Arbor, Michigan.

Unger-Hamilton, R., 1991, Natufian Plant Husbandry in the Southern Levant and Comparison with That of the Neolithic Periods: The Lithic Perspective. In *The Natufian Culture in the Levant,* edited by O. Bar-Yosef and F. R. Valla, pp. 483–520. International Monographs in Prehistory, Ann Arbor, Michigan.

Valla, F. R., 1981, Les etablissements Natoufiens dans le Nord d'Israel. In *Prehistoire de Levant,* edited by J. Cauvin and P. Sanlaville, pp. 409–419. CNRS, Paris.

Valla, F. R., 1987, Chronologie Absolue et Chronologie Relative dans le Natoufien. In *Chronologies in the Near East,* edited by O. Aurenche, J. Evin and F. Hours, pp. 367–394. British Archaeological Reports International Series 379, Oxford.

Valla, F. R., 1991, Les Natoufiens de Mallaha et l'espace. In *The Natufian Culture in the Levant,* edited by O. Bar-Yosef and F. R. Valla, pp. 111–122. International Monographs in Prehistory, Ann Arbor, Michigan.

Watkins, T., 1992, The Beginning of the Neolithic: Searching for Meaning in Material Culture Change. *Paléorient* 18: 63–76.

Watkins, T., Baird, D., and Betts, A., 1989, Qermez Dere and the Early Aceramic Neolithic of Northern Iraq. *Paléorient* 15: 19–24.

Watkins, T., Betts, A., Dobney, K., Nesbitt, M., Gale, R. and Molleson, T., 1991, *Qermez Dere, Tell Afar—Interim Report No. 2.* Project Paper No. 13, Department of Archaeology, University of Edinburgh, Edinburgh.

Chapter 3

Hallan Çemi and Early Village Organization in Eastern Anatolia

Michael Rosenberg and Richard W. Redding

INTRODUCTION

However one chooses to define cultural complexity (e.g., Flannery 1972a; Service 1978; McGuire 1983), there was a time not so long ago when expectations concerning Neolithic lifeways were that they were anything but complex. Thus, the large size of the then newly discovered Neolithic sites of Jericho (Kenyon 1957) and Çatal Höyük (Mellaart 1967) was genuinely surprising, as were the public structures they contained and the elaborate ritual life they evidenced. Since then there has developed an increasingly sophisticated understanding of the factors producing cultural complexity in hitherto simpler societies (e.g., Price and Brown 1985; Upham 1990). This has produced the expectation that at least some minimal level of social complexity is present in virtually any sedentary society. Now, even the possible

MICHAEL ROSENBERG • Department of Anthropology, University Parallel Program, University of Delaware, Newark, Delaware 19716. **RICHARD W. REDDING** • Museum of Anthropology, University of Michigan, Ann Arbor, Michigan 48109.

Life in Neolithic Farming Communities: Social Organization, Identity, and Differentiation, edited by Ian Kuijt. Kluwer Academic/Plenum Publishers, New York, 2000.

existence of hereditary status distinctions in the Levantine Natufian is the subject of legitimate debate (pro: Henry 1989; Wright 1978; con: Byrd and Monahan 1995; Kuijt 1996; Olszewski 1991).

General expectations, however, are not the same as a clear understanding of the specific configuration social complexity took at a given site or in a given prehistoric culture. This is no minor point, given recent suggestions that social forces—themselves implicitly predicated on the existence of a specific sociopolitical and economic milieu—were a factor in the development of both sedentary lifeways and food production (e.g., Bender 1978, 1985; Blanton and Taylor 1995; Hayden 1990, 1992; see also Marquardt 1985). Unfortunately, such a clear understanding of social specifics is generally lacking for Neolithic sites in southwestern Asia, both individually and collectively. This is because, with only a few very recent exceptions (e.g., Byrd 1994; Kuijt 1995), archaeologists investigating Neolithic social structure have tended to focus primarily on evidence for hereditary status distinctions and similar proxies for emergent political complexity (e.g., Henry 1989; Wright 1978). In keeping with the general theme of this volume, this chapter focuses on the general structure and organization of the society that once inhabited Hallan Çemi, a proto-Neolithic site in the Taurus foothills of eastern Anatolia.

Hallan Çemi is the oldest fully settled village site thus far known from eastern Anatolia. It represents the remains of an essentially sedentary hunter–gatherer society, albeit one on the threshold of animal domestication, that inhabited the upper reaches of the Tigris toward the end of the eleventh millennium bp (uncalibrated). More to the point, Hallan Çemi has yielded an interesting array of material remains relating to the possible structure and organization of the society that once inhabited it (see Rosenberg and Davis 1992; Rosenberg 1994a, 1994c; Rosenberg et al., 1995).

Of particular interest are the elements strongly suggesting that the site's inhabitants engaged in recurrent activities involving the conspicuous, formalized preparation and consumption of food—in other words, feasting (see Hayden 1995). Of even greater interest are the other elements suggesting that any such feasting was less a medium of competition than one for establishing cooperative relationships in an otherwise highly competitive environment (cf. Rosenberg n.d.). That is, they suggest that any such feasting was more in the Yanomamo mold (see Chagnon 1983) than in the form proposed by Hayden (1990, 1992).

Finally, the site data suggest that the community that once inhabited this site already exhibited at least rudimentary forms of the organizational features that characterize the later, fully developed Neolithic societies of the region. This is despite Hallan Çemi's relatively small size and generally proto-Neolithic economy. That, in turn, suggests that the basic structure of

Neolithic societies emerged during the very earliest stages of the trend toward the development of food producing economies (i.e., with sedentism), not differentially in tandem with a growing dependence on food production (see Rosenberg 1994b). The reason for this is apparently that such organizational features are necessary for sedentary life, whatever type of subsistence base it is predicated on.

ORGANIZATIONAL COMPLEXITY AND SEDENTARY SOCIETIES

The increased complexity typically evident in the social systems of sedentary hunter–gatherer and simple food-producing societies is largely due to two departures from typical mobile hunter–gatherer norms. The first is surplus collection or production, minimally on a seasonal basis, in the context of some form of delayed return system (cf. Woodburn 1980, 1982, 1988; Southall 1988; see also Byrd 1994). These seasonal surpluses, when coupled with storage, are what typically permit year-round occupation of a site to begin with (see Cohen 1985; Price and Brown 1985; Redding 1988; Testart 1982). Such temporary surpluses, to the degree that they are not fully consumed within the annual subsistence cycle that created them, have the potential to become true surpluses. Thus, they become potentially available for other socioeconomic and sociopolitical purposes, such as those suggested by Hayden (1990, 1992).

The second departure is the institutionalization of (subsistence) resource-related ownership systems, a prerequisite for the establishment of delayed return systems, such as necessary for storage, not to mention food production (Rosenberg 1990). The initial establishment of such (ownership based) delayed return systems will tend to precipitate the collapse of the social systems that make a mobile hunting–gathering adaptation functionally possible (Chapter 2, this volume; Rosenberg 1990, 1994b; Yellen 1985: 47–48)—ones characterized by high group and individual mobility, immediate returns, and resource sharing based on generalized reciprocity, etc. This collapse perforce leads to the rapid development of new socioeconomic and sociopolitical structures that can accommodate the concept of subsistence resources as property and incidentally make a sedentary lifeway functionally possible. As ably summarized by Byrd (1994:642), these would of necessity include new, more formal socioeconomic structures, presumably based on at least some degree of balanced reciprocity. These new institutions would function to integrate the multiple, new, individually more restrictive social networks for sharing production and consumption that replace the now-defunct single, all-embracing generalized reciprocal system characteristic of mobile hunter–gatherer groups (cf. Flannery 1972b;

Netting 1990; Plog 1990; Wilson 1988; Winterhalder 1990). Also included would be new sociopolitical structures that function to more effectively resolve conflicts within the typically larger and, of necessity, generally more stable groups that characterize village systems (cf. Adler and Wilshusen 1990; Flannery 1972b; Wilson 1988). Lastly, implicit in the existence of any such new structures is the existence of new superstructural elements that symbolize and legitimize the new structures, as well as function to further integrate the various social elements of the community as a whole. The territoriality/ownership statements implicit in many mortuary practices (cf. Saxe 1970; Goldstein 1976) are just one potential case in point.

HALLAN ÇEMI: AN OVERVIEW

Hallan Çemi is situated at an altitude of *ca.* 640 m on the west bank of the Sason Çayi, a tributary of the Batman Çayi and the Tigris, respectively. It is a small, roughly 4.3-m high mound, about 0.7 ha in area, of which less than 0.5 ha is covered by an aceramic occupation dating to the last few hundred years of the eleventh millennium bp (uncalibrated). To date, approximately 750 m^2 of this aceramic occupation have been exposed to depths of between about 0.5 and 3 m. Though at least four building levels are known to exist, architectural remains from only the uppermost three have thus far been excavated to any meaningful degree (Figure 1). The site was apparently occupied year-round (Rosenberg 1994a:130).

The economy of the site's inhabitants was based primarily on hunting and gathering. Almonds, pistachios, and pulses were apparently the most intensively utilized wild plant resources. Wild sheep and deer were the most intensively utilized animal resources, constituting *ca.* 36% and 27%, respectively, of the mammalian remains in the faunal assemblage. Sheep/goat remains collectively constitute *ca.* 43% of the mammalian bone. Among those bones types where the difference between sheep and goats can be distinguished, sheep outnumber goats by about 6:1. On the assumption that the ratio of sheep to goats in these particular categories of bone is representative of the larger body of sheep/goat bone, sheep constitute *ca.* 36% of the mammalian bone in the faunal assemblage. Pig was a significantly smaller component of their meat diet, constituting only *ca.* 12% of the mammalian remains. However, judging from the molar sizes (cf. Flannery 1982), butchering patterns, sex ratios, and survivorship curves, the site's inhabitants engaged in some degree of pig husbandry by at least the last building level (see Redding n.d.; Rosenberg et al. 1995:5).

In the three uppermost building levels (and very likely in the fourth as well) the spatial layout of the community consisted of a variety of structures

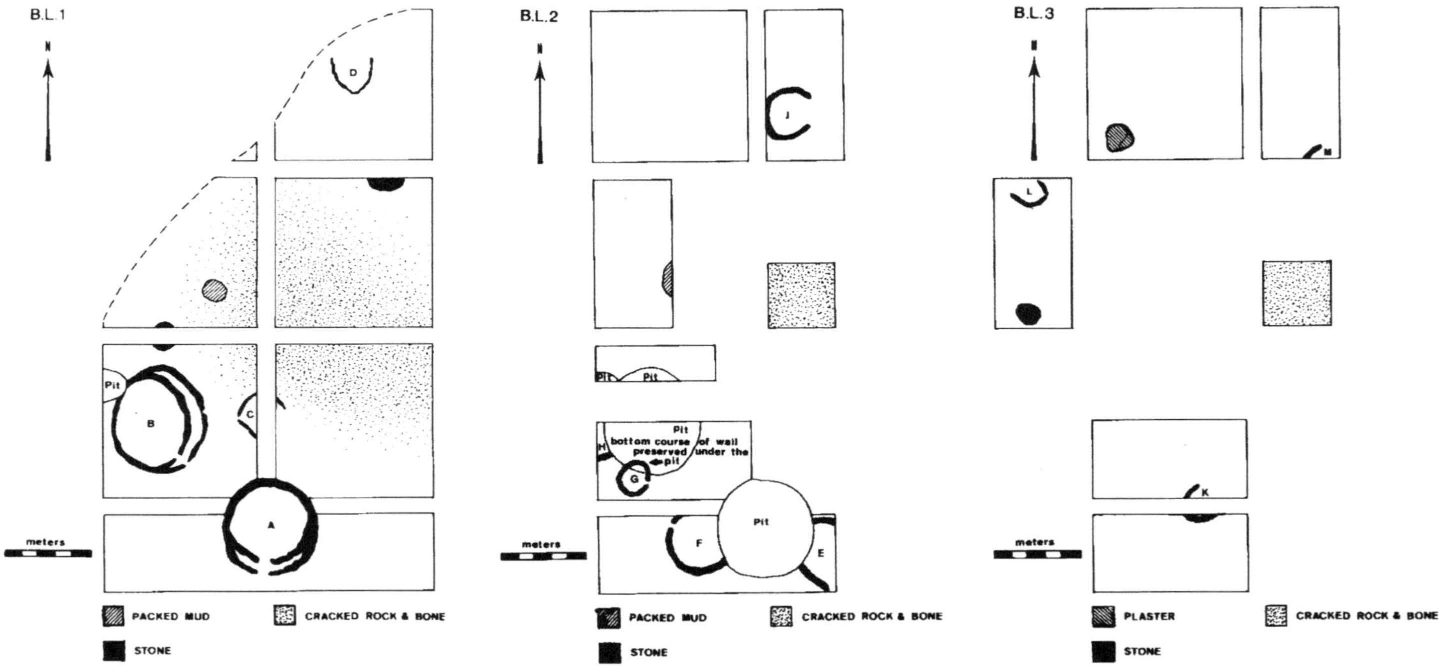

Figure 1. Sketch plans of uppermost three building levels with structures and platforms indicated.

and features arranged around an open central activity area over 15 m in diameter. This central area is devoid of architectural features. It is also generally devoid of large-scale food preparation equipment such as stone mortars and grinding slabs, while such hand-held grinding stones as do occur in this area tend to be fire-cracked. The deposits in this open central area are generally characterized by inordinately dense concentrations of animal bone and fire-cracked river pebbles/cobbles. The animal bone in this area is often in the form of large, still-articulated portions of animal carcasses and includes a linear arrangement of three sheep crania on one of the surfaces associated with the uppermost building level. What fragmentation the bones from this area generally exhibit is more a product of postdepositional processes (most commonly excavation) than of the processes that lead to their deposition. This contrasts with the generally more fragmentary nature of the bone found in the surrounding areas. The central activity area is surrounded by a variety of features, including curvilinear structures and circular platforms. The recognizable (i.e., stone) structures vary in size and construction, in some respects by building level. Storage pits are apparently absent in all levels of the site, as are human burials.

A total of four recognizable stone structures is thus far known from the uppermost building level (1) and all utilized sandstone slabs to one degree or another in their construction. Two of them are relatively small surface structures (Figure 1C, D), about 2.5 m in diameter, and U- or C-shaped in configuration. In the case of both these two small stone structures, all that remains is one course of sandstone slab orthostats enclosing an area that, in at least the case of one, contained a poorly preserved plaster feature that is presumably a hearth, such as are known from better preserved structures in the lower building levels. A feature that may represent a small, clay-walled surface structure is also known from this level, but will be omitted from further discussion at this time due to its uncertain status. It is, however, similar in size and configuration to these smaller stone structures.

The remaining two stone structures attributable to this building level are larger (between 5 and 6 m in diameter), fully circular, and semisubterranean in their construction (Figure 1A, B). In both cases, walls constructed of sandstone slabs lined the pit portion of the house and extended up beyond it. Within the pit portion, these walls were constructed of flat-lying coursed slabs and orthostats, alone or in combination. The free-standing above-ground portions of the walls were uniformly of coursed slab construction. Regularly spaced gaps in the walls presumably held roof supports; a small stone feature in the center of each floor may have been a foundation for a central support. Other interior features included plaster hearths on the floors and semicircular stone benches/platforms against the walls. In contrast to the smaller structures, the floor surfaces within these two structures were finished with a thin sand and plaster mixture and were each resurfaced multiple times (Figure 2).

Both plant and animal remains (skulls and antlers aside) were very sparse within the two larger structures, as were objects that could reasonably be interpreted as ground stone food processing equipment. On the other hand, copper ore (presumably used for pigment) does occur at the site, and virtually all the fragments of copper ore thus far discovered were found in and around these two semisubterranean structures. Also, the two largest obsidian blade cores thus far found were discovered on a surface within one of these structures (Figure 1A) in association with the only clearly identifiable obsidian knapping area thus far identified in the entire site. The copper ore and obsidian were imported materials, something that limited their availability but made them relatively valuable (Rosenberg 1994c).

Also found within the same (i.e., Figure 1A) semisubterranean structure as the relatively large obsidian cores was a complete aurochs skull (sans mandible) that appears to have once hung on its north wall, facing the entrance (see Rosenberg 1994a). This skull aside, bovid remains are apparently absent among the faunal remains (see Rosenberg et al. 1995), meaning that for whatever reason bovids were very rarely if ever actually used as a food animal. This, coupled with its position, strongly suggests that the above-

Figure 2. Large semisubterranean structure (A) in building level 1.

mentioned aurochs skull had symbolic significance. Several partially preserved sheep skulls and several deer antlers were found on surfaces in the other semisubterranean structure; unfortunately, it is not clear whether they too once hung on the walls of that building.

Four recognizable structures are thus far known from building level 2. The upper part of a possible fifth (Figure 1E) structure has also been excavated, but little can yet be said about it beyond the fact that its walls are constructed in the same fashion as the other four. The four structures that have been excavated down to their floors are all surface structures, with walls constructed of river pebbles/cobbles cemented in courses with a white plaster-type substance. Three of these four level 2 structures are cut by the level 1 semisubterranean structures, making it unclear whether they were fully circular or C-shaped (Figure 1F, G, H). All three of these structures had floors paved with closely fitted sandstone slabs. These three structures varied in diameter from *ca.* 4 to *ca.* 2 m and none appeared to contain raised plaster hearths. In the case of the two smaller ones (Figure 1G, H), this may be a product of preservation, as the floors were only partially preserved. In the case of the largest (Figure 1F), this may be due to the obscuring effect of the plaster–stone debris that covered the paved surface. It was produced when this structure's walls were leveled—it appears purposefully—preparatory to the construction of the level 1 semisubterranean structures. This largest paved structure, however, did have a plaster-lined depression at its center. No other clearly recognizable features were evident on the floors of these three structures. The fourth building level 2 structure (Figure 1J) is clearly C-shaped and *ca.* 3 m in diameter. It was located across the central activity area from the three above-described buildings. It did not have a stone-paved floor and it did contain a raised plaster hearth.

Three recognizable structures attributable to building level 3 are known thus far (Figure 1K, L, M), though only one fell completely within the excavation and building L was very poorly preserved and may be associated with level 2. All three are surface structures, about 2 m in diameter, with walls constructed in the same fashion as the level 2 structures. All three are apparently C-shaped and their floors are unpaved. No plaster hearths were noted, but this could be a product of partial excavation or poor preservation. In addition to the recognizable stone structures and exterior to them are several expanses of thick plaster that are clearly surfaces. They occur in all levels and, as found, are of varying sizes and configurations. Some are associated with one or more recognizable postholes; others are not. Also, some clearly abut the known stone structures and may represent sheltered extensions or exterior activity areas; others are just as clearly not associated with any of the stone structures. These latter surfaces may or may not represent the interiors of freestanding structures made of materials less durable than stone.

The structures aside, circular platforms also occur at the site (Figure 3). These platforms average about 40 cm in preserved height and vary in diameter from slightly under 1 m to almost 2 m. They are constructed of stone, packed mud, or a plaster like material. The stone examples are often mud-plastered and constructed of either solid stone or a stone exterior enclosing earth fill. The function of these circular platforms remains unclear. However, Hallan Çemi was clearly occupied year-round and, as noted, storage pits are absent. Year-round occupation of the site almost certainly required some kind of storage facilities. Thus, for the moment, the most plausible explanation for these platforms is that they are the foundations for silos or similar such above-ground storage facilities. If, as it appears, pigs were kept at the site, the penchant pigs have for rooting around in the ground makes such above-ground storage facilities a very sensible alternative to storage pits.

COMMUNITY ORGANIZATION

Hallan Çemi was clearly not a particularly large community, as is perhaps to be expected of a society not far removed from a mobile hunting–gathering lifeway. The site, at under 0.5 ha, is not particularly large and the number of

Figure 3. Stone platform (building level 1).

clearly identifiable structures, at no more than five per several hundred square excavated meters in each level, is also relatively small. Even if we make the highly debatable assumption that all of the exterior plaster expanses that do not abut stone structures are in fact the traces of other freestanding structures, it is still very hard to justify the premise that the site encompassed a total number of structures much higher than the teens at any one time. Thus, it seems unlikely that Hallan Çemi was inhabited by a community very much larger than that thought to characterize mobile hunting–gathering bands (cf. Lee and DeVore 1968). However, despite the community's relatively small size and essentially hunter–gatherer subsistence base, certain key elements that characterize the organization of later, larger, Neolithic sites (e.g., Çayönü, Beidha) are already evident at Hallan Çemi.

At the most basic organizational level, the entrances to structures in all levels of the site generally face away from the central (i.e., communal) activity area (Figure 1). As noted by Yellen (1985, 1990), with respect to similar layouts in the camps of some !Kung groups that have recently begun to practice pastoralism, this is a departure from the layout characteristic of mobile hunter–gatherer camps. Yellen suggests that it is adopted to enhance household privacy and associated with the abandonment of generalized reciprocal sharing of the type that characterizes mobile hunting–gathering lifeways and its replacement by the more restrictive networks for sharing (see also Byrd 1994:649) that develop in response to the institutionalization of private economic property.

At the community level, public buildings and the new social structures they represent are a consistent feature of later Anatolian Neolithic sites, such as Çayönü (Özdogan and Özdogan 1989; Schirmer 1990) and Nevali Çori (Hauptmann 1993). In the Levant they are also present, not just at later sites, such as Beidha (Byrd 1994), but at very early sites, such as 'Ain Mallaha (Perrot 1966) and the PPNA tower of Jericho, which Bar-Yosef (1986) and Kuijt (1996) view as a communal constructions probably employed for ritual. Therefore, it is not surprising that public structures are also apparently present at Hallan Çemi. While the case for the large level 2 paved structure (Figure 1F) being a public building is relatively weak, revolving primarily around its size (but see Byrd 1994:646), the case for the two large semisubterranean buildings (Figure 1A, B) in level 1 being public buildings is much stronger (see Rosenberg 1994a; Rosenberg et al. 1995). First, they are approximately three times the size of the contemporary smaller stone structures and similar in area to what Byrd (1994:646) proposes to be public buildings at Beidha. Second, the interiors are also distinctive, with frequent resurfacing of the floor and distinctive architectural features such as platforms/benches. There is also the general dearth of food-processing equipment in these structures and the presence within one of the only clearly identifiable obsidian-processing area in the site. The occurrence in the vicinity of these two structures of virtually all the copper ore fragments found at the site is also a point of distinction and a further indication that the activities carried out in and around them were not domestic in nature. Lastly, the obviously symbolic nature of the aurochs skull that once hung opposite the entrance to one suggests that the group using this building was sufficiently important to be symbolized. The presence within the other semisubterranean structure of several sheep skulls (whether or not they once hung on its walls) is also intriguing. As noted, three other such skulls were found neatly aligned in the central activity area (Figure 4). Whatever the meaning of this open-air skull alignment, it was produced by a purposeful public act, raising the possibility that the sheep skulls found in this second public structure also reflect some facet of the public domain (see Goring-Morris, Kuijt, and Rollefson, all this volume).

Byrd (1994) suggests that public buildings served as elements of the more formal sociopolitical structures that emerged with the beginning of settled village life. They developed to facilitate conflict resolution and group-level decision making, as well as to promote group cohesion within the larger groups produced by sedentary lifeways. Byrd (1994:643) also suggests that the suprahousehold corporate groups represented by such public

Figure 4. Linear arrangement of three sheep crania in central activity area.

buildings may have controlled rights to either or both of the community's real property (communal lands) and intellectual property (practical, sacred, and supernatural knowledge). In that vein, several points regarding the Hallan Çemi public buildings are worth noting.

First, in the only level (1) for which we have reasonably good evidence for the existence of public buildings, there are apparently two such buildings. Thus, there may very well be two suprahousehold groups in concurrent existence, despite the community's relatively small size. This differs from the apparent case at PPNB Beidha (see Byrd 1994:656) and another PPNB era site in eastern Anatolia—Nevali Çori (cf. Hauptmann 1993)—at both of which only one public building is in use at a time. The Çayönü data (Özdogan and Özdogan 1993), on the other hand, hint that this point of contrast may be indicative of a trend; there are two or more public buildings in concurrent use during the earlier subphases and only a single one (the terrazzo building) by the cell-plan subphase. Second, there is some evidence for control over practical knowledge, if not real property, by the group(s) using these public buildings. Both the copper ore and the obsidian were materials imported into Hallan Çemi from some distance (see Rosenberg 1994a; Rosenberg and Davis 1992). Thus, the presence of otherwise rare obsidian cores in one of these structures and the general association of copper ore finds in and around both these structures suggest that, even if these suprahousehold groups did not control outright the long-distance trade in these relatively valuable commodities at Hallan Çemi, they played a central role in that trade. Lastly, there is some evidence for control of sacred/supernatural knowledge by these suprahousehold groups. This takes the form of the aurochs skull within one of these structures and the strong association between the symbolization of aurochs and the sacred domain in later Anatolian cultures (e.g., Mellaart 1967).

The public buildings aside, the material culture assemblage from the site contains several categories of objects that very likely relate to the structural aspects of the inhabitant's culture. Included within this group are the numerous stone bowl fragments found at the site. These are made of a gray/green-black chloritic stone and white limestone, with the chloritic examples often elaborately decorated with incised designs in geometric and naturalistic motifs (Figure 5). Preliminary analysis suggests that the frequency of elaborately decorated bowls increases over time. Similar type stone bowls also occur in limited numbers at the nearby and somewhat later site of Çayönü (Özdogan and Özdogan 1993) and other, as yet undated, aceramic sites along the upper Tigris in eastern Turkey (e.g., Özdogan and Özdogan 1993:91). However, they do not seem to occur at roughly contemporary aceramic sites within the Tigris drainage in northern Iraq (e.g., Nemrik 9 and Qermez Dere).

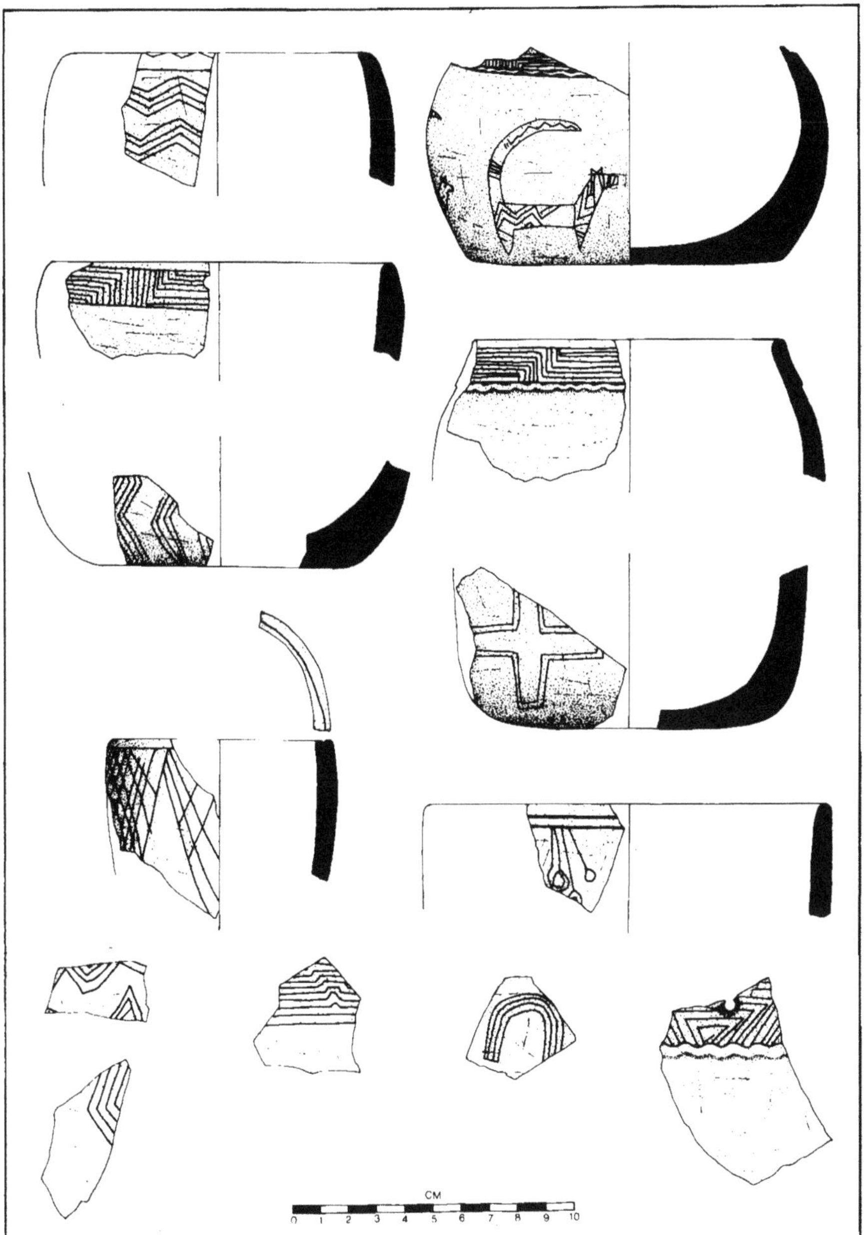

Figure 5. Stone bowls, showing selection of design motifs.

Also included are a series of sculpted fancy pestles (Figure 6:1–7). These are most commonly made of the same chloritic stone as the bowls, but occasionally sandstone was used. The majority of both kinds have handles sculpted into a variety of naturalistic forms, stylized to varying degrees. Recognizable motifs include goats' heads (Figure 6:7), paired laterally projecting straight or curved flanges that may represent bovid horns (Figure 6:3–6), and one or more kinds of mammalian animal devoid of horns (Figure 6:2). As a type, these objects are clearly related to the sculptures found at the slightly later site of Nemrik 9 in northern Iraq (see Kempisty and Kozlowski 1990; Kozlowski,1989). The chloritic examples, particularly the sculpted examples, were extensively conserved, with new working ends continuously fashioned on the remaining broken handle portion, until such time as the pestle was too small for further use. The stone bowls were also highly conserved, though less extensively so than the sculpted pestles. Lastly, it should be noted that the stone bowls and sculpted fancy pestles, particularly those made of the chloritic stone, are typically made to the same approximate scale and were thus quite capable of being used together.

Inasmuch as the aforementioned stone bowls and sculpted pestles were clearly labor intensive to produce and extensively conserved, they were clearly valuable objects. Moreover, the large majority of the pestles and a significant number of the bowls were decorated with motifs that, by virtue of their recurrence, presumably had symbolic significance. Their apparent value and their apparent status as vehicles for symbolization suggest usage in the ritualized preparation and consumption of food or drink, presumably in the context of special occasions. Given the extremely high frequency of stone bowl fragments in the overall site assemblage, the behavioral context in which stone bowl (and possibly associated sculpted pestle) usage occurred was apparently a relatively frequent occurrence. In that vein, the presence of extremely high concentrations of animal bone, including still-articulated portions of animal carcasses, and fire-cracked stone in the central activity area is worth repeating because Hayden (e.g., 1995:296) considers them to be indicative of feasting.

There is yet another type of object in the assemblage that likely reflects some aspect of social structure or organization. This type encompasses a series of small notched stone batons made of a relatively soft micaceous (perhaps schistic) stone (Figure 6:8–11). These objects are fairly standardized in form (see Rosenberg 1994a). They are generally lenticular in vertical section and taper longitudinally to end in a plain (e.g., Figure 6:9) or dimpled (e.g., Figure 6:11) tip. The dimpling aside, with the exception of one example made of orange mudstone that is triangular in vertical section, these objects differ from each other primarily in the number of notches incised on their edge(s). The notches are generally very neat, clean cut, and show no

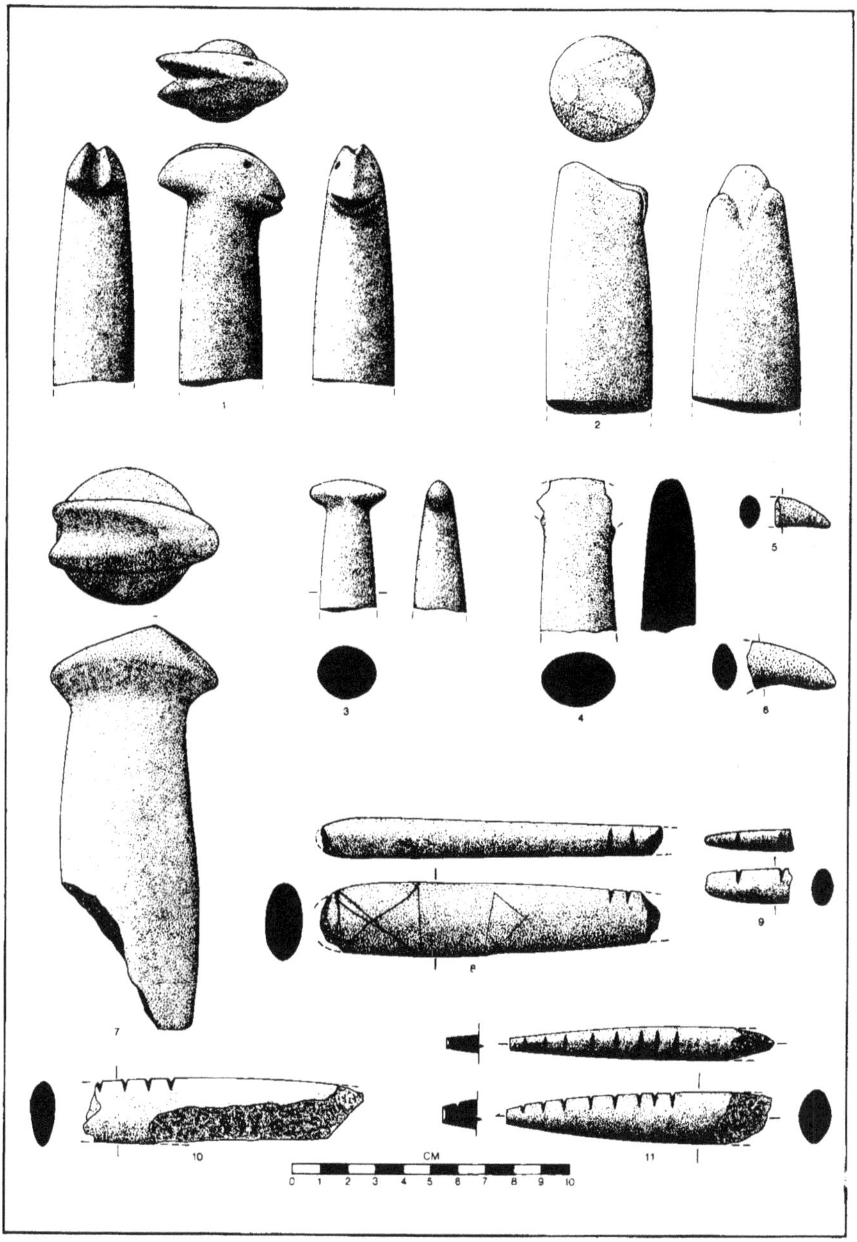

Figure 6. Sculpted pestles (1–7) and notched stone batons (8–11).

signs of wear within them, nor is wear apparent on any other part of these relatively soft objects. The single reasonably intact example (Figure 6:8) is decorated with an incised hourglass design. Objects of this type seem thus far to be known only from Hallan Çemi and can most readily be construed to represent formal tallies of some sort (see Rosenberg et al. 1995).

If so, the tally system's very formality implies that these records are of events having sociocultural significance. The specific things being tallied are probably not knowable archaeologically. However, socially significant events can be broadly categorized as falling under the general heading of either things done or things given. The third possibility is that these tallies represent things owned. It is also the most remote, given that the ethnographic record indicates that amassed wealth is socially significant in only the most complex societies. Needless to say, if this is what these tallies represent, the behaviors they imply are a most dramatic departure from the mobile hunter–gatherer norm. If these records represent things done, then, to the degree that individual actions or achievements in any given cultural domain are recorded, they are noteworthy and thus a potential source of political power (e.g., see Lee 1969). This is a clear departure from the fiercely defended egalitarian ethos of mobile hunting–gathering societies, with its strong tendencies toward publicly playing down individual achievement for precisely that reason. On the other hand, if these tallies represent things given, then, to the degree that giving or receiving is being recorded, this is a departure of the generalized reciprocal sharing that dominates mobile hunter–gather exchange. Thus, they would constitute evidence for the existence of at least some balanced reciprocal exchange systems at Hallan Çemi. In either case, these notched stone batons, to the degree they can be considered tallies, are indicative of profound departures from the sociopolitical and socioeconomic systems that characterize mobile hunting–gathering groups.

SYMBOLISM AND ORGANIZATION

Symbols can, among other things, stand for the ideological elements that characterize a given group or they can stand for the groups themselves. Unfortunately, it is not clear whether the exclusive use of bird motifs to decorate the sculpted fancy pestles from Nemrik 9 (see Kempisty and Kozlowski 1990: Figures 64–68; Kozlowski 1989:29), versus the use of mammalian motifs to decorate the ones from Hallan Çemi is one or the other. Burials are also symbolic actions with meaning at a number of potential levels, most of them purely ideological. However, at the level of territorial claim (cf. Charles and Buikstra 1983; Goldstein 1976; Saxe1970) they are suggested to stand for the group that is exercising the claim. In that context, the apparent absence of burials within the site implies certain things about community organization at Hallan Çemi.

As noted earlier, no human burials have yet been found at Hallan Çemi, and, given the current size and depth of the excavation exposure at this site, it has become reasonable to conclude that they simply do not occur within at the site proper. This strongly suggests that human remains were consistently disposed of away from the site in one fashion or another, perhaps in a cemetery. It should be noted that there is apparently a complete absence of human bone at Hallan Çemi. Thus, a central cache within the site can probably be ruled out on grounds that, if it exists, at least some human bone would have worked its way into the surrounding general deposits due to postdepositional processes. A cemetery, thought to be associated with the occupation at Zawi Chemi, does exist in the proto-Neolithic levels at Shanidar Cave (Ferembach 1970), a slightly earlier site in the same general region as Hallan Çemi, and the use of cemeteries by the broadly contemporary (terminal Natufian) cultures in the Levant is well documented (e.g., see Chapter 2 of this volume; Henry 1989). Along the Tigris, as in the Levant, intramural burials do occur at tenth millennium sites, such as Nemrik 9 (Borkowski 1992) and Çayönü (Özdogan and Özdogan 1989).

The use of cemeteries departs from the mobile hunting–gathering pattern for dealing with the dead. It is also different from the pattern of intramural internment that characterizes the later, fully developed Neolithic cultures of southwestern Asia. In that vein, Charles and Buikstra's (1983) conclusion concerning the organization of archaic central Mississippi drainage (i.e., settled hunter–gatherer) communities is particularly interesting. Following Goldstein's (1976) general suggestion that the use of cemeteries correlates with the corporate lineal inheritance of restricted resources, Charles and Buikstra (1983) suggest that the presence of bluff top cemeteries in the archaic corresponds to corporate (resource controlling) groups consisting of all the residents of the associated small villages, who are also probably all members of single extended families. If so, and if we make the further assumption that the shift to intramural burial later in the Neolithic is indicative of a change in the symbolizing, kin-based, resource-controlling group engaging in the activity (from the community to the household), then the social changes implicit in changing burial practices are also consistent with the changing socioeconomic patterns suggested by Flannery (1972b) to be indicated by changes in architectural design during the Neolithic.

Thus, the interesting point about the possibility of cemetery internment at Hallan Çemi is that it, like the small number of structures (see above), implies a relatively small size for Hallan Çemi's community. Single kin group communities are also consistent with the potential effects of the latent intergroup hostility proposed to arise from initial attempts to assert ownership in a mobile hunting–gathering society (see Rosenberg 1994b: 327–328). Both of these, in turn, are consistent with the suggested practice of kin group endogamy at Natufian sites (Henry 1989:208; see also Smith 1973). Single-kin-group communities also help to explain how the community acting as a whole could be the basic unit of production and consumption for valuable imported commodities such as copper ore and obsidian. On the other hand, such a hypothetical form of community organization at Hallan Çemi makes puzzling the presence of two contemporary public structures at the site.

In any case, such a hypothetical form of community organization raises an interesting possibility concerning the socioeconomic/sociopolitical role played by the feasting suggested by Hayden (1995) to have occurred at Hallan Çemi. Specifically, if the Hallan Çemi community consisted of a single extended family group, then political power within the community would have been delegated primarily along familial lines. Thus, power relationships within the community would likely have been primarily kinship defined, tradition based, and, consequently, more or less static over the course of adult lifetimes. However, competitive feasting of the sort proposed by Hayden (1990, 1992), implies the existence of a much more fluid sociopolitical structure than this, for otherwise there would be no point to it.

On the other hand, such single kin group community organization is theoretically compatible with a very different kind of feasting. This second possible form is reciprocal feasting on the Yanomamo model as a mechanism to overcome any latent (if not overt) hostility and promote cooperative ties between groups (Chagnon 1983). The sociopolitical ties so established would serve as replacements for those dissolved by the institutionalization of territoriality, to facilitate trade, and conceivably to also facilitate some movement of people between groups through marriage. Given the small size and social restrictiveness of community-equals-single-kin-group equation, such ties would certainly be useful. Moreover, the evidence for long-distance trade, implicitly indicating the existence of some form of ties to other groups, is certainly there at Hallan Çemi.

Lastly, it was suggested earlier that the organizational changes that distinguish early settled village societies from mobile hunting–gathering ones are not directly rooted in new food procurement techniques (e.g., food production) but in the changing patterns of property ownership that are a prerequisite for the institutionalization of these procurement techniques. Theoretically, at the earliest stages of the transition from mobile to settled lifeways, these social changes could conceivably still be just behaviorally (i.e., expedience) based and not yet be formally legitimized by prevailing beliefs (Rosenberg 1994b). However, it is noteworthy that at Hallan Çemi this does not appear to be the case, judging from the association of symbolic elements (decorative elements) with evidence for behavioral change (artifact types). That is, the socioeconomic and sociopolitical changes alluded to by the abovementioned aspects of the material culture assemblage already appear to be substantially reconciled with the belief system, and *vice versa*. Given the absence of evidence for the existence of settled village communities along the Taurus flanks prior to the establishment of Hallan Çemi, it would appear that the sedentary cultural system represented by Hallan Çemi coalesced very rapidly. This is expected on theoretical grounds (Rosenberg 1994b) and is consistent with the suggested punctuational pace of similar type changes in the Levant (e.g., Bar-Yosef and Belfer-Cohen 1991, 1992).

SUMMARY

Hallan Çemi represents the remains of a fully sedentary group of hunter–gatherers on the threshold of food production. It also exhibits the basic socioeconomic and sociopolitical characteristics of a fully settled village society. This indicates that the basic structure of Neolithic society coalesced with the very beginnings of sedentary lifeways and did not develop in tandem with either the gradually increased reliance on food production or with the gradual elaboration of culture during the Neolithic.

Specifically, Hallan Çemi exhibits a community layout that provides for some degree of individual privacy. This indirectly implies a significant departure from the generalized reciprocal sharing that characterizes mobile hunter–gatherer societies. Also, the formal tallies represented by the notched batons, whether they stand for things done or things given, indicate the formal recognition of individual social action and thus constitute another significant departure from the mobile hunting–gathering norm. In addition, Hallan Çemi appears to contain public buildings representing the existence of sociopolitical groupings at the suprahousehold level. These represent still another departure from the mobile hunter–gatherer norm and would have functioned to, among other things, resolve conflicts and otherwise promote group cohesion in the context of a fully sedentary lifeway. Lastly, while the precise size of the Hallan Çemi community is debatable, as was its precise sociopolitical organization, there is little doubt that it was a relatively small community, probably not much larger than what is thought to be typical of mobile hunting–gathering bands. This means that the above-described aspects of community organization were not a product of community size, suggesting that they were instead a direct product of sedentism.

Finally, there is evidence for public feasting. This evidence takes the form of the central activity area deposits themselves, with their bone and firecracked stone concentrations. It also takes the form of the formalized food preparation and consumption implicit in the stone bowls and sculpted pestles. Whether this feasting was for purposes of sociopolitical competition, building socioeconomic and sociopolitical ties with other neighboring communities, or some combination of the two is not clear. However, the latent hostility suggested to characterize the earliest stages of the shift to settled village life (cf. Rosenberg 1994, n.d), coupled with Hallan Çemi's apparent small size, the evidence for trade, etc., favor a cooperation fostering rather than competitive role for such feasting.

ACKNOWLEDGMENTS

The Hallan Çemi material presented here is derived from four seasons of excavation (1991–1994). Major funding for this project was provided by the National Geographic Society, the National Science Foundation, the University of Delaware Research Foundation, and the American Research Institute in Turkey. Additional funding and other types of support were provided by Mobil Exploration Mediterranean Inc. (Turkey), various institutions within the Turkish Ministry of Culture, the University of Delaware, University of Pennsylvania, and the University of Michigan. We would like to thank Ian Kuijt and Mitchell S. Rothman for their thoughtful critiques of earlier drafts of this chapter. All of their suggestions could not be acted on within the constraints of the present format, but those that were acted on greatly improved the chapter. Their efforts are deeply appreciated.

REFERENCES

Adler, M., and Wilshusen, R. H., 1990, Large-Scale Integrative Facilities in Tribal Societies: Cross-Cultural and Southwestern U.S. Examples. *World Archaeology* 22: 133–146.

Bar-Yosef, O., 1986, The Walls of Jericho: An Alternative Interpretation. *Current Anthropology* 27: 157–162.

Bar-Yosef, O., and Belfer-Cohen, A., 1991, From Sedentary Hunter-Gatherers to Territorial Farmers in the Levant. In *Between Bands and States*, edited by S. A. Gregg, pp. 181–201. University Center for Archaeological Investigations, Occasional Paper No. 9. Southern Illinois, Carbondale.

Bar-Yosef, O., and Belfer-Cohen, A., 1992, From Foraging to Farming in the Mediterranean Levant. In *Transitions to Agriculture in Prehistory*, edited by A. B. Gebauer and T. D. Price, pp. 21–41. Monographs in World Archaeology No. 4. Prehistory Press, Madison, Wisconsin.

Bender, B., 1978, Gatherer-Hunter to Farmer: A Social Perspective. *World Archaeology* 10: 204–222.

Bender, B., 1985, Prehistoric Developments in the American Midcontinent and in Brittany,

Northwest France. In *Prehistoric Hunter-Gatherers: The Emergence of Cultural Complexity*, edited by T. D. Price and J. A. Brown, pp. 21–57. Academic Press, Orlando, Florida.

Blanton, R. E., and Taylor, J., 1995, Patterns of Exchange and Social Production of Pigs in Highland New Guinea: Their Relevance to Questions about the Origins and Evolution of Agriculture. *Journal of Archaeological Research* 3: 113–145.

Borkowski, W., 1992, Graves. In *Nemrik 9: Pre-Pottery Neolithic Site in Iraq, Volume 2: House No 1/1A/1B*, edited by S. K. Kozlowski, pp. 35–42. Warsaw University Press, Warsaw, Poland.

Byrd, B. F., 1994, Public and Private, Domestic and Corporate: The Emergence of the Southwest Asian Village. *American Antiquity* 59: 639–666.

Byrd, B. F., and Monahan C. M., 1995, Death Mortuary Ritual, and Natufian Social Structure. *Journal of Anthropological Archaeology* 14: 251–287.

Chagnon, N. A., 1983, *Yanomamo: The Fierce People*. Holt, Rinehart and Winston, New York.

Charles, D. K., and Buikstra, J. E., 1983, Archaic Mortuary Sites in the Central Mississippi Drainage: Distribution, Structure, and Behavioral Implications. In *Archaic Hunters and Gatherers in the American Midwest*, edited by J. L. Phillips and J. A. Brown, pp. 117–145. Academic Press, New York.

Cohen, M. N., 1985, Prehistoric Hunter-Gatherers: The Meaning of Social Complexity. In *Prehistoric Hunter-Gatherers: The Emergence of Cultural Complexity*, edited by T. D. Price and J. A. Brown, pp. 99–119. Academic Press, Orlando, Florida.

Ferembach, D., 1970, Etude Anthropologique des Ossements Humains Proto-Neolithiques de Zawi Chemi Shanidar. *Sumer* 26: 21–64.

Flannery, K. V., 1972a, The Cultural Evolution of Civilizations. *Annual Review of Ecology and Systematics* 3: 399–426.

Flannery, K. V., 1972b, The Origin of the Village as a Settlement Type in Mesoamerica and the Near East. In *Man, Settlement and Urbanism*, edited by P. J. Ucko, R. Tringham, and G. W. Dimbleby, pp. 23–53. Duckworth, London.

Flannery, K. V., 1982, Early Pig Domestication in the Fertile Crescent: A Retrospective Look. In *The Hilly Flanks and Beyond: Essays on the Prehistory of Southwestern Asia*, edited by T. C. Young, Jr., P. E. L. Smith, and P. Mortensen, pp. 163–188. Studies in Ancient Oriental Civilization 36, University of Chicago Oriental Institute, Chicago, Illinois.

Goldstein, L., 1976, *Spatial Structure and Social Organization*. Ph.D. dissertation, Northwestern University. University Microfilms, Ann Arbor, Michigan.

Hauptmann, H., 1993, Ein Kultgebäude in Nevali Çori. In *Between the Rivers and over the Mountains: Archaeologica Anatolica et Mesopotamica Alba Palmieri Dedicata*, edited by M. Frangiapane, H. Hauptmann, M. Liverani, P. Matthiae, and M. Mellink, pp. 37–69. University of Rome, Rome.

Hayden, B., 1990, Nimrods, Piscators, Pluckers and Planters: The Emergence of Food Production. *Journal of Anthropological Archaeology* 9: 31–69.

Hayden, B., 1992, Models of Domestication. In *Transitions to Agriculture in Prehistory*, edited by A. B. Gebauer and T. D. Price, pp. 11–19. Prehistory Press. Madison, Wisconsin.

Hayden, B., 1995, A New Overview of Domestication. In *Last Hunters–First Farmers*, edited by T. D. Price and A. B. Gebauer, pp. 273–299. School of American Research Press, Santa Fe, New Mexico.

Henry, D. O., 1989, *From Foraging to Agriculture: The Levant at the End of the Ice Age*. University of Pennsylvania Press, Philadelphia, Pennsylvania.

Kempisty, A., and Kozlowski, S. K., 1990, Objects of Art. In *Nemrik 9: Pre-Pottery Neolithic Site in Iraq (General Report–Seasons 1985–1986)*, edited by S. K. Kozlowski, pp. 155–162. Warsaw University Press, Warsaw, Poland.

Kenyon, K., 1957, *Digging up Jericho*. Ernest Benn, London.

Kozlowski, S. K., 1989, *Nemrik 9, A PPN Neolithic Site in Northern Iraq. Paléorient* 15(1): 25–31.

Kuijt, I., 1995, Pre-Pottery Neolithic A Settlement Variability: Evidence for Sociopolitical Developments in the Southern Levant. *Journal of Mediterranean Archaeology* 7: 165–192.

Kuijt, I., 1996, Negotiating Equality through Ritual: A Consideration of Late Natufian and Prepottery Neolithic A Period Mortuary Practices. *Journal of Anthropological Archaeology* 15: 313–336.

Lee, R. B., 1969, Eating Christmas in the Kalahari. *Natural History* 78: 14–22.

Lee, R. B., and DeVore, I., 1968, Problems in the Study of Hunter-Gatherers. In *Man the Hunter*, edited by R. B. Lee and I. DeVore, pp. 3–12. Aldine, Chicago, Illinois.

Marquardt, W. H., 1985, Complexity and Scale in the Study of Fisher-Gatherer-Hunters: An Example from the Eastern United States. In *Prehistoric Hunter–Gatherers: The Emergence of Cultural Complexity*, edited by T. D. Price and J. A. Brown, pp. 59–98. Academic Press, Orlando, Florida.

McGuire, R., 1983, Breaking down Complexity: Inequality and Heterogeneity. *Advances in Archaeological Method and Theory* 6: 91–142.

Mellaart, J., 1967, *Çatal Höyük: A Neolithic Town in Anatolia*. McGraw-Hill, New York.

Netting, R. McC., 1990, Population, Permanent Agriculture, and Polities: Unpacking the Evolutionary Portmanteau. In *The Evolution of Political Systems: Sociopolitics in Small Scale Sedentary Societies*, edited by S. Upham, pp. 21–61. Cambridge University Press, New York.

Olszewski, D. I., 1991, Social Complexity in the Natufian? Assessing the Relationship of Ideas and Data. In *Perspectives on the Past: Theoretical Biases in Mediterranean Hunter-Gatherer Research*, edited by G. A. Clark, pp. 322–350. University of Pennsylvania, Philadelphia, Pennsylvania.

Özdogan, M., and Özdogan, A., 1989, Çayönü: A Conspectus of Recent Work. *Paléorient* 15: 65–74.

Özdogan, M., and Özdogan, A., 1993, Pre-Halafian Pottery of Southeastern Anatolia: With Special Reference to the Çayönü Sequence. In *Between the Rivers and over the Mountains: Archaeologica Anatolica et Mesopotamica Alba Palmieri Dedicata*, edited by M. Frangiapane et al. pp. 87–103. University of Rome, Rome.

Perrot, J., 1966, Le Gisement Natoufien de Mallaha (Eynan), Israël. *L'Anthropologie* 70: 437–484.

Plog, S., 1990, Agriculture, Sedentism, and Environment in the Evolution of Political Systems. In *The Evolution of Political Systems: Sociopolitics in Small Scale Sedentary Societies*, edited by S. Upham, pp. 177–202. Cambridge University Press, New York.

Price, T. D., and Brown J. A., 1985, Aspects of Hunter-Gatherer Complexity. In *Prehistoric Hunter-Gatherers: The Emergence of Cultural Complexity*, edited by T. D. Price and J. A. Brown, pp. 3–20. Academic Press, Orlando, Florida.

Redding, R. W., 1988, A General Explanation of Subsistence Change: From Hunting and Gathering to Food Production. *Journal of Anthropological Archaeology* 7: 59–97.

Redding, R. W., n.d., The Origin of Food Production in Anatolia: A Test of 'Foreign' Ideas. Paper presented at the annual meeting of the Society for American Archaeology, April 20–24, 1994.

Rosenberg, M., 1990, The Mother of Invention: Evolutionary Theory, Territoriality, and the Origins of Agriculture. *American Anthropologist* 92:399–415.

Rosenberg, M., 1994a, Hallan Çemi Tepesi: Some Further Observations Concerning Stratigraphy and Material Culture. *Anatolica* 20: 121–140.

Rosenberg, M., 1994b, Pattern, Process, and Hierarchy in the Evolution of Culture. *Journal of Anthropological Archaeology* 13: 307–340.

Rosenberg, M., 1994c, A Preliminary Description of the Lithic Industry from Hallan Çemi. In *Neolithic Chipped Stone Industries of the Fertile Crescent: Proceedings of the First Workshop on PPN Chipped Stone Industries*, edited by H. G. Gebel and S. K. Kozlowski, pp. 223–238. Studies in Early Near Eastern Production, Subsistence, and Environment 1. Freie Universität, Berlin.

Rosenberg, M., n.d., Cheating at Musical Chairs: Sedentism in an Evolutionary Context. Paper presented at the World Archaeological Congress 3 (New Delhi, India), December 4–11, 1994.

Rosenberg, M., and Davis M., 1992, Hallan Çemi Tepesi, an Early Aceramic Neolithic Site in Eastern Anatolia: Some Preliminary Observations Concerning Material Culture. *Anatolica* 18:1–18.

Rosenberg, M., Nesbitt, R. M. A., Redding, R. W., and Strasser, T. F., 1995, Hallan Çemi Tepesi: Some Preliminary Observations Concerning Early Neolithic Subsistence Behaviors in Eastern Anatolia. *Anatolica* 21:1–12.

Saxe, A. A., 1970, *Social Dimensions of Mortuary Practices*. Ph.D. dissertation, University of Michigan. University Microfilms, Ann Arbor, Michigan.

Schirmer, W., 1990, Some Aspects of Building at the "Aceramic-Neolithic" Settlement of Çayönü Tepesi. *World Archaeology* 21:363–383.

Service, E. R., 1978, *Profiles in Ethnology*, Third Edition. New York: Harper and Row.

Smith, P., 1973, Family Burials at Hayonim. *Paléorient* 1:69–72.

Southall, A., 1988, On Mode of Production Theory: The Foraging Mode of Production and the Kinship Mode of Production. *Dialectical Anthropology* 12:165–192.

Testart, A., 1982, The Significance of Food Storage among Hunter-Gatherers: Residence Patterns, Population Densities, and Social Inequalities. *Current Anthropology* 23:523–537.

Upham, S., 1990, Decoupling the Process of Political Evolution. In *The Evolution of Political Systems: Sociopolitics in Small Scale Sedentary Societies*, edited by S. Upham, pp. 1–17. Cambridge University Press, New York.

Wilson, P., 1988, *The Domestication of the Human Species*. Yale University Press, New Haven, Connecticut.

Winterhalder, B., 1990, Open Field, Common Pot: Harvest Variability and Risk Avoidance in Agricultural and Foraging Societies. In *Risk and Uncertainty in Tribal and Peasant Economies*, edited by E. Cashdan, pp. 67–88. Westview, Boulder, Colorado.

Woodburn, J., 1980, Hunter-Gatherers Today and Reconstruction of the Past. In *Soviet and Western Anthropology*, edited by E. Gellner, pp. 95–117. Duckworth, London.

Woodburn, J., 1982, Egalitarian Societies. *Man* 17: 431–451.

Woodburn, J., 1988, African Hunter-Gatherer Social Organization: Is it Best Understood as a Product Encapsulation? In *Hunters and Gatherers Volume 1: History, Evolution and Social Change*, edited by T. Ingold, D. Riches, and J. Woodburn, pp. 31–64. Berg, Oxford.

Wright, G. A., 1978, Social Differentiation in the Early Natufian. In *Social Archaeology*, edited by C. L. Redman, M. J. Berman, E. V. Curtin, W. T. Langehorne, Jr., N. H. Versaggi, and J. C. Wanser, pp. 201–223. Academic Press, New York.

Yellen, J., 1985, Bushmen. *Science* 85: 41–48.

Yellen, J., 1990, The Transformation of the Kalahari !Kung. *Scientific American* 262(4): 96–105.

Chapter 4

Households in Transition

Neolithic Social Organization within Southwest Asia

BRIAN F. BYRD

INTRODUCTION

The Late Pleistocene and Early Holocene was a period of unprecedented economic change in portions of southwest Asia as semimobile hunters and gatherers became sedentary agriculturalists. Significant changes in social organization and ideology are predicted to be associated with the emergence of these novel community forms, and discussion has generally focused on new patterns of social interaction, increased social complexity, and the emergence of nascent elites (e.g., Bender 1978, 1990; Hayden 1990, 1995a; Hodder 1990; Hole 1984; Price and Feinman 1995). Several factors, based on ethnographic observations, are considered to have played a role in fundamentally altering social interaction (Flannery 1972, 1993; Southall 1988; Upham 1990). Mobile foraging economies are by nature extensive, while economies of sedentary agriculturalists are characterized by intensive exploitation of local, often highly circumscribed resources. With highly circumscribed resources, there is greater potential for varied access between

BRIAN F. BYRD • ASM Affiliates, Encinitas, California 92024, and Department of Anthropology, University of California, San Diego, La Jolla, California 92093.

Life in Neolithic Farming Communities: Social Organization, Identity, and Differentiation, edited by Ian Kuijt. Kluwer Academic/Plenum Publishers, New York, 2000.

members of a single community and for competition for those resources (Hayden 1990; Netting 1990; Wilk and Netting 1984:11; Wilson 1988). Rights to many resources among sedentary ethnographic populations are typically well defined and noncommunal. A number of scholars have argued that with sedentism and agricultural production the household became the basic social unit within communities that maintained and transmitted inheritance rights and resource access (Flannery 1972:48; Netting 1990:60; Wilson 1988).

Research on community changes associated with the emergence of sedentism and food production within southwest Asia has generally focused on social complexity and ritual behavior, particularly associated with mortuary practices, with minimal attention given to unraveling the nature of household organization (Byrd and Monahan 1995; Cauvin 1972, 1994; Henry 1985; Kuijt 1995, 1996; Wright 1978). Recently I have suggested that early agricultural villages in southwest Asia were characterized by a restriction in the social network for sharing production and consumption activities, and the development of more institutionalized mechanisms for integrating the community as a whole (Byrd 1994a). Typically in southwest Asia, examination of household size and organization has been limited to brief statements focused on the Early Neolithic. For example, Redman stated that during the early Neolithic "each building housed a family having from five to eight members" (Redman 1978:143). Others have offered contradictory reconstructions that early Neolithic households were representative of nuclear families or extended families or the potential for the latter (Aurenche 1981; Bar-Yosef and Meadow 1995; Hole 1987:82; Moore 1985: 19; Voigt 1990:3; Watson 1978:156). These assertions were rarely explicitly linked to empirical evidence. Discussion of social organization for direct Epi-Paleolithic antecedents of the Early Neolithic is more infrequent (yet see Goring-Morris 1987).

There has been only one specific attempt to reconstruct household organization in southwest Asia during this time period. Flannery's (1972) ambitious reconstruction of social structure and the nature of households focused on the transition to agricultural villages. He identified two site types in the archaeological literature—circular hut compounds and rectangular house villages—and examined the ethnographic literature for insights into the social structure of these settlement types. Based on analogy from modern African pastoralists, early clusters or compounds of circular buildings during the Natufian and Pre-Pottery Neolithic A (PPNA) were hypothesized as representative of patrilineal, polygynous extended families. Each building was typically occupied not by a family but by one person or sometimes two people (typically a man or one of his wives). At later Neolithic villages, in contrast, "the rectangular buildings are designed to accommodate *families*, rather than individuals" (Flannery 1972:39, emphasis original). These buildings of the Pre-Pottery Neolithic B (PPNB) were hypothesized as being occupied by families of three or four people. Only during subsequent periods were such structures combined to form extended households. Thus, the transition from round to square architecture coincided with a shift in residential unit from a polygynous extended household to a monogamous nuclear household (Flannery 1972:42). Recently this study has been criticized (Saidel 1993), although the critique has been quite thoroughly rebutted by Flannery (1993), and their debate was not focused on household size or its organization (since Saidel endorses Flannery's household hypothesis) but on causality and the shape of buildings.

This chapter explicitly examines two questions: (1) Can changes be discerned in the size and organization of households during the transition to settled food-producing villages in southwest Asia? (2) At what point during this extended transition did such developments occur? In examining the available archaeological evidence for changes in households during the transition to sedentary villages, earlier discussions that these communities comprised nuclear households or extended households are examined. This is done by first discussing how such inferences can be made by linking variation in social organization to specific archaeological correlates. The study area is restricted to one portion of southwest Asia, the southern Levant, since the data set for this area is the strongest. The time period under consideration includes the Epi-Paleolithic through the Middle Pre-Pottery Neolithic B. I argue in this chapter that, although there are changes in the size of domestic structures during this time period, they remained primarily the residences of nuclear families. No major alterations occurred in the numerical composition of the residential unit. Developments in the organization of interior domestic space did take place as facilities and compartmentalization increased. These latter changes are interpreted as reflecting increased household autonomy and alterations in how nuclear households interacted with each other, particularly during the Middle PPNB. These results provide new insights into how households, a fundamental social unit within these communities, dealt with a novel reorientation in settlement pattern and subsistence strategies.

DEFINITIONS AND APPROACH OF THE STUDY

Any study of prehistoric households requires constructing an approach for operationalizing their elucidation in the archaeological record. Households have been defined in a myriad of ways by social scientists. Depending on their definition and resulting social behavior, more than one household can be in a single building and one household can have several buildings (Wilk and Netting 1984; Wilk and Rathje 1982:620). For the purposes of this study, a widely used functional definition is employed. A household is defined as a task-oriented residence unit that shared a combination of production, coresidence, reproductive tasks, and consumptive tasks (Netting 1982; Netting et al. 1984). Nuclear households, or simple households, generally comprise spouses and their offspring while extended families or complex households include two or more coresident married siblings or additional adult members (Blanton 1994:5; Hammel and Laslett 1974).

The difficulty lies in how to identify households in the archaeological record. Architecture or the built environment provides a framework for the spatial analysis of prehistoric social organization (Kent 1990; Lawrence and Low 1990; Netting et al. 1984). Architecture fulfills a variety of social demands, organizing, regulating, and delimiting contact between individuals and households (Wilson 1988). The tangible structure of the built environment provides a focus for spatial analysis aimed at gaining insights into social organization, how social relations are reiterated, and how community organization changes over time. A major analytical problem, however, is that spatial organization is not directly related to social organization or ideology (Lawrence and Low 1990; Wilk 1990). Instead, it is imperfectly reflected in built form (for example, buildings and households are not necessarily identical), and other factors, including the specific social institutions, past historical events, and ideology play a role in the character of the built environment. In addition, the dynamic nature of architecture adds to the complexity of the issue. Buildings can be changed and modified over time to meet new or changing community needs (Banning and Byrd 1987; Goody 1958).

The initial step in this process entails making inferences regarding the probable function of individual buildings and isolating those that are best interpreted as domestic units. To do so, one must examine the arrangement of space within building interiors, focusing on the location and function of internal structural features, the nature and spatial distribution of associated artifacts, and, if present, circulation paths within and between buildings. A number of discrete and continuous variables (particularly hearths and artifacts) have been correlated with different building functions in ethnoarchaeological investigations within southwest Asia (Horne 1980:23, 1994; Kramer 1982, 1983:349; Watson 1979:295). Features, for the purpose of this discussion, are considered to be the fixed construction elements or furniture within the built environment that do not have an intrinsic structural purpose (such as the walls and primary posts needed to enclose the space and support the roof). They are critical in understanding the function of buildings (Watson 1979: 295). Moreover, features tend to either restrict or provide a focus for particular activities to take place, including food processing, preparation, and cooking, artifact manufacture and maintenance, and storage (Whitelaw 1991; Wiessner 1982).

The function of prehistoric buildings, however, is more difficult to reconstruct. Over time building function may change, and the archaeological evidence may imply only the ultimate use of the building or include a compilation of the signatures from a sequence of functions. The nature of building abandonment creates a variety of different situations as in situ artifacts are typically removed upon planned abandonment and left in place with rapid or catastrophic abandonment (Cameron and Tonka 1993). Furthermore, some economic and social functions are differentially preserved. It is not sufficient, for example, to assume that artifacts resting on or near floors reflect activities carried out while the building was in use. In dealing with this issue, it is useful to keep in mind the distinction made by Carr (1984:114) between depositional sets and areas (in the archaeological record) and activity sets and areas (in the behavioral past), in order to consider the effect that formation processes had in creating floor assemblages (Schiffer 1983, 1987). Activity area research has clearly demonstrated the role that formation processes play in the creation of depositional sets (e.g., Brooks and Yellen 1987; Cameron 1990; Carr 1987; Hayden and Cannon 1983; Seymour and Schiffer 1987). This understanding allows one to better understand and characterize the processes that contributed to the creation of floor associated assemblages. In other words, what formation processes contributed significantly to the correlation of particular sets of artifacts associated with building floors? For example, four types of floor artifact assemblages were distinguished at Beidha, each characterized by a different form of formation process and sometimes disparate archaeological signatures (Byrd n.d.). These categories of floor associated assemblages include portable and nonportable in situ artifacts left as the result of sudden abandonment, nonportable in situ artifacts whose abandonment probably was not the result of a sudden event, stockpiled in situ caches of discarded raw materials and artifacts, and, finally, artifacts deposited on the floor as the result of trash dumping after building abandonment. In certain circumstances, several of these processes may have contributed to the creation of floor context artifact assemblages requiring researchers to examine these independently.

Based on recent research at Beidha (Byrd 1994a, n.d.), three classes of buildings were defined: domestic dwellings, storage facilities, and nondomestic buildings. Domestic dwellings were distinguished based on the nature of internal features, the presence of in situ artifacts for food processing and preparation, and the production and maintenance of tools and other domestic equipment. Domestic dwellings may be overrepresented, because unless there was sufficient evidence to the contrary, buildings of similar size and form were interpreted as domestic buildings. A number of criteria can be employed to discern building use: variation in building size, patterns in building construction style and morphology, the function of associated structural features, the range of activities represented by in situ artifacts on building floors, and the nature of building abandonment. Given that disasters, primarily the burning of buildings, typically provided the greatest insight into building function, interpretive emphasis can be placed on evidence derived from burned buildings and generalized to analogous buildings. Repeated patterning, be it in the presence or absence of evidence, also plays a role in inferring function.

Ideally, to address the issues at hand, a restricted number of sites would be selected that were occupied during the transition to food producing villages: from the onset of Early Natufian through the Pre-Pottery Neolithic B (12,500 to 8,000 bp). These sites would have undergone extensive excavations, exposing many buildings with well-preserved architecture and burned buildings with floor context artifacts, and would be fully published as monographs. Thus, a study of primary contexts would provide precise information on the size of buildings, the nature of internal features, and the range of activities carried out. Each site would have been occupied over the entire time period under consideration, have multiple phases of occupation (allowing for the potential for intrasite diachronic changes), relatively well preserved buildings, and occur within a relatively restricted portion of southwest Asia. In this situation, inferences into changes in building function and the number and nature of the household could be quite strong. Unfortunately, such a data set is lacking. This study examines published architectural data from sites covering the transition, focusing on those sites with the most detailed information available in published reports. Monographs are rare, and full information on features within buildings and the nature of artifacts associated with floors is often lacking. For example, it is often difficult to discern whether hearths were present, absent, or uncertain, along with the precise nature of floor-associated artifacts. In general, the sample sizes for each time period are small, particularly for the earlier portion of the sequence. This is not an exhaustive study since some sites with architecture were excluded owing to too small a sample or the lack of suitable published information. The samples from any period are often from different environmental settings, and, where possible, these subsamples are considered separately. In addition, the end of the sequence is treated in a varied manner.

This study primarily considers sites within the time period from 20,000 to 8,500 bp. This includes the following cultural complexes in the southern Levant: the Pre-Natufian Epi-Paleolithic, of which a number of terms have been utilized (20,000–12,800/12,500 bp); the Natufian (12,800/12,500–10,300 bp); the Pre-Pottery Neolithic A or PPNA (10,300–9300/9200 bp); the Harifian

(10,600–10,100 bp); and the Middle Pre-Pottery Neolithic B or PPNB (9,300/9200–8,500 bp) (Bar-Yosef and Meadow 1995; Byrd 1994b; Goring-Morris 1991; Kuijt 1995; Rollefson et al. 1992). The Natufian has been subdivided into the early (12,800/12,500–11,000 bp) and late (11,000–10,300 bp) phases, with the latter subsuming the Final Natufian (Valla 1987). The Epi-Paleolithic Harifian, which is limited in areal extent to the Negev and Sinai, is generally considered contemporaneous with the end of the Natufian and overlaps slightly with the onset of the PPNA (Goring-Morris 1991). Note that the Middle PPNB (from now on referred to as the PPNB) is temporally contiguous with the PPNA (Kuijt 1995). For comparative continuity in the Negev, I have also included two sites that chronologically fall within the Late PPNB. However, Late PPNB and Final PPNB sites in the highlands and Jordan Valley are not discussed since consideration of the social implications of subsequent changes in domestic architectures at these sites is outside the scope of this study.

The Early Epi-Paleolithic sample includes only four buildings from two sites in the upper Jordan Valley (Fig. 8.1). A time gap exists between these and the next sample set. The Natufian period sample includes structures from five sites: one in the western highlands, one in the Jordan Valley, and three (including two Harifian sites) in the Negev. The former two sites are primarily Early Natufian, with less evidence for the Late Natufian, while the latter three are Late Natufian or Harifian in age. The PPNA sample consists of structures from five sites, three in the Jordan Valley and two in the western highlands. Finally, the PPNB sample of structures is from a more dispersed region, including two in the Jordan Valley, one in the western hills, two in the eastern hills, and two in the Negev.

Only three of the twenty sites in this sample were occupied for an extensive period of time that encompassed more than one period or witnessed strong changes in the built environment. These key sites for inferring the nature of diachronic trends are 'Ain Mallaha, Jericho, and Beidha, and each is discussed in more detail. The open-air Natufian site of 'Ain Mallaha lies in the Huleh Basin of the upper Jordan Valley. This extensive site contains thick occupation deposits and substantial stone architecture in the form of oval domestic structures and has been interpreted as a permanent Natufian settlement (Perrot 1966:477). The excavations by Jean Perrot have identified four geological units (Beds I–IV, with IV the oldest), with four Natufian occupation phases termed Ancienne (Early), Moyenne (Middle), Recente (Late), and Finale (Final). These occupation phases span a considerable period of time from the Early Natufian into the Late Natufian (Perrot and Ladiray 1988; Valla 1991). Jericho, Tell es-Sultan, is situated adjacent to a perennial spring in the lower Jordan Valley. At an elevation of 825 ft below sea level, the Dead Sea lies 7 miles to the south. Jericho was occu-

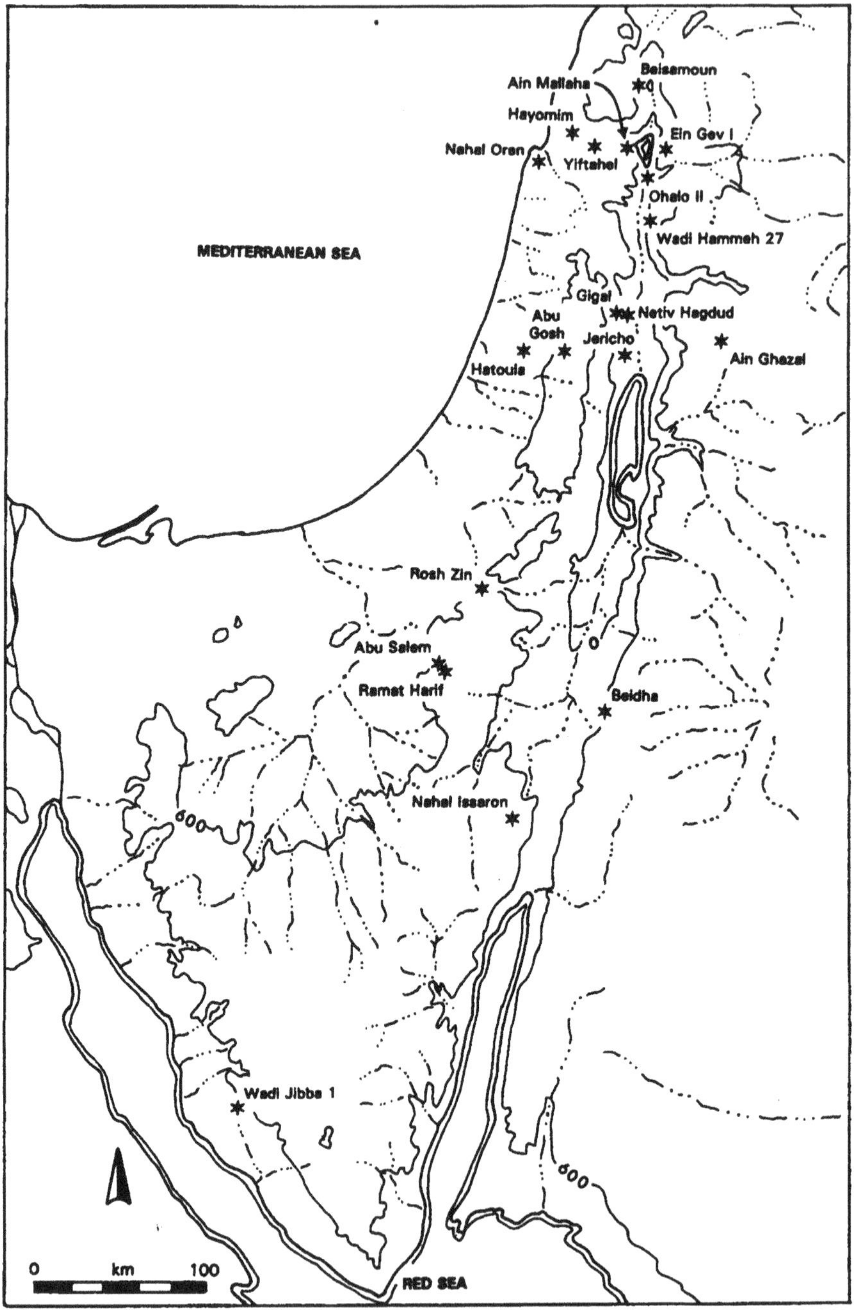

Figure 1. Location of sites within southern Levant used in the study.

pied briefly during the Early Natufian and then reoccupied at the onset of the Neolithic (Kenyon 1981). Three phases of aceramic Neolithic occupation are generally recognized: Proto-Neolithic, Pre-Pottery Neolithic A, and Pre-Pottery Neolithic B. Based on a series of conventional radiocarbon dates, the aceramic Neolithic occupation appears to have spanned the period from 10,300 to 8,600 bp (Kuijt and Bar-Yosef 1994; Walterbolk 1987). Mud-brick buildings were round during the Proto-Neolithic and PPNA and rectangular during the PPNB. Beidha is situated in the southern Jordanian highlands, just north of Petra. Prehistoric occupation includes a Natufian encampment (primarily during the thirteenth millennium bp) and a PPNB Neolithic village (during the ninth millennium bp) (Byrd n.d.; Kirkbride 1966). The Neolithic settlement consists of a small, low tell over 3 m thick, and sixty-five buildings have been excavated. Final stratigraphic analysis distinguished three phases of occupation labeled from earliest (A) to latest (C) (Byrd 1994a).

ARCHAEOLOGICAL DATA

Although it is not the goal of this study to review the history of domestic architecture for this area (see Aurenche 1981), some salient trends and developments are described in summarizing the archaeological evidence. The Early Epi-Paleolithic sample consists of three structures from Ohalo II and one partial structure from Ein Gev I in the upper Jordan Valley with a mean interior structure size of 10.9 m^2 (Table 1; see Fig. 1). Dated to the twentieth millennium bp, the moderately sized semicircular hut structures from Ohalo II are oriented in a linear manner (Nadel 1991; Nadel et al. 1995). They are built of perishable construction material, opening to the east, with no internal features present. These structures have multiple floors with a range of debris and trash built within them. Hearths are present outside and adjacent to the structures and indicate spatial patterns in the processing and possibly discard of plant and animal remains. In contrast, the Ein Gev I structure was built into the side of the hill, partially using stone construction and possibly open on the downslope side (Arensberg and Bar-Yosef 1973; Bar-Yosef 1970:109–111). The hut was rebuilt six times and internal features were limited. A hearth, a pit of unknown function, and a stone paved area were present in layer 4. A stone mortar and pestles were associated with the former, and horn cores with the latter.

The three Natufian sites with architectural assemblages occur in different settings (the upper Jordan Valley, the western highlands, and the arid Negev) and are highly varied. At 'Ain Mallaha, although the sample size of buildings where interior area can be estimated is small, remnants reaffirm

Table 1. Summary of Interior Area for Domestic Structures in the Southern Levant[1]

Period	N	Minimum	Maximum	Median	Mean
Early Epi-Paleolithic	4	7.5	15.75	9.5/10.5	10.8
Natufian	14	2.0	29.4	5.6/7.5	10.1
'Ain Mallaha Early	3	9.25	29.4	22.5	20.4
'Ain Mallaha Middle	3	7.5	24.6	8.4	13.5
Hayomim Cave	6	2.0	5.6	3.75/4.1	3.9
Rosh Zin	2	5.0	11.5	—	8.2
Harifian—Negev & Sinai	10	1.9	12.4	9.0	6.9
Abu Salem	5	1.9	12.4	9.0	7.5
Ramat Harif	5	3.0	10.5	6.5	6.4
PPNA—Jordan Valley	13	5.25	35	—	15.8
Gigal	2	11.5	18.75	—	15.1
Jericho	5	9	20.1	17.5	14.7
Netiv Hagdud	6	5.25	35	7.25/23	16.9
PPNA—Western Hills	15	3.8	19	10.2	10.6
Hatoula	3	15	19	16	16.7
Nahal Oren	12	3.8	16.6	8.0/9.6	9.0
PPNB—Jordan Valley and Hills	6	28.6	44.5	32.1/34.4	35.25
PPNB Beidha					
Phase A	11	7.4	14.9	10.7	10.6
Phase B	8	3.8	14.2	5.9/6.1	6.9
Phase C	11	13.3 (5.8)	48.3 (21.0)	27.4 (11.9)	29.4 (12.8)
Estimate (basements only)					
PPNB—Negev & Sinai	14	2.8	8.6	—	5.1
Wadi Jibba 1	6	4.5	8.6	5.7/6.2	6.0
Nahal Issaron	8	2.8	7.5	4.2/4.5	4.5

[1]These values are taken from plans provided in published reports. Only those structures that were complete or whose full circumference could be reliably estimated are included. Interior area does not include walls.

the following trends. The early phase consists of a series of oval and semicircular structures with a considerable size range and a mean interior area of 20.4 m^2 (Perrot 1960, 1966; Valla 1991, 1994). The buildings were initially aligned along the side of the hill and built into the hillside. Constructed of stone, these structures generally had mud floors, with one example of a red-painted crushed limestone surface. The latter may represent the only nondomestic structure at the site. In contrast to the small sample of Early Epi-Paleolithic structures, internal features are well documented and include hearths, stone bins, and stone paving/slabs. Nonportable *in situ* artifacts such as mortars and stone vessels are also reported. New structures were often built almost directly overlying earlier ones. The small sample includes a large semicircular structure that is interpreted as only two-thirds enclosed (Valla 1991, 1994: 86). It contains three hearths each with different associated floor artifacts: one, near the front, lacked associated artifacts; lithic reduction material was associated with the second; and the third had an adjacent area with pestles and a nearby area with material indicative of weapon regearing (Valla 1994:186). Thus, a wide range of activities appears to be spatially segregated within this structure. Partially analogous (in architecture form, features, and possible open ramada-like style) structure remnants are documented at nearby Early Natufian Wadi al-Hammah 27 (Edwards 1991). In the middle phase at 'Ain Mallaha, semisubterranean buildings are more numerous and building size decreased to a mean area of 13.4 m^2. The largest of these buildings (No. 26) is rebuilt, each time decreasing in size. These more circular buildings of the middle phase at 'Ain Mallaha include stone lined hearths, stone bins, and at least one example of a stone partition. Although the late phase has no evidence of domestic structures, new research on the final phase has documented a series of oval stone structures that have many similarities with middle phase buildings (Valla et al. 1999). Burials pits and smaller feaatures such as a plaster-lined storage facility are also well documented in the late and final phases.

In contrast, the architecture of the Early Natufian of Hayonim Cave Layer B was dominated by a series of small, honeycomb-stye subcircular stone structures with a mean interior area of only 3.9 sq m (Bar-Yosef 1991; Belfer-Cohen 1988). These structures were organized in two parallel rows across the width of the cave, and five main phases of construction discerned. The floors were constructed of stone paving. Probable stone hearths were documented in half the structures, typically near the walls, in contrast to their prominent central location in structures at 'Ain Mallaha. No other features are reported on these floors. Multiple floors were occasionally distinguished, in situ floor artifacts uncommon, and debris built up within the structures. Earlier structures are only partially exposed, and Later Natufian structures are more ephemeral (both within the cave and on the terrace) (Valla et al. 1991).

The Late Natufian of Rosh Zin in the Negev includes four slightly semisubterranean, interlocking stone structures that represent the final phase of occupation (Henry 1976). Mean interior area of the two complete structures is 8.2 m^2. The largest structure included the remnants of a stone pillar or monolith with a possible dedicatory offering at its base. No other features, such as hearths, were documented, and the structures were filled with occupational debris. The subsequent Harifian structures of Abu Salem and Ramat Harif in the Negev are well discussed by Goring-Morris (1991) (see also Goring-Morris 1987; Scott 1977). Site layout was generally in a linear pattern with one dwelling per spatially distinct architectural unit (with occasionally some smaller associated structures). These structures range considerably in size with a mean of 6.9 m^2. They are semisubterranean, subcircular stone structures with an occasional shelf or bench. Stone slabs and tables or bedrock with shallow-ground cup marks and cut marks and one or more deep stationary mortars occur within interiors. No storage facilities were noted. Makeshift interior hearths were present in some buildings and were situated against walls. The smaller structures adjacent to larger ones also typically had mortars and cup marks. Considerable trash occurred within the buildings.

The PPNA sample from the Jordan Valley includes structures at Gigal, Jericho, and Netiv Hagdud. The sample includes various size structures, with a mean interior area of 15.8 m^2. In general, these are independent, freestanding structures with separate entrances. The small published sample of buildings from Gigal includes two burned subrectangular buildings with walls made of small stones and mud or daub (Noy 1989). Hearths (burnt gravel and ash surrounded by clay) and silos were documented inside some houses. Floor-associated material included stored plant remains, bitumen baskets, stone cup-hole artifacts, mortars, stone bowls, and stone slabs. The radiocarbon dates indicate these structures may be slightly earlier than Netiv Hagdud.

The Netiv Hagdud sample includes large and small oval structures (Bar-Yosef et al. 1980; 1991; Bar Yossef and Gopher 1997). Floors were typically of mud plaster with wall foundations of upright limestone slabs and walls of unbaked, planoconvex mud bricks. A partition is present in only one structure. Building features are uncommon and include a few bins and silos and concave oval-shaped cobble areas that are interpreted as interior hearths. Slabs with cup holes or cup marks occur on some floors, occasionally along with other artifacts.

The mud-brick structures at Jericho are generally oval or almost circular (Kenyon 1981). Multiple clay floors were common, as was multiple reuse of walls. Interior hearths were noted only occasionally, and only one other clearly identified feature documented (a 1.5 m diameter circular mud platform). Partitions are absent, and in only one instance was there direct access between houses. Floor-associated artifacts were not reported. A notable aspect of Jericho, of course, is the presence of a series of nondomestic constructions, including the wall, tower, and a series of adjacent, apparently special-function structures.

The PPNA sample from the western highlands includes a small sequential sample at Hatoula (Lechevallier and Ronen 1985) and a large sample from Nahal Oren (Stekelis and Yizraely 1963). The Hatoula architecture includes an early phase oval structure of baked brick and two subsequent phase semisubterranean oval structures made of stone. Mean interior area is 16.7 m^2. No features or artifacts were noted. The Nahal Oren Stratum II sample includes a series of oval and subcircular structures built of stone with a mean interior area of 9.0 m^2. The structures are situated on four artificial terraces along the hillside, and some buildings on adjacent terraces shared walls. A regularity in internal plan was asserted, although not documented in the short preliminary article. Internal features set into the mud plaster floors included hearths lined with stones situated in various locations, and thick and thin cup-marked stones. An underlying Natufian phase included a sequence of poorly preserved oval shaped structures with small installations, including silos and hearths.

The two PPNB sites from the Negev are characterized by clusters of stone structures. At Nahal Issaron Layer C four phases of architectural construction represent at least eight structures (Gopher et al. 1994; Goring-Morris and Gopher 1983). These interlocked stone structures have well-preserved stone walls and include a wide range of oval to subrectangular forms clustered in a beehive form. Floors were not discernible, yet hearths were distinguished at sequential levels within structures (these included types that had stone lined with ash in larger structures and types with small stones and charcoal). Entrances are uncommon. Although not discussed in the text, the site plan reveals stone partitions and querns in several structures (Gopher et al. 1994:Figure 1), and possible sleeping platforms were mentioned (Goring-Morris 1993:70). Shallow slabbed-lined pits were also present, including one within a structure. Two small structures (less than 1.5 m^2 each) were classified as nondomestic structures and not included in the floor area sample. In contrast, Wadi Jibba included six attached structures in a linear arrangement (Bar-Yosef 1984). Entrances all opened to one courtyard area, and no structural features are mentioned. Goring-Morris (1993) interprets the former site as a possible winter/spring settlement, while the latter is considered a winter settlement.

The architecture of PPNB Beidha represents the largest available sample, and hence is discussed in more detail. It was continuously occupied, and an indigenous architectural progression took place from clusters of oval posthouses (phase A), through individual oval and subrectangular buildings (phase B), and ultimately to full rectangular buildings with two stories (phase C) (Byrd 1994a, n.d.; Kirkbride 1966). A series of large and medium sized nondomestic structures and small storage facilities was also documented throughout the sequence. Initially, the layout of the community consisted of open courtyard spaces and small aggregates of buildings that were often constructed together using shared walls with entrances opening adjacent to each other. The phase A buildings were semisubterranean with very thin, plaster-coated floors with a mean interior area of 10.6 m^2. These structures typically had two wide entrances and a simple internal structural organization. The initial subphase A1 buildings lacked internal features, while almost half of the buildings had interior features in subphase A2, typically either plastered hearths or small stone platforms. No built storage features were documented. Burned buildings of this phase produced *in situ* portable artifacts (including bone tools, flaked stone tools, ground stone axes, pestles, stone grinders, and raw materials) along with nonportable artifacts, including querns, stone work slabs, and stone blocks on building floors. Variation within and between buildings is documented.

Phase B included above-ground curvilinear construction and semisubterranean construction and the first subrectangular and rectangular buildings. Buildings were freestanding with single, narrow entrances via stone steps in the later portion of this phase. Internal structural features (particularly plastered hearths and stone platforms) became more widespread, and almost one-quarter had more than one feature. Based on the presence of nonportable artifacts (querns and a stone bowl) and hearths, medium/small buildings were primarily interpreted as domestic units. Mean interior area was 6.9 m^2. Similar to phase A, some adjacent buildings lacked features and had no in situ portable artifacts.

Phase C buildings were generally built directly against each other, and primary multiple-construction events occasionally shared side walls. The interior area of medium sized buildings increased significantly and is represented by two-story corridor buildings with an upper story set slightly above ground level and a basement. The upper stories, although poorly preserved, were open in plan and contained plastered floors. In contrast, the basements had earthen floors and included up to seven very small rooms (~1.5 m by 1.0 m). The mean area of the basements was 12.8 m^2. The corridor building upper stories were estimated, based on preserved remnants, to be approximately 1.3 times (16.6 m^2) the area of the lower story, yielding an overall mean floor area of 29.4 m^2.

The internal spatial organization of two-story buildings became more diverse and complex with each story generally having a separate exterior entrance. The diversity of interior building features increased and seven types are represented. Internal partitions, wall niches, and monoliths first occurred in this phase, and partitions predominated. The relative frequency of different features varied between the upper and lower stories of corridor buildings. Corridor building upper stories (where preservation permits determination) and all but one single story building contained hearths, whereas corridor building basements did not. Partition walls were most prevalent in corridor building basements. The function of the corridor building lower and upper stories diverged markedly. Stone partitions across the entrances to small basement rooms may have functioned as dividing walls for storage set on clusters of stone slabs. During phase C, individual stone slabs only occurred in corridor building basements and primarily functioned as work surfaces for the production of various items, including beads and bone tools. Food-processing equipment, typically querns set into the floor, were consistently situated in rooms near the front of the basements.

Rectangular buildings typified the PPNB structures of Jericho (Garstang and Garstang 1948; Kenyon 1981), Beisamoun (Lechevallier 1978), Yiftahel (Garfinkle 1987), and 'Ain Ghazal (Banning and Byrd 1987; Rollefson and Simmons 1988; Rollefson et al. 1992). A distinct regional style of architecture termed *pier houses* characterizes these sites in the western hills, Jordan Valley, and the eastern hills (Byrd and Banning 1988). These rectangular structures were usually freestanding, and typically had an entrance at one of the short ends and a series of stone piers, mud brick piers, or wooden posts situated symmetrically along the long axis to support the roof. The area between piers was often sealed off, either fully or partly, to create work spaces or storage areas. Formal hearths constructed as part of the plaster floor were common, occurring most often in the center of the large back room. Low stone partitions occur occasionally as well. Unfortunately very few structures are preserved sufficiently that their full size can be calculated. The sample includes six buildings (three from Jericho, one each from 'Ain Ghazal, Beisamoun, and Yiftahel) with a mean interior area of 35.5 m^2.

DIACHRONIC TRENDS IN THE ARCHAEOLOGICAL RECORD

With respect to the interior size of domestic structures, no clear cut pan-southern Levantine trends are discernible between cultural complexes from the Early Epi-Paleolithic through the PPNA. Subsequently, a major increase in size occurs during the PPNB. However, overall sample sizes are low. Regional or site-specific trends appear to be a major factor contributing to the difficulty of discerning broad diachronic trends. Site setting (such as open air versus cave or hillside versus flat terrain) is potentially a major factor influencing the size and layout of structures. For example, size differences such as those between 'Ain Mallaha and Hayonim Cave during the Early Natufian may be due in part to natural factors. In addition, more regionally specific patterns are discernible particularly within the Negev highlands contrasted with the Jordan Valley. Notably for the Negev, the size of domestic structures remains statistically constant from the Late Natufian through the PPNB, with means from 4.5 m^2 to 7.5 m^2 (in fact, the lowest means occur during the PPNB), possibly due to more consistency in settlement practices through time. In contrast, structure size varies considerably between time segments at sites in the Jordan Valley. The largest pre-PPNB structures occur during the early phase at 'Ain Mallaha. These structures, however, were not all fully enclosed, as some appear to have been ramada-like (with a similar situation possibly at Wadi al-Hammeh 27 and earlier Ein Gev I). During the middle phase at 'Ain Mallaha they decrease in size, comparable to PPNA structures in the Jordan Valley. One of the interesting developments is the presence of a bimodal size difference in domestic structures during the Harifian of the Negev and Sinai and the PPNA in the Jordan Valley. Such a pattern may also be present during the Early and Final Natufian of 'Ain Mallaha, although the sample size is too small for confirmation. This variation in intrasite structure size may imply either functional or social organizational differences. The PPNA in the western hills is much more varied and includes one site (Hatoula) with structures of equivalent size and shape to those in the PPNA of the Jordan Valley and one site (Nahal Oren) with a series of much smaller structures.

The major and significant change in interior area occurs between the PPNA and the PPNB of the Jordan Valley and adjacent highlands. Mean values for the PPNB are 2.2 times those of the PPNA of the Jordan Valley (where the largest structures of the PPNA occur). They are also considerably larger than the mean values for the Early Natufian of 'Ain Mallaha (the next largest mean floor area). A similar trend occurs during the PPNB at Beidha. The domestic structures of the initial two phases are considerably smaller than the PPNA of the Jordan Valley (closer to building size in the Negev). In phase C, however, building size increases dramatically, reaching a value almost comparable to that of other PPNB sites and considerably larger than the PPNA of the Jordan Valley. As mentioned, the PPNB structures of the Negev remain small.

The second major trend entails the formal allocation and organization of interior space, and this changes dramatically over time. Although the small sample of structures predating the Natufian does not allow strong inferences to be made, it does reveal that structures were constructed and reused prior to the Natufian (sometimes with more permanent building materials) and occasionally included interior features. Hearths appear to occur primarily outside structures. Unfortunately a considerable time gap separates this sample from that of the Early Natufian.

The structures of the Early Natufian in the Jordan Valley differ from their predecessors only to the extent that more permanent building material was used, and the relative frequency of structures and features within them. During the Early Natufian, the presence of hearths (both makeshift and more formally constructed) and other interior features (including stone bins) occur within many structures at sites in the Jordan Valley and western highlands. Notably, the large ramada-like structures included multiple hearths or multiple stone bins. At Hayonim Cave, hearths were discerned in half the small structures. The buildup of debris in structures is well documented at a number of sites and floors noted within this debris accumulation. Whether this indicates reuse after a period of abandonment and trash dumping, or relatively constant accumulation of debris in a living context (with occasional surfaces discerned due to the presence of features), or both is uncertain. Large stone mortars, stone slabs, and cup-mark stones within a number of structures are interpreted as *in situ* nonportable artifacts that provide evidence for recurrent processing of food resources within houses. Notably, many of the Harifian structures have stone slabs with cup marks. Throughout the Epi-Paleolithic there are examples where spatial patterns in the distribution of artifacts, including portable artifacts, within a single structure are suggestive of spatial patterning of domestic activities including food production and tool maintenance and manufacture. These activities are best documented at 'Ain Mallaha and Wadi al-Hammeh 27. One of the major questions that needs to be further addressed is, what behavioral processes created these artifact sets on the floors of buildings that lack evidence of catastrophic abandonment. For apparent caches of larger portable artifacts such as pestles, why did they not continue to be used? For smaller artifact classes and those that represent debris or by-products of activities, floor clusters may represent gradual accumulation of residue from the same events carried out in the same general locations that were not subjected to cleaning and secondary disposal.

No major changes in the organization of interior space are discernible when the Early Natufian is contrasted with the PPNA. Hearths are documented in some structures but not all. In addition, they are not always formally constructed features but instead are makeshift. Other features, notably stone bins and large stones with cup hole or mortars, occur in some structures. This is a trend that appears to have begun earlier, particularly based on the Harifian evidence (although we lack Late Natufian structures in the Jordan Valley area).

As with building size, major changes are most notable during the PPNB. Unfortunately we lack evidence from the very onset of the PPNB to determine whether these changes are gradual or dramatic. Given the associated shift from round to rectangular building form, this change may not be gradual. No extensive use of formal subdivisions within structures occurs until the PPNB. Prior examples are infrequent (e.g., Netiv Hagdud building 8). PPNB pier buildings of the highlands and Jordan Valley are characterized by a range of facilities and small compartments. Thus, the syntax is more complex than it was previously (depicted graphically as tree diagrams versus the simple paths of previous structures). The small subdivided interior spaces within pier houses were presumably for processing or storage. Hearths are almost always present within these structures and are an integral part of the interior plaster floors. At Beidha these changes occur during the PPNB as basements in the final phase were used for a variety of activities, including storage, food processing, and artifact production. These endeavors were very compartmentalized. Cooking and eating, entertaining, and sleeping presumably occurred primarily in the upper stories, based on the presence of hearths and a more open plan. One of the interesting regional trends is that in the Negev, despite the presence of small round structures in the PPNB, the structures are more elaborate than previous ones in the same area (although not as elaborate as the PPNB rectangular structures of the highlands and Jordan Valley in terms of the number of internal features, including the presence of nonportable artifacts).

INTERPRETATION

Domestic Building Sizes and Implications for Household Sizes

Examination of diachronic trends in prehistoric domestic structure size and organization have revealed that the most prominent changes occurred between the PPNA and the PPNB. No major changes are well documented prior to this period (except for the site-specific exception of the Early Natufian at 'Ain Mallaha). Thus, the fundamental questions are: What was the nature of social organization that characterized pre-PPNB settlements? What do the changes from the PPNA to PPNB imply in terms of social organization. Does household size increase? Does it reflect a change from loose extended families with buildings inhabited by individuals to nuclear families residing in single buildings as Flannery (1972) hypothesized? Does it represent a shift from nuclear family residences to extended family residences? Are both periods characterized primarily by nuclear family or multifamily residences, or are other interpretations more plausible?

The size of the residential unit is one aspect of household reconstruction that has been the subject of considerable discussion. Cross-cultural studies of modern communities have asserted a correlation between the total floor area of dwellings and settlement population, although mean values vary considerably (e.g., Casselberry 1974; Cook and Heizer 1968; Kolb 1985; Naroll 1962; LeBlanc 1971). The most prominent study is that of Naroll (1962), who argued that a mean of 10 m^2 per individual could be effectively used to estimate prehistoric populations. Naroll's (1962) results in particular have been widely used by archaeologists (e.g., Flannery 1972) to estimate the number of individuals residing in a structure. Ethnoarchaeological studies of agricultural villages in southwest Asia provided further insight into the correlation between floor area and household size (Horne 1994; Kramer 1982; Watson 1978). Kramer's research at Shahabad, Iran, indicated that residential floor area values per individual, when those under the age of two were excluded, were consistent with Naroll and LeBlanc's findings (Kramer 1979:155). In contrast, the average dwelling area was 7.3 m^2 per person at Hasanabad, Iran (Watson 1978:137), lower than many cross-cultural findings (yet see Kolb 1985). These studies have found that variation in household size, rather than wealth, was positively correlated with the amount of roofed dwelling space (Kramer 1979:154–155; Watson 1978:133–137). This is, however, a subject of which there is some disagreement (e.g., Wilk 1983:111). One complicating factor revealed in studies of modern southwest Asian villages is deciding what portions of the habitation area to include in the calculation of floor area (see Jacobs 1979). The values given for Shahabad and Hasanabad in Iran only include the roofed living area, not associated storage facilities, stables, and enclosed formal courtyards. When portions of the latter are considered, then obviously these values would increase.

If Naroll's (1962) average floor area estimate is applied to the sites in this study, the number of individuals exceeds two per domestic structure only during the Early Natufian of 'Ain Mallaha and then much later in the PPNB. This suggests that nuclear families were only present in these contexts, as argued by Flannery (1972). If so, then we are required to explain why dwellings inhabited by nuclear families appeared in the Early Natufian at one site, were absent elsewhere, and disappeared from the Middle Natufian onward at 'Ain Mallaha until they reappeared in the PPNB. At PPNB Beidha, the primary aspect of the increase in building size during phase C was the addition of a basement with small storage and work rooms. If one contrasts the estimated size for phase C upper stories with that of the residential space for the previous phases, there is only a more modest increase to 16.9 m^2 of floor area (Table 1). Applying Naroll's (1962) floor area estimate for these Beidha phase C upper stories, there are still fewer than two people inhabiting each phase C building, too low to represent a nuclear family. Thus, following Naroll's formula, houses large enough for nuclear family residences occur primarily only during the PPNB and only at some sites (Table 1). Prior to that, structures were not large enough to contain complete households.

Although these ethnographic and ethnoarchaeological studies of modern communities are useful in revealing links between the floor area of dwellings and household size, one should be cautious in trying to directly apply these results to the interpretation of prehistoric sites (Kolb 1985). Other factors, such as community views on crowding, economic basis, residential patterns, degree of nucleation, and overall population size, can have significant effects on the mean amount of floor area per person (Casselberry 1974; Dohm 1990). The prehistoric sites in this southern Levantine study have quite different village layout and organization, lacking the large dwelling compounds with formal internal courtyards and multiple residential rooms within them that typify modern Iranian villages. Nor are there compelling reasons to expect that prehistoric household organization was analogous to modern Iranian villages, which are characterized by extended or joint families with typically more than one nuclear family residing within a compound. For example, modern Levantine villages, which have been subjected to less ethnoarchaeological study, often lack the courtyard or compound organization of Iranian villages (Antoun 1972; Aurenche and Desfarges 1982, 1985; Khammash 1986). Moreover, inheritance may have been different than in modern Iran and in historic Mesopotamia where compound rooms, including their associated storage units and stables, often become partitioned over time, and one family may own rooms of varying functions in different parts of the village (e.g., Horne 1994; Yoffee 1988).

Given this situation, it is worthwhile to consider whether the summary value stated by Naroll (1962) may be too high for this prehistoric context. Kolb (1985) and Wiesner (1982) have pointed out that Naroll s relationship was nonlinear and that there was a wide standard deviation. Ten of the eighteen ethnographic societies reported by Naroll (1962:Table 1) have mean floor areas of less than 7 m^2 per individual, and some values are considerably less. Kolb (1985:594) noted that the distribution was essentially bimodal and skewed due to a few examples with large means. Naroll (1962) also used the largest settlement within a society and incorporated many nonliving areas, such as wall bases, patios, courtyards, and alleys into his calculation. Watson (1978, 1979:296–297) noted that prehistoric Neolithic room and building size (in southwest Asia and elsewhere) were typically considerably smaller than in ethnographic settings such as Hasanabad. She suggested that this may not indicate a smaller number of individuals per room or buildings, but rather the amount of floor area per individual within living quarters was less than today. Given that early prehistoric settlements in the southern Levant were part of the transition from a long-enduring, relatively mobile hunting and gathering lifestyle to a radically new sedentary food producing economy, some persistence of earlier organizational traditions would not be surprising. Therefore, ethnoarchaeological results in hunter–gatherer use of space should be considered as well (e.g., Yellen 1977; Whitelaw 1991; Wiesner 1982).

Cook and Heizer's (1968) analysis of ethnographic California hunter–gatherers provides further support for nuclear families' use of much smaller structures. Of the thirty California Native American groups reviewed, none had mean floor areas approaching 10 m^2 per individual. Mean floor area per individual was always less than 8 m^2, with a range from 1.26 to 7.7 m^2 (Cook and Heizer 1968:Table 2). In addition, the !Kung of Africa, with nuclear households living in individual huts, had floor areas of 5.9 to 10.5 m^2 per person depending on the size of the camp (Wiesner 1974; Yellen 1977). Based on these results and the previous observations, Naroll's values are considered too high for the prehistoric hunter–gatherer and founder agricultural settlements under consideration.

I suggest that nuclear family households (most typically two adults and their offspring) best characterize domestic dwellings throughout this time frame in the southern Levant, from the Early Natufian through the PPNA and into the PPNB. Thus, no change in household size occurred during this period. This interpretation is supported by the correlation of ethnographic patterns in the size of houses and the floor area per individual of many hunters and gatherers and small agricultural communities and the prehistoric structures under consideration (Figure 2), along with several other lines of evidence. The latter include the widespread presence of hearths within prehistoric structures, the presence of other facilities such as bins and storage units within houses, and, where preservation and documentation permits, the presence of artifacts on the floors of buildings indicative of a range of domestic activities.

The interpretation that nuclear family households persisted from at least the Natufian through the PPNB differs considerably from Flannery's (1972) interpretation of individual structures prior to the PPNB being primarily inhabited by single individuals of polygamous households (see also Kuijt n.d.; Cauvin, Kuijt, Rollefson, all this volume). Community organizational patterns further support the nuclear family hypothesis. Structures at these southern Levantine sites were generally organized as clusters, often linear and along terraces. They were predominantly freestanding even when situated close together, with the only exceptions being sites in the Negev and Beidha phase A. Often, such as during the PPNA in the Jordan Valley, freestanding structures with considerable open space from one to the next predominate. Settlements were never organized similar to Flannery's (1972) ethnographic analog of pastoralist kraal-like compounds arranged around a large central space and contained by a wall or fence. These early settlements were less syntactically complex than kraal structures (Banning and Byrd 1989; Hillier et al. 1976). They were generally characterized by a three-syntax path represented by a single cell, which may have internal subdivisions, linked to an open space, and the two linked by a doorway. In contrast, a kraal compound is a six-syntax path where aggregates of closed cells are unified not by attachment but by inclusion.

The structures prior to the PPNB and during the PPNB are not considered to comprise multiple, extended families. If these structures housed multiple families, then they should be considerably larger. For example, the mean interior area for the small sample of multifamily dwellings presented by Cook and Heizer (1968) and Casselberry (1974) is never less than 75 m^2.

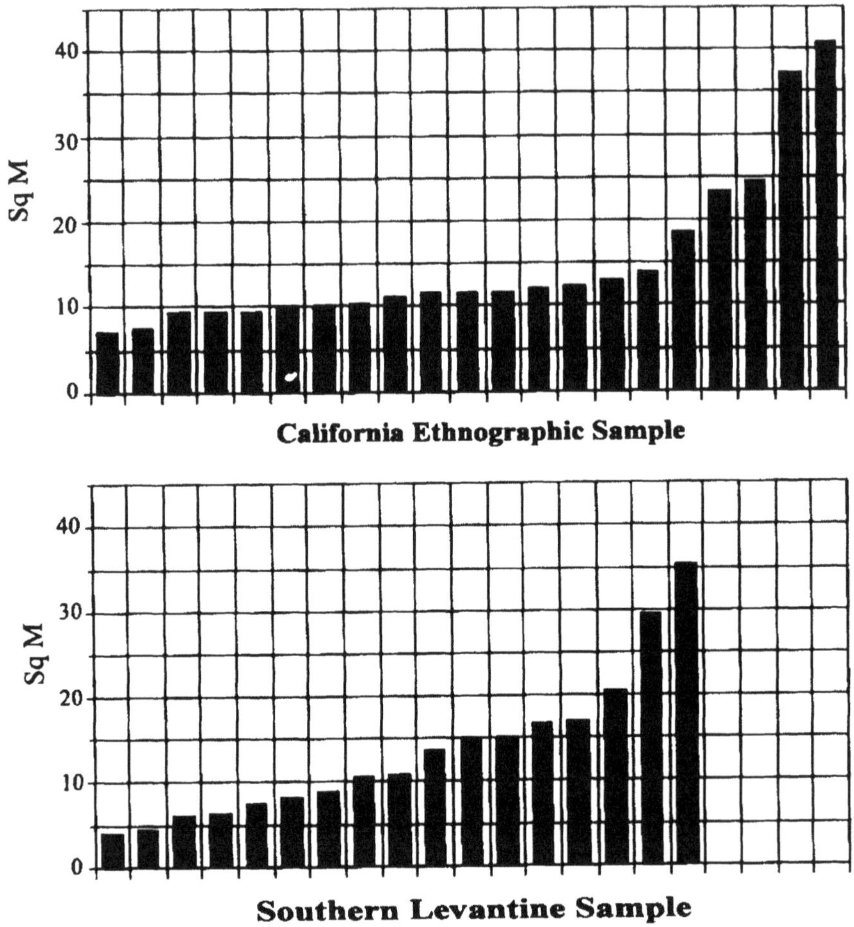

Figure 2. Distribution of mean interior areas of domestic structures for California ethnographic hunter–gatherer sample (Cook and Heizer 1968:Table 2) and southern Levantine Epi-Paleolithic and Early Neolithic Sample (from Table 1).

The buildings prior to the PPNB are rarely multiroom (which could be expected if individual families resided together), and the multiroom PPNB structures generally have a single large room with a hearth along with a corridor and very small storage or work rooms. These are also dissimilar to the modern Iranian village houses previously discussed. Nor are there other indicators of multiple families residing in a single structure, such as multiple hearths (hearths are never situated within the small PPNB rooms). The only possible exception to this pattern may be during the Early Natufian if the ramada-like structures at 'Ain Mallaha (and possibly Wadi Hammeh 27) housed multiple, perhaps extended, families. This latter possibility is based on the presence of multiple hearths within one structure, the lack of information regarding direct antecedents, and the small sample size for these Jordan Valley Natufian sites.

Other Factors Affecting Domestic Building Size and Organization

Thus, if household size did not change in a profound way during this time period (including from the PPNA to the PPNB), what accounts for the temporal and spatial patterned variation in the size of structures? I suggest that a series of factors influenced the size, internal organization, and sitewide spatial distribution of structures in this sample, and these factors varied in importance during this transition. These factors included natural aspects of individual site settings (essentially as a limiting factor, such as the use of caves or terraces), the degree of settlement permanence, the extent of reliance on domesticates (particularly cereals and legumes), the degree of household autonomy, and the total settlement population.

The expansion in the size of structures from the PPNA to the PPNB is interpreted as primarily due to a trend toward greater use of internal space for domestic activities, storage, and production, and possibly the increased size of settlement populations. A number of studies have noted an allometric relationship between settlement population and floor space (Cook and Heizer 1968; Wiesner 1974), and increasing floor space within larger settlements provided a mechanism for coping with crowding. One of the major changes asserted for the PPNB is an increase in the population of individual settlements (Bar-Yosef and Belfer-Cohen 1991; Bar-Yosef and Meadow 1995; Kuijt 1995). In addition, jealousy and interhousehold disagreements may have increased from the Natufian onward with greater settlement permanence and more circumscribed resources, particularly if subsistence plots were subject to household ownership (Byrd 1994a). This facilitated limiting information on the amount of foodstuffs and material within households and access to them, and provided an impetus for more production and processing activities to be carried out within structures. Hence, more interior space within structures was required to carry out these tasks and to store resources (Byrd 1994a; Redman 1983). Ultimately, this led to the formalization of interior spatial structure and an increase in building size in the PPNB. The size increase entailed the introduction of internal compartments either within pier houses or, as at Beidha, in subdivided basements that formalized the spatial discreteness of household production activities and storage. Thus, household autonomy increased over time. This formalization of internal space also may have aided in reinforcing the existing social order within households (Blanton 1995).

Greater household autonomy is often considered to be advantageous for facilitating increased productivity, particularly since it is easier to single out noncontributing individuals. If greater household autonomy was reflected in the ability of some families to mobilize more labor by getting unattached individuals into their families (Netting et al. 1984), then we would expect that some structures would be considerably larger than others. However, during the PPNB there was considerable uniformity in size and style of buildings at specific sites in the core agricultural portion of the southern Levant (Byrd and Banning 1988). This shared external appearance may reflect aspects of a shared ideological tradition that aimed at promoting an egalitarian ethos by limiting external variation.

With respect to variation between portions of the southern Levant, Natufian and PPNA settlements in the Jordan Valley were typically larger than their counterparts in more arid portions of the southern Levant as a result of some of the aforementioned factors, most notably greater settlement permanence and reliance on high-cost cereals and legumes, and possibly larger total settlement population. More interior space may have been needed and dedicated for drying, storing, and processing these gathered products, particularly during the PPNA. The trend toward larger interior spaces in the Natufian and PPNA of the Jordan Valley and subsequent PPNB throughout the agricultural zone did not occur in more arid southern and eastern areas and took place slightly later in time at Beidha (on the fringe of the dry farming zone today). The key factors for why settlements in these areas retained smaller structures during the PPNB entails seasonal use of settlements, with less reliance on domesticated plants and, hence, less need for storage and processing localities. Ethnographic studies of hunter–gatherers have noted that the degree of seasonality and anticipated mobility are important factors in determining the size of hunter–gatherer buildings (Kelly 1983; Kent 1990). Thus, the built environment of hunter–gatherers and small scale agriculturalists in the Negev (and other arid areas such as the eastern steppe/desert area) had a slightly different trajectory. This process included the introduction of more internal facilities for processing and production, and more activities may have been carried out within structures, but it did not entail an increase in the size of structures.

A major unresolved question entails the extent of household economic autonomy and whether it changed over time. At what point, if any, did nuclear households become the primary unit of production? Two possibilities can be offered. First, nuclear families were the key economic unit of production throughout this time sequence. Alternatively, related nuclear families (presumably extended families) shared resources and many processing and production tasks for portions of this time segment, particularly before the PPNB. The key lines of evidence for determining the degree of household economic autonomy are the location of storage facilities, processing areas, and production areas (see also Rosenberg, this volume). If nuclear families were the primary economic unit, then these activities should be associated with each household in a redundant manner. If extended families shared these economic activities, then one may expect that adjacent structures (particularly those that are built close to or abutted to each other) were occupied by related households. These households may have shared storage facilities, situated either within only one building or as smaller adjacent storage structures. One would also predict variation in the range of in situ artifacts between such structures, indicating specialized locations where tasks were carried out. For example, one building may have a number of ground stone artifacts and the adjacent one have none. Two potential factors that are difficult to identify archaeologically complicate this issue: the potential presence of specialized structures only for processing foodstuffs for community-wide events or festivals (Hayden 1990); and localities where independent households share, for socializing reasons, a work or processing area but do not share resources.

In general, the published material in the southern Levant lacks the level of detail needed to rigorously evaluate these hypotheses. Examination of the evidence from PPNB Beidha provides preliminary insights into this question and highlights the complexity of the issue and the importance of formation processes in creating floor-associated artifact assemblages (Byrd 1994a, n.d.). Some sharing of activities between adjacent structures may have occurred at Beidha, particularly during the initial subphase (A1) of occupation. There is much less evidence for sharing during the final phase (C) of occupation, although sudden abandonments (ideal for gaining such insights) rarely occurred. Beidha Phase A is characterized by a somewhat higher frequency of in situ floor assemblages (62% versus 52%) and a considerably higher frequency of in situ portable artifacts (38% versus 6%) than phase C. Not surprisingly, phase A has more burned occupation episodes than phase C (57% versus 4%).

Phase A buildings included clusters of portable artifacts in one or, sometimes, two internal areas, and this implies that certain activities were conducted in discrete areas or stored there for later use. A portion of each floor was devoid of any artifacts or features, and these open areas may have constituted sleeping locals. Amid the artifact clusters, a suite of related artifacts was situated directly adjacent to each other: these included polishing stones, hand-held ground stone, and pestles all in different buildings. This pattern may indicate that certain activities were group events where several individuals, perhaps from different households, participated. Unfortunately, comparable information is lacking for phase C.

Nonportable artifacts occurred both separately and in arrays on building floors in both phases. In some phase A situations, nonportable stone slab artifacts clustered on floors slightly separated from portable artifacts. In other phase A cases, as with the stone blocks on one floor and three querns on another, portable and nonportable artifacts co-occurred in a cluster. In both phases, querns occurred separately and within a cluster of portable artifacts. At least three querns were present in several buildings (two in phase A and one in phase C, a noncorridor building). Querns were also absent in two phase A burned buildings despite a sudden abandonment and large numbers of portable artifacts. This variability implies that associated plant food processing was not carried out in every domestic building but only in some buildings and perhaps on a larger scale. These trends correspond to architectural and spatial organizational patterns that indicate household autonomy increased over time. It does not preclude, however, that some sharing occurred between related households, particularly with respect to acquisition events. In addition, it appears that some processing and production activities (such as flint knapping and large-scale roasting) occurred in site areas that were much more public.

It is possible that similar patterns may occur at other sites (see Goring-Morris, Kuijt, Rosenberg, and Rollefson, this volume). At present, sample sizes are too small to clearly articulate the full range of patterns in domestic spatial organization. There is also a dearth of detailed studies of floor artifacts with a healthy consideration of the role that abandonment process and postabandonment events played in forming artifact assemblages on the floors of buildings. Indications entail the presence of a bimodal distribution of structures at some Harifian and Jordan Valley PPNA sites, the absence of hearths in some structures, and the occasional direct access between buildings. Potentially, some of these buildings may have served as interhousehold storage or task specific areas. However, much more information is needed to rigorously address this question, and this should be considered an important and potentially profitable line of investigation (Flannery 1993).

In sum, the domestic structures from the Natufian through the Middle PPNB are interpreted as the residences of nuclear families. No major changes in the numerical composition of the primary unit of residence occurred, and extended family residences were not present. The only possible exception entails the Early Natufian of the Jordan Valley. It is conceivable that the initial ramada-like structures were constructed to house multiple families. If so, this may reflect continuity with earlier, poorly documented social strategies that were characterized by more extensive community sharing behavior. Unfortunately our knowledge of pre-Natufian settlements in this area is extremely limited, and hence it is unclear if the Early Natufian spatial organization represents continuity or a major departure from the prior time segment (Goring-Morris 1995).

Transformation did take place in how nuclear households interacted with each other and household economic autonomy increased over time. The present evidence is indicative of gradual changes prior to the PPNB with possible indications of economic interdependence between related and adjacent nuclear households. The onset of the PPNB witnessed substantial change in the nature of domestic space. Increases in the sizes of structures and their interior elaboration (more formalized areas for storing and processing) is interpreted as evidence of increased nuclear family household autonomy. Nondomestic public buildings are also well documented during this time period (and to a limited extent before that). These changes in domestic structures occurred during a period of rich ideological development and expression that functioned on the household and suprahousehold level (Bar-Yosef and Meadow 1995 Byrd 1994a; Cauvin 1972; Kuijt 1995; Rollefson 1983, 1986; Rollefson and Simmons 1987).

Increases in the sizes of domestic structures, their permanence, and elaboration in the organization of interior space occurred initially in the area that includes the Jordan Valley and adjacent highlands. This area is considered to represent the heartland of early plant domestication (Bar-Yosef and Meadow 1995), and these organizational changes appear correlated with intensification of these resources and settlement permanence. A range of different expressions in local architecture, economic strategies, and the use of storage space occur in the more arid east and south.

DISCUSSION

Consideration of broader cross-cultural patterns in the anthropological literature provides a context for strengthening insights into why the prehistoric communities in the southern Levant were organized in the manner proposed. There has been much discussion regarding the conditions that favor nuclear versus extended family social organizations. Wilk and Rathje (1982) noted that variation within categories of household function (production, distribution, transmission, and reproduction) produce different types of households. For example, different sizes of households are more efficient at performing certain key tasks with respect to scheduling labor for production. Large households are better suited for coping with many simultaneous tasks. Pasternak et al. argue that extended families are predicted in situations where there are incompatible activity requirements that adults cannot avoid and where hired or coerced labor is not available (Pasternak, et al. 1976). Based on ethnographic surveys, these are typically women's economic tasks and child care, along with men's subsistence tasks that require overnight stays outside the home. An extending strategy, of which extended families are included within, is a way of coping with labor shortages and abundant land (Netting et al. 1984; Reyna 1976). As such, it is a way of intensifying resource procurement. Hayden and Cannon (1982) have observed that there must be strong factors (typically economic, environmental, or defensive) for groups of families to live closely together, and control over resources is the glue that keeps them together (although see Goring-Morris, Kuijt, Rollefson, Rosenberg, and Voigt, all in this volume, for consideration of the means by which community identities were structured within Near Eastern Neolithic contexts).

Nuclear families are more common in situations where these factors are not applicable, since extended families are harder to maintain due to conflict and jealousy. Smaller households are considered best suited for situations where mobility is important or where linear scheduling of spatially restricted resources takes place (Wilk and Rathje 1982). Small households are also considered the most effective way of passing resources for generation to generation since there is less conflict over inheritance. This point is reiterated by Hayden (1995b), who argued that ownership is correlated with high resource density and reliability. Ownership over resources is also more likely in areas where stored foodstuffs are used during lean seasons, and considerable effort is needed to prepare them for storage (Hayden 1995b:28). It is not surprising to produce surpluses in these situations.

Based on these cross-cultural generalized differences between different sizes of households, the transition from hunting and gathering to a strong reliance on domestic products in the southern Levantine is a situation best suited for the perpetuation of nuclear households. Key factors include the utilization of relatively abundant but spatially restricted resources that can be effectively exploited as small plots of land, a limited need for long stays outside the community by adult members of the family, a lack of multiple simultaneous tasks, and a lean season during which stored resources could be utilized. Thus, nuclear families appear to have had an adaptive advantage over extended households in this context.

Explaining why communities and the households that constitute them followed particular trajectories is a more difficult problem to unravel. Blanton (1994, 1995) has recently presented a model that attempts to explain cross-cultural patterns in the nature and organization of households. Households perpetuate themselves by controlling economic actions, marriage, and postmarital residence. However, households can change over time if it is considered in the best interest for all the individuals in the group. Blanton (1995:112–116) argues that examination of household idealogy (rituals and habits) provides a context for understanding how households change, particularly away from an egalitarian social organization. Habits perpetuate households and are the means for a household reproduction strategy. Material media messages such as gender use of space, shrines, and layout based on cosmology are reflections of household social structure. Symbolic behavior can create and perpetuate household inequality, and more rigid use of space typically occurs in less egalitarian societies. Blanton (1995) identified a statistical correlation between household centralization, pooling of resources, and arranged marriages. Nuclear families occur most frequently in more egalitarian situations since there is less household control over members, and individuals can marry early on and establish their own residences. In contrast, a continuity strategy that keeps adults part of an extended household is more often associated with increased standardization, more ideological reiteration of household values, and control of social habits.

In a related argument, Feinman (1995) has recently discussed alternative pathways to cultural complexity (see also Blanton et al. 1996). He asserts that there are two mutually opposing courses: a corporate based and network based pathway to inequality (yet see Hayden 1990, 1995b). A corporate based pathway emphasizes few overt differences between houses (in terms of size and value), minimal economic differentiation and display of individual wealth, integrated social segments, kin affiliation importance, and a focus on collective ritual, group space and ritual areas, and public construction. In contrast, network-based pathways are characterized by individual wealth, craft production, long-distance exchange, and personal networks of power (see also Renfrew 1974). Societies can go from one to another over time since both aspects are always in play.

These theoretical constructions provide additional insight into understanding how changes in domestic structures and the households that inhabited them correspond to developments on the community level during the Middle PPNB. The transition to food production in the southern Levant appears to be characterized by a corporate pathway that included public construction, group rituals and areas to conduct them, and little evidence of variation in individual or household wealth. If the nuclear families that composed these early villages controlled or owned plots of land that were the focus of plant resource exploitation, then these were inherently unequal in their yield. In addition, the more restricted sharing of resources between households and the greater household control over access and information increased the probability of jealousy and conflict between households. At the same time community and/or lineage level power and authority may have grown. This may have been the impetus for the similarity in size and outward appearance of PPNB domestic structures and uniformity in mortuary practices which reiterated a community-wide egalitarian ethos (Kuijt 1995, 1996). Standardization and elaboration of internal domestic structures in the PPNB aided household autonomy but also reinforced social order within and between households. Elders may have controlled prestige goods, postmarital residence choices, and other items, including marriage costs (Blanton 1995). Thus, the pathway that hunter–gatherers in the southern Levant took in becoming some of the earliest food producers was both novel and conservative. There was a tendency to try and reinforce community social order through the efforts of community leadership while the fundamental social units—households—became more autonomous and more unequal in their ability to perpetuate themselves.

ACKNOWLEDGMENTS

This chapter has benefited from discussions with Ofer Bar-Yosef, Ted Banning, Gary Rollefson, and, especially, Seetha Reddy. Thanks go to Ian Kuijt for asking me to contribute and for his comments on an earlier draft. Much of this chapter was written while holding a National Endowment for the Humanities University Teacher's Fellowship.

REFERENCES

Antoun, R. T., 1972, *Arab Village*. Indiana University, Bloomington, Indiana.

Arensburg, B. and Bar-Yosef, O., 1973, Human Remains from Ein Gev I, Jordan Valley, Israel. *Paléorient* 1: 201–206.

Aurenche, O., 1981, *La Maison Orientale. L'Archifecure du Proche-Orient Ancien des Origines au Milieu du Quatrième Millènaire*. Geuthner, Paris.

Aurenche, O., Desfarges, P., 1982, Utilisation et Transformation de l'Espace Architectural a El Kowm (Palmyre, Syrie). *Cahiers de L'Euphrates* 3: 99–113.

Aurenche, O., and Desfarges, P.,1985, Enquete Ethnoarchitecturale a Smakieh (Jordanie). Rapport Preliminaire. In *Studies in the History and Archaeology of Jordan, II*, edited by A. Hadidi, pp. 331–345. Routledge and Kegan Paul, London.

Banning, E. B. and Byrd, B. F., 1987, Houses and the Changing Residential Unit: Domestic Architecture at PPNB 'Ain Ghazal. *Proceedings of the Prehistoric Society* 53: 309–325.

Banning, E. B., and Byrd, B. F., 1989, Alernative approaches for exploring Levantine Neolithic architectire. *Paléorient* 15(1): 153–159.

Bar-Yosef, O., 1970, *The Epi-Paleolithic Cultures of Palestine*. Hebrew University, Jerusalem.

Bar-Yosef, O., 1984 Seasonality among Neolithic Hunter-Gatherers in Southern Sinai. In *Animals and Archaeology. 3. Early Herders and Their Flocks*, edited by Clutton-Brock, J. and C. Grigson, pp. 145–160, BAR International Series 202, Oxford.

Bar-Yosef, O., 1991, The Archaeology of the Natufian Layer at Hayonim Cave. In *The Natufian Culture in the Levant*, edited by Bar-Yosef, O., and F. Valla, pp. 81–92. International Monographs in Prehistory, Ann Arbor, Michigan.

Bar-Yosef, O., and Belfer-Cohen, A., 1991. From Sedentary Hunter-Gatherers to Territorial Farmers in the Levant. In *Between Bands and States*, edited by S. A. Gregg, pp. 181–202. Center for Archaeological Investigations Occasional Paper 9. Southern Illinois University, Carbondale, Illinois.

Bar-Yosef, O., and Gopher, A., 1997, *An Early Neolithic Village in the Jordan Valley. Part I: The Archaeology of Netiv Hagdud.* American School of Prehistoric Research Bulletin 43: Cambridge, Massachusetts.

Bar-Yosef, O., Gopher, A., and Goring-Morris, A. N., 1980, Netiv Hagdud: A "Sultanian" Mound in the Lower Jordan Valley. *Paléorient* 6: 201–206.

Bar-Yosef, O., Gopher, A., Tchernov, A., and Kislev, M. E., 1991, Netiv Hagdud: An Early Neolithic Village Site in the Jordan Valley. *Journal of Field Archaeology* 18: 405–424.

Bar-Yosef, O. and Meadow, R. H., 1995, The Origins of Agriculture in the Near East. In *Last Hunters–First Farmer,* edited by T. Douglas Price and A. Birgitte Gebaure, pp. 39–94. School of American Research,Santa Fe, New Mexico.

Belfer-Cohen, A. 1988, *The Natufian Settlement at Hayonim Cave.* Ph.D. dissertation, Hebrew University, Jerusalem.

Blanton, R. E., 1994, *Houses and Households. A Comparative Study.* Plenum Press, New York.

Blanton, R. E., 1995, The Cultural Foundations of Inequality in Households. In *Foundations of Social Inequality,* edited by T. E. Price and G. M. Feinman, pp. 105–128. Plenum Press, New York.

Blanton, R. E., Feinman, G. M., Kowalewski, S. A., and Peregrine, P. N., 1996, A Dual-Processual Theory for the Evolution of Mesoamerican Civilization. *Current Anthropology* 37(1): 1–14.

Brooks, A. and Yellen, J., 1987, The Preservation of Activity Areas in the Archaeological Record: Ethnoarchaeological and Archaeological Work in Northwest Ngamiland, Botswana. In *Method and Theory for Activity Area Research,* edited by S. Kent. pp. 63–106. Columbia University Press, New York.

Bender, B., 1978, Gatherer-Hunter to Farmer: A Social Perspective. *World Archaeology* 10(2): 204–237.

Bender, B., 1990, The Dynamics of Nonhierarchial Societies. In *The Evolution of Political Systems,* edited by S. Upham, pp. 247–263. Cambridge University Press, Cambridge.

Byrd, B. F., 1994a, Public and Private, Domestic and Corporate: The Emergence of the Southwest Asian Village. *American Antiquity* 58(4): 639–666.

Byrd, B. F., 1994b, Late Quaternary Hunter-Gatherer Complexes in the Levant between 20,000 and 10,000 BP. In *Late Quaternarty Chronology and Paleclimates of the Eastern Mediterranean,* edited by O. Bar-Yosef and R. Kra. pp. 205–226. Radiocarbon Publications, Tucson, Arizona.

Byrd, B. F., n.d., The Neolithic Village of Beidha: Architecture, Occupation History and Spatial Organization. In *The Excavations of Mrs. Diana Kirkbride-Helbeck,* by British Academy Monographs in Archaeology. Oxford University Press, Oxford.

Byrd, B. F. and Banning, E. B., 1988. Southern Levantine Pier Houses: Intersite Architectural Patterning during the Pre-Pottery Neolithic B. *Paléorient* 14(1): 65–72.

Byrd, B. F., and Monahan, C., 1995, Death, mortuary ritual, and Natufian social structure. *Journal of Anthropological Archaeology,* 14: 251–287.

Cameron, C. M., 1990, Pit Structure Abandonment in the Four Corners Region of the American Southwest: Late Basketmaker III and Pueblo I Periods. *Journal of Field Archaeology* 17(1): 27–37.

Cameron, C. M. and S. A. Tonka, 1993, *Abandonment of Settlements and Regions: Ethnoarchaeological and Archaeological Approaches.* Cambridge University Press, Cambridge.

Carr, C. 1984, The Nature of Organization of Intrasite Archaeological Records and Spatial Analytic Approaches to Their Investigation. In *Advances in Archaeological Method and Theory,* Volume 7, edited by M. B. Schiffer, pp. 103–222. Academic Press, New York.

Carr, C., 1987, Dissecting Intrasite Artifact Palimpsests. In *Method and Theory for Activity Area Analysis,* edited by S. Kent , pp. 236–294. Columbia University Press, New York.

Casselberry, S., 1974, Further Refinement of Formulae for Determining Population from Floor Area. *World Archaeology* 6: 118–122

Cauvin, J., 1972, *Religions Néolithiques de Syro-Palestine.* Centre de Recherches d'Ecologie et de Prehistore. Maisonneane, Paris.

Cauvin, J., 1994, *Naissance des divinités, Naissance de l agriculture.* CNRS, Paris.

Cook, S. F., and Heizer, R. F., 1968, Relationship among Houses, Settlement Areas, and Population in Aboriginal California. In *Settlement Archaeology,* edited by K. C. Chang, pp. 79–116. National Press Books, Palo Alto, California.

Dohm, K., 1990, Effect of Population Nucleation on House Size for Pueblos in the American Southwest. *Journal of Anthropological Archaeology* 9: 201–239.

Edwards, P., 1991, Wadi Hammeh 27: An Early Natufian Site at Pella, Jordan. In *The Natufian Culture in the Levant,* edited by Bar-Yosef, O. and F. Valla, pp. 123–148. International Monographs in Prehistory, Ann Arbor, Michigan.

Feinman, G. M., 1995, The Emergence of Inequality: A Focus on Strategies and Processes. In *Foundations of Social Inequality,* edited by T. D. Price and G. M. Feinman, pp. 255–280. Plenum Press, New York.

Flannery, K. V., 1972, The Origin of the Village as a Settlement Type in Mesoamerica and the Near East: A Comparative Study. In *Man, Settlement and Urbanization,* edited by P. Ucko, R. Tringham, and G. Dimbleby, pp. 23–53. Duckworth, London.

Flannery, K. V., 1993, Will the Real Model Please Stand Up: Comments on Saidel's "Round House or Square?" *Journal of Mediterranean Archaeology* 6(1): 109–117.

Garfinkel, Y. 1987 Yiftahel: A Neolithic Village from the Seventh Millennium B.C. in Lower Galilee, Israel. *Journal of Field Archaeology* 14: 199–212.

Garstang, J. and Garstang, J. B. E., 1948, *The Story of Jericho,* 2nd edition. Marshal, Morgan and Scott, London.

Goody, J. (ed.) 1958, *The Developmental Cycle in Domestic Groups.* Cambridge University Press, Cambridge.

Gopher, A., Goring-Morris, A. N., and Gordon, D., 1994, Nahal Issaron: The Lithics of the Late PPNB Occupation. In *Neolithic Chipped Stone Industries of the Fertile Crescent,* edited by H. G. Gebel and S. K. Kozolowski, pp. 479–494. Ex Oriente, Berlin.

Goring-Morris, A. N., 1987, *At the Edge: Terminal Pleistocene Hunter-Gatherers in the Negev and Sinai.* BAR International Series 361, Great Britain.

Goring-Morris, A. N., 1991, The Harifian of the Southern Levant. In *The Natufian Culture in the Levant,* O. Bar-Yosef, and F. R. Valla , pp. 173–216. International Monographs in Prehistory, Ann Arbor, Michigan.

Goring-Morris, A. N., 1993, *From Foraging to Herding in the Negev and Sinai: The Early to Late Neolithic Transition. Paléorient* 19(1): 65–89.

Goring-Morris, A. N., 1995, Complex Hunter/Gatherers at the End of the Paleolithic (20,000–10,000 BP). In *The Archaeology of Society in the Holy Land,* edited by T. E. Levy , pp. 141–168. Leicester University Press, London.

Goring-Morris, A. N. and Gopher, A., 1983, Nahal Issaron: A Neolithic Settlement in the Southern Negev. *Israel Exploration Journal* 33(3): 149–162.

Hammel, E. A. and Laslett, P., 1974, Comparing Household Structure over Time and between Cultures. *Comparative Studies in Society and History* 16: 73–109.

Hayden, B., 1990 Nimrods, Piscators, Pluckers, and Planters: The Emergence of Food Production. *Journal of Anthropological Archaeology* 9: 31–69.

Hayden, B., 1995a, An Overview of Domestication. In *Last Hunters–First Farmer,* edited by T. D. Price and A. Birgitte Gebaure, pp. 273–300. School of American Research, Sante Fe, New Mexico.

Hayden, B., 1995b, Pathways to Power: Principles for Creating Socioeconomic Inequalities. In *Foundations of Social Inequality,* edited by T. D. Price and G. M. Feinman, pp. 15–78. Plenum Press, New York .

Hayden, B., and Cannon, A., 1982, The Corporate group as an archaeological unit. *Journal of Anthropological Archaeology* 1: 132–158.

Hayden, B., and Cannon, A., 1983 Where the Garbage Goes: Refuse Disposal in the Maya Highlands. *Journal of Anthropological Archaeology* 2: 117–163.

Henry, D. O., 1976, Rosh Zin: A Natufian Settlement near Ein Avdat. In *Prehistory and Paleoenvironments in the Central Negev, Israel,* Volume 1, edited by A. E. Marks, pp. 317–348. Southern Methodist University, Dallas, Texas.

Henry, D. O., 1985, Preagricultural sedentism: The Natufian example. In *Prehistoric Hunter-Gatherers: The Emergence of Cultural Complexity,* edited by T. D. Price and J. A. Brown, pp. 365–381. Academic Press, New York.

Hiller, B., Leaman, A., Stansall, P., and Bedford, M., 1976, Space syntax. *Environment and Planning B* 3:147–185.

Hodder, I., 1990, *The Domestication of Europe: Structure and Contingency in Neolithic Societies.* Basil Blackwell, Oxford.

Hole, F., 1984, A Reassessment of the Neolithic Revolution. *Paléorient* 10(2):49–60.

Hole, F., 1987 Archaeology of the Village Period. In *The Archaeology of Western Iran,* edited by F. Hole, pp. 29–78. Smithsonian Institution Press, Washington, D.C.

Horne, L., 1980, Village Morphology. *Expedition* 22(4):18–23.

Horne, L.,1994, *Village Spaces: Settlement and Society in Northeastern Iran.* Smithsonian Institution Press, Washington D.C.

Jacobs, L., 1979, Tell-i-Nun: Archaeological Implications of a Village in Transition. In *Ethnoarchaeology: Implications of Ethnography for Archaeology,* edited by C. Kramer, pp. 175–191. Columbia University, New York.

Kelly, R. L., 1983, Hunter-Gatherer Mobility Strategies. *Journal of Anthropological Research* 39: 277–306.

Kent, S., 1990, *Domestic Architecture and the Use of Space: An Interdisciplinary Cross-Cultural Study.* Cambridge University, Cambridge.

Kenyon, K.,1981, *Excavations at Jericho, Volume 3, The Architecture and Stratigraphy of the Tell.* London: The British School of Archaeology in Jerusalem, London.

Khammash, A., 1986, *Note on Village Architecture in Jordan.* University of Southwestern Louisiana, Lafayette, Louisiana.

Kirkbride, D., 1966, Five Seasons at the Prepottery Neolithic Village of Beidha in Jordan. *Palestine Exploration Quarterly* 98:5–61.

Kolb, C. C., 1985, Demographic estimates in Archaeology: Contributions for Ethnoarchaeology on Mesoamerican Peasants. *Current Anthropology* 26(5):581–599.

Kramer, C., 1979, An Archeological View of a Contemporary Kurdish Village: Domestic Architecture, Household Size and Wealth. In *Ethnoarchaeology: Implications of Ethnography for Archaeology,* C. Kramer, pp. 139–163. Columbia University, New York.

Kramer, C., 1982, *Village Ethnoarchaeology, Rural Iran in Archaeological Perspective.* Academic Press, New York.

Kramer, C., 1983 Spatial Organization in Contemporary Southwest Asian Villages and Archaeological Sampling. In *The Hilly Flanks and Beyond,* edited by T. C. Young, P. E. L. Smith, and P. Mortensen, pp. 347–368. Studies in Ancient Oriental Civilizations No. 36. Oriental Institute, Chicago, Illinois.

Kuijt, I., 1995, *New Perspectives on old Territories: Ritual Practices and the Emergence of Social Complexity in the Levantine Neolithic.* Ph.D. dissertation. Harvard University. University Microfilms, Ann Arbor, Michigan.

Kuijt, I., 1996, Negotiating Equality through Ritual: A Consideration of Late Natufian and Pre-

Pottery Neolithic A Period Mortuary Practices. *Journal of Anthropological Archaeology* 15(4):313–336.

Kuijt, I. and Bar-Yosef , O., 1994, Radiocarbon Chronology for the Levantine Neolithic: Observations and Data. In *Late Quaternary Chronology and Paleoclimates of the Eastern Mediterranean,* edited by O. Bar-Yosef and R. S. Kra, pp. 227–245. Radiocarbon, Tucson, Arizona.

Lawrence, D. and Low, S., 1990, The Built Environment and Spatial Form. *Annual Reviews in Anthropology* 19:453–505.

LeBlanc, S., 1971, An Addition to Naroll's Suggested Floor Area and Settlement Population Relationship. *American Antiquity* 36: 210–211.

Lechevallier, M., 1978, *Abou Gosh et Beisamoun, Deux Gisements du VIIe Millenaire Avant L'ere Chretienne en Israel.* Memoires et Travaux du Centre de Recherches Prehistoriques Francaise de Jerusalem, No. 2. Association Paléorient, Paris.

Lechevallier, M., and A. Ronen 1985, *Le Site Natoufien-Khiamien de Hatoula, Pres de Latroun, Israel. Jerusalem.* Centre de Recherche Francais de Jerusalem, 1.

Moore, A. M. T., 1985, The Development of Neolithic Societies in the Near East. *Advances in World Archaeology* 4:1–69.

Nadel, D., 1991, Ohalo II—The third season. *Mitekufat Haeven* 23:48–59.

Nadel, D., Carmi, I., and Sega, D., 1995, Radiocarbon Dating of Ohalo II: Archaeological and Methodological Implications. *Journal of Archaeological Science* 22:811–822.

Naroll, R., 1962, Floor Area and Settlement Population. *American Antiquity* 27:587–588.

Netting, R. M., 1982, Some Home Truths on Household Size and Wealth. *American Behavioral Scientist* 25:641–662.

Netting, R. M., 1990, Population, Permanent Agriculture, and Polities: Unpacking the Evolutionary Portmanteau. In *The Evolution of Political Systems,* edited by S. Upham pp. 21–61. Cambridge University Press, Cambridge.

Netting, R. Mc., Wilk, R. R., and Arnould, E. J., 1984, *Households: Comparative and Historical Studies of the Domestic Group.* University of California Press, Berkeley, California.

Noy, T., 1989, Gigal I., A pre-Pottery Neolithic Site, Israel. The 1985–1987 Seasons. *Paléorient* 15(1): 15–22.

Pasternak, B., Ember, C. R., and Ember, M., 1976. On conditions favoring extended family households. *Journal of Anthropological Anthropological Research* 32: 109–123.

Perrot, J., 1960, Excavation at Eynan (Ein Mallaha), 1959 Season: Preliminary Report. *Israel Exploration Journal 10*: 14–22.

Perrot, J., 1966, Le Gisement Natoufien de Mallaha (Eynan), Israël. *L'Anthropologie* 70(5): 437–484.

Perrot, J. and Ladiray, D., 1988, *Les Hommes de Mallaha (Eynan) Israël, les Sepulchres.* Memoires et Travaux du Centre de Recherche Francaise de Jerusalem, Number 7. Association Paléorient, Paris.

Price, T. D., and Feinman, G. M., 1995, *Foundations of Social Inequality.* Plenum Press, New York.

Redman, C. L., 1978, *The Rise of Civilization.* Freeman, San Francisco, California.

Redman, C., L., 1983 Regularity and Change in the Architecture of an Early Village. In *The Hilly Flanks and Beyond,* edited by T. C. Young, P. E. L. Smith, and P. Mortensen, pp. 189–196. Studies in Ancient Oriental Civilization No. 36. Oriental Institute, Chicago, Illinois.

Renfrew, C. 1974, Beyond a Subsistence Economy: The Evolution of Social Organization in Prehistoric Europe. In *Reconstructing Complex Socieites,* edited by C. B. Moore, pp. 69–85. Supplement to the Bulletin of the American Schools of Oriental Research No. 20, American Schools of Oriental Research, Cambridge, Massachusetts.

Reyna, S. P., 1976, The Extending Strategy: Regulation of the Household Dependency Ratio. *Journal of Anthropological Research* 32:182–198.

Rollefson, G. O., 1983, Ritual and Ceremony at Neolithic 'Ain Ghazal (Jordan). *Paléorient* 9(2):29–37.

Rollefson, G. O., 1986, Neolithic 'Ain Ghazal (Jordan): Ritual and Ceremony, II. *Paléorient* 12(1):45–52.

Rollefson, G. O., and Simmons, A. H., 1987, The Life and Death of 'Ain Ghazal. *Archaeology* 40(6):38–45.

Rollefson, G. O., and Simmons, A. H., 1988, The Neolithic Settlement at 'Ain Ghazal. In *The Prehistory of Jordan,* edited by A. N. Garrard and H. G. Gebel, pp. 393–421. BAR International Series 396, Oxford.

Rollefson, G. O., Simmons, A. H., and Kafafi, Z., 1992 Neolithic Cultures at 'Ain Ghazal, Jordan. *Journal of Field Archaeology* 19: 443–470.

Saidel, B. A., 1993, Round House or Square? Architectural Form and Socio-Economic Organization in the PPNB. *Journal of Mediterranean Archaeology* 6(1): 65–108.

Seymour, D., and Schiffer, M., 1987, A Preliminary Analysis of Pithouse Assemblages from Snaketown, Arizona. In *Method and Theory for Activity Area Research,* edited by S. Kent, pp. 549–603.Columbia University Press, New York.

Schiffer, M. B., 1987, *Formation Processes of the Archaeological Record.* University of New Mexico Press, Albuquerque, New Mexico.

Schiffer, M. B.,1983, Toward the Identification of Formation Processes. *American Antiquity* 48: 675–706.

Scott, T. R., 1977, The Harifian of the Central Negev. In *Prehistory and Paleoenvironments in the Central Negev, Israel,* Volume 2, edited by A. E. Marks, pp. 271–322. Southern Methodist University, Dallas, Texas.

Southall, A., 1988, On Mode of Production Theory: The Foraging Mode of Production and the Kinship Mode of Production. *Dialectical Anthropology* 12: 165–192.

Stekelis, M., and Yizraely, T., 1963, Excavations at Nahal Oren: Preliminary Report. *Israel Exploration Journal* 13(1): 1–12.

Upham, S. 1990, Decoupling the Process of Political Evolution. In *The Evolution of Political Systems,* edited by S. Upham, pp. 1–20. Cambridge University, New York.

Valla, F., 1987, Chronologie Absolue et Chronologies Relatives Dans le Natoufien. In *Chronologies of the Near East,* edited by O. Aurenche, J. Ervin, F. Hours, pp. 267–294. BAR International Series 379. Oxford.

Valla, F., 1991, Les Natoufiens de Mallaha et L'espace. In *The Natufian Culture in the Levant,* edited by O. Bar-Yosef and F. Valla, pp. 111–122. International Monographs in Prehistory, Ann Arbor, Michigan.

Valla, F., 1994, The First Settled Societies—Natufian (12,500–10,200 BP). In *The Archaeology of Society in the Holy Land,* edited by T. E. Levy, pp. 170–187. Leicester University, London.

Valla, F. R., Khalaily, H., Samuelian, N., Bocquentin, F., Delage, C., Valentin, B., Plisson, H., Rabinovich, R., and Belfer-Cohen, A., 1999, Le Natiufien Final et les Nouvelles fouilles a Mallaha (Eynan), Israel 1996–1997. *Journal of the Israel Prehistoric Society* 28:105–176.

Valla, F. R., Le Mort, F., and Plisson, H., 1991, Les Fouilles en Cours sur la Terrasse D'Hayonim. In *The Natufian Culture in the Levant,* edited by O. Bar-Yosef and F. Valla, pp. 930–1100. International Monographs in Prehistory, Ann Arbor, Michigan.

Voigt, M. M., 1990, Reconstructing Neolithic Societies and Economies in the Middle East: An Essay. *Archaeomaterials* 4(1): 1–14.

Waterbolk, H. T., 1987, Working with Radiocarbon Dates in Southwestern Asia. In *Chronologies in the Near East,* edited by O. Aurenche, J. Evin, and F. Hours, pp. 39–60. British Archaeological Reports International Series 379, Oxford.

Watson, P. J., 1978, Architectural Differentiation in Some Near Eastern Communities, Prehistoric and Contemporary. In *Social Archaeology*, edited by C. Redman, M. J. Berman, E. V. Curtin, W. T. Langhorne Jr., N. M. Versaggi, and J. C. Wanser, pp. 131–157. Academic Press, New York.

Watson, P. J., 1979, *Archaeological Ethnography in Western Iran*. Viking Fund Publications in Anthropology, No. 57. University of Arizona, Tucson, Arizona .

Whitelaw, T., 1991, Some Dimensions of Variability in the Social Organization of Community Space among Foragers. In *Ethnoarchaeological Approaches to Mobile Campsites*, edited by C. S. Gamble and W. A. Boismier, pp. 139–188. International Monographs in Prehistory, Ann Arbor, Michigan.

Wiesner, P,. 1974, A Functional Estimator of Population from Floor Area. *American Antiquity* 39(2): 343–350.

Wiesner, P., 1982, Beyond Willow Smoke and Dog's Tails: a Comment on Binford's Analysis of Hunter-Gatherer Settlement Systems. *American Antiquity* 47: 171–177.

Wilk, R. R. 1983, Little House in the Jungle: The Causes of Variation in House Size among Modern Kekchi Maya. *Journal of Anthropological Archaeology* 2(2): 99–116.

Wilk, R. R., 1990, The Built Environment and Consumer Decisions. In *Domestic Architecture and the Use of Space*, edited by S. Kent, pp. 34–42. Cambridge University Press, New York.

Wilk, R. R., and Netting, R., 1984, Households: Changing Forms and Functions. In *Households: Comparative and Historical Studies of the Domestic Group*, edited by R. Netting, R. Wilk, and E. Arnould, pp. 1–28. University of California, Berkeley, California.

Wilk, R. R., and Rathje, W. L., 1982, Household Archaeology. *American Behavioral Scientist* 25:(6) 617–639.

Wilson, P., 1988, *The Domestication of the Human Species*. Yale University Press, New Haven, Connecticut.

Wright, G. A., 1978, Social differentiation in the Early Natufian. In *Social Archaeology, Beyond Subsistence and Dating*, edited by C. Redman, M. J. Berman, E. V. Curtin, W. T. Langhorne Jr., N. M. Versaggi and J. C. Wanser, pp. 201–223. Academic Press, New York.

Yellen, J. E., 1977, *Archaeological Approaches to the Present: Models for Reconstructing the Past*. Academic Press, New York.

Yoffee, N., 1988, Aspects of Mesopotamian Land Sales. *American Anthropologist* 90(1): 119–130.

Part III

Organization of Social Relations

Mortuary Ritual, Social Differentiation, and Settlement Systems

While providing explorations of a range of material realms of the Neolithic, including mortuary practices, ritual, architectural systems, and settlement patterns, the chapters in this section share a common interest: examining the possible ways in which different forms of material culture reflect growing social differentiation and the social use of ritual practices in Pre-Pottery Neolithic contexts. Traditional arguments for the emergence of social differentiation in the Neolithic have often failed to provide detailed considerations of archaeological correlates and data sets. Through in-depth exploration of mortuary practices, architectural systems, the organization of ritual, and consideration of how Neolithic site and regional settlement patterns might reflect past social organization and change through time, the authors address the nature of Neolithic social systems and the emergence of social inequality in general.

In Chapter 5 Nigel Goring-Morris examines the archaeological evidence for variability in mortuary practices seen at the Middle Pre-Pottery Neolithic settlement of Kfar HaHoresh and places ideological, ritual, and mortuary practices within a wider developmental context of the Levantine Pre-Pottery Neolithic. He argues that mortuary and architectural data, such as differential treatment of the dead through the elaborate painting and plastering of human skulls in the MPPNB, indicate that incipient social hierarchies and ritual ideologies would have been necessary to regulate and edify increasingly complex interpersonal and intragroup relationships, particularly in the control over economic resources, land tenure, ownership, and inheritance.

In light of the variety of mortuary practices seen at Kfar HaHoresh, as well as those noted from other settlements, Goring-Morris argues that this site may have served as a secluded regional mortuary/ritual center, potentially visited by kin from surrounding settlements for common rituals associated with funerary rites. Moreover, the author argues that the codification of funerary tasks are linked to increases in the size of communities, adjustments to new social and economic systems, and the breakdown of egalitarian lifeways. This is believed to be expressed through the emergence of differential wealth, ownership, and the development of differential inherited status in the MPPNB.

Working on the assumption that the power of household ritual is based, in part, on the realization that mortuary practices are a form of public action rather than a direct reflection of the status, authority, and importance of the dead, Ian Kuijt, in Chapter 6, examines mortuary practices during the Middle Pre-Pottery Neolithic B period in the south-central Levant. The author argues that mortuary rituals were organized by a series of ordering principles based on the age, and possibly status, of the deceased. Kuijt explores three dimensions of regional mortuary practices: the employment of decapitation as a common theme, the numeric organization within human skull caches, and the social impact of secondary mortuary practices and skull caching on community integration and cohesion. It is argued that the development of specific ritual practices were linked to the need to maintain existing household political, economic, and social ties during times of social, resource, and environmental stress. Thus, Kuijt envisions Neolithic ritual and mortuary practices as having been intentionally employed as a means of consolidating community membership, and that similarities in these practices across much of the Near East illustrate the importance these practices played as physical and symbolic vehicles for the creation of social meaning and community identity.

In a detailed case study of data from 'Ain Ghazal, Gary Rollefson outlines how the nature and location of human burial practices, the use of small animal and human figurines, as well as stylistic changes of large statuary from the Middle Pre-Pottery Neolithic B period to the Pottery Neolithic period inform us as to changes through time in the employment of specific ritual practices at an individual settlement. Available evidence from 'Ain Ghazal indicates that in the MPPNB different burial systems reflect—either symbolically or physically—some form of differential status of individuals at the household and community level. Reflecting upon the architectural evidence at 'Ain Ghazal throughout time, Rollefson notes that as with other Pre-Pottery Neolithic settlements in the Near East, there is evidence to suggest that rare buildings were designed for community rituals. He directs attention to the past existence of different forms of ritual behavior that may reflect increasingly inclusive social groups, from individual household usage of small figurines to the use of large human statues and busts as part of a collective or individual representation of membership within communities. In discussing the transition from the Pre-Pottery to Pottery Neolithic, Rollefson notes that at some point between 8,000 and 7,750 bp, residents at 'Ain Ghazal ceased to practice skull removal and to regularly employ small and large statues in ritual and, increasingly, buried people in groups as secondary disarticulated bundles. He argues that this transition reflects a reduction in ancestor veneration and increased residential mobility and part-time residence at 'Ain Ghazal.

In Chapter 8 Frank Hole critically examines the argument that the size of Neolithic settlements across the Near East reflects relative social complexity, the existence of a social hierarchy within individual communities, and regional site hierarchies. Citing ethnographic sources, Hole notes that there are many reasons why select Neolithic settlements may have been larger than others: they were political, economic, religious centers, or they were located near rich environmental locations. He argues that variation in Neolithic settlement size, however, can also be a by-product of the sequential use of settlements, wide spacing of dwelling, and the multiple use of individual structures. Comparing PPNB settlements from the Near East, Hole examines the relationship between settlement size, the presence or absence of monumental, public, or cultic structures, and evidence for elite status within social systems. While noting there was a tendency for community members from larger settlements to have more exotic, nonlocal goods, Hole argues that the concept of regional centers and settlement hierarchy may not be applicable to PPNB. Moreover, the overall size of individual settlements is not linked to the presence of nonresidential community buildings, and he concludes that the size of settlements does not appear to be linked to increased complexity in social relationships or a determining factor in the development of specialization.

In a further illustration of a renewed attention to the possible links between regional settlement systems and Neolithic social organization, Alan Simmons, in Chapter 9, explores demographic shifts in the Pre-Pottery Neolithic B of the south-central Levant, as well as the process of abandonment of large aggregate villages at the beginning of the eighth millennium. He argues that this abandonment may have been related to deteriorating climatic conditions and an environmental crisis brought about by expanding population levels, intensive agriculture and herding, and deforestation for cooking fuel. Noting that the majority of existing PPNA archaeological data appear to be located west of the Jordan River and that the majority of large Late PPNB agricultural communities appear to be located along the environmentally marginal zones of the hills along the Mediterranean and eastern desert ecotone, Simmons presents a three-stage model outlining this demographic shift. He argues that increased subsistence specialization would have necessitated more hierarchical social systems and some degree of central authority. As part of a broader consideration of the collapse of Pre-Pottery Neolithic social systems, Simmons emphasizes regional climatic change and ecological degradation as a major dimension of the fragmentation of large regional communities and a return to smaller-scale social units. In this context, as with that of other global areas, the explosive growth of Middle and Late Pre-Pottery Neolithic B period communities appears to be linked to subsistence specialization starting in the MPPNB and culminating in the increased dependence upon select plant and animal species in the LPPNB.

Chapter 5

The Quick and the Dead

The Social Context of Aceramic Neolithic Mortuary Practices as Seen from Kfar HaHoresh

NIGEL GORING-MORRIS

INTRODUCTION

The end of the tenth and the entire ninth millennia bp (uncalibrated) represent a period of major transition in lifeways, from mobile hunter–gatherers to food production in the Mediterranean Levant (Bar-Yosef and Belfer-Cohen 1992; Bar-Yosef and Meadow 1995; Horwitz 1993; Köhler-Rollefson 1989; Mellaart 1975; Moore 1985; Redman 1978; Rollefson and Köhler-Rollefson 1989). This transformation is accompanied by a marked increase in overall population densities in the area and recolonization of adjacent semiarid regions, and a hierarchy in community sizes ranging from hamlets to large villages. The mosaic of ecological settings in the Levant is accompanied by a wide array of socioeconomic lifeways: there are thus farmers and herders, farmers and hunters, specialized hunters and gatherers, and probably fishers as well (Figure 1). At the end of the ninth and beginning of the eighth millennia bp a probable combination of socioeconomic and environmental factors caused the larger communities in the southern and central Levant to

---

NIGEL GORING-MORRIS • Department of Prehistory, Institute of Archaeology, Hebrew University, Jerusalem, Israel 91905.

Life in Neolithic Farming Communities: Social Organization, Identity, and Differentiation, edited by Ian Kuijt. Kluwer Academic/Plenum Publishers, New York, 2000.

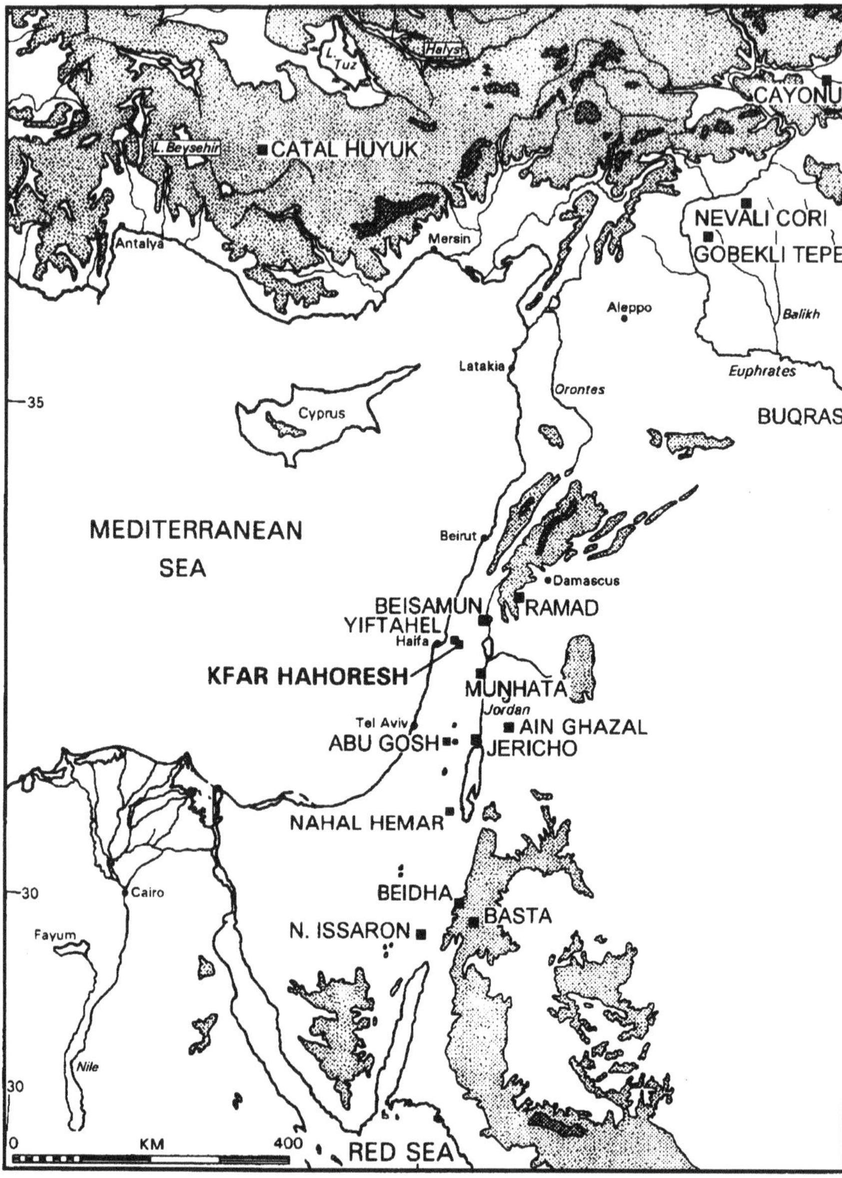

Figure 1. Map of south-central Levant showing the location of Kfar HaHoresh and several other contemporary Middle and Late PPNB sites.

gradually collapse and disintegrate, at about which time smaller, more mobile and dispersed agropastoralist societies first seem to appear in both the Mediterranean areas, as well as in the semiarid periphery (Baird 1992; Banning 1994; Betts 1988; Gopher and Gophna 1993; Goring-Morris 1994a).

At one end of the scale are large permanently occupied villages with populations numbering in the hundreds, if not thousands, especially in the Rift Valley and in western Transjordan (e.g., 'Ain Ghazal, Kharaysin, Wadi Shueib, Beisamoun, Jericho, and Basta) as well as smaller settlements (e.g. Munhata, Wadi Ghwair, Beidha) (Edwards and Thorpe 1986; Kenyon 1979; Kenyon and Holland 1983; Kirkbride 1966, 1967, 1968; Lechevallier 1978; Nissen et al. 1991, 1992; Rollefson et al. 1992; Simmons et al. 1988, 1989). Smaller villages (e.g. Yiftahel, Abu Gosh) and hamlets (e.g. Nahal Oren, Horvat Galil, Nahal Betzet) are more characteristic in Cisjordan (Garfinkel 1987; Gopher 1989; Lechevallier 1978; Noy et al. 1973). Coevally, and not too far away, in the more arid periphery, small bands continued what was essentially an Epi-Paleolithic-type mobile foraging mode of subsistence, albeit probably more sophisticated, at small seasonally occupied sites (Bar-Yosef 1981, 1984, 1985; Betts 1990; Gopher et al. 1995; Goring-Morris 1994a; Goring-Morris and Gopher 1983).

Notwithstanding this diversity, various aspects of the material culture remains demonstrate a high degree of homogeneity throughout the Levant, from southeast and central Anatolia in the north, to southern Sinai in the south, and from the Mediterranean east to the Saudi Arabian Desert. Information and exotic material exchange networks are well developed during the PPNB. Accordingly this period throughout the entire Levant has widely come to be termed the PPNB koine after Kenyon's original terminology, Neolithic 2, or the PPNB "interaction sphere" (Bar-Yosef and Belfer-Cohen 1989; Cauvin 1994; Moore 1985; Rollefson 1990b). It has become increasingly clear in recent years that by at least the beginning of the PPNB, if not earlier, the hub of some of the demonstrated innovation in material culture had shifted from the south-central to the central-northern Levant (the middle to upper Euphrates), as opposed to the situation earlier during the Epi-Paleolithic. This is perhaps most graphically illustrated by chipped-stone lithic technologies and typologies (Gopher 1990).

Yet, as noted, considerable regional variability is discernible, in part at least, reflecting the diversity in local adaptations and environmental settings, as well as social alliances. Thus, for example, large, individual rectangular and lime-plaster-floored habitations are common features in the larger, permanently occupied villages in the Mediterranean zone of the central and southern Levant, as opposed to circular, beehive plans, and more flimsy, coarsely constructed architecture in the more arid regions (Banning and

Byrd 1987; Bar-Yosef 1984; Goring-Morris 1994a). Other local architectural traditions are also apparent on the Euphrates and further to the north, as are different iconographic themes (Aurenche and Calley 1988). There are also subtle, regional stylistic differences in such aspects as basic lithic technology and projectile point morphologies (Gopher 1994; Nishiaki 1994; Schmidt and Beile-Bohn \1996). These and similar stylistic phenomena presumably reflect more localized intercommunity interactions and ties, in addition to chronological vectors.

Until lately it was widely believed that both farming and herding were already well established throughout the Mediterranean zone by the beginning of the PPNB. However, while cultivation of domesticated plants had been initiated by the later stages of the PPNA in the south-central Levantine Corridor, goat husbandry appears only during the course of the MPPNB, as documented at 'Ain Ghazal, and perhaps also at Yiftahel and Kfar HaHoresh (Köhler-Rollefson, Gillespie, and Metzger 1988; Horwitz 1987, 1993; Köhler-Rollefson 1989).

The stresses, dislocations, and impact of such rapid changes in lifestyles, with increasingly larger populations in permanently occupied settlements, must have been considerable. The emergence of incipient social hierarchies and ritual ideologies were necessary to regulate and codify increasingly complex interpersonal, intragroup and intergroup relationships (see Flannery 1995). This would be reflected in such aspects as control over, and access to, a variety of resources, including land tenure, ownership, and inheritance, as well as communal ventures, kinship relations, changing gender roles, etc. Yet the seemingly profane aspects of material culture appear to indicate that PPNB society was basically egalitarian in outlook (notwithstanding Byrd and Monahan 1995). Indeed to date the only obvious evidence for the emergence of some form of social hierarchy is in the fluorescence of ritual and ceremonial aspects of the material culture, often associated with mortuary practices, which surely represents only a fraction of the PPNB ideological belief system.

The major role of ritual and ideology in the PPNB has long been documented in the southern Levant from the excavations at Jericho in the Jordan Valley, and at Ramad in southwest Syria, as well as more recently especially from excavations at 'Ain Ghazal near Amman, Nahal Hemar Cave in the Judean Desert, with plastered human skulls, stone masks, and near-life-size human sculptures (Bar-Yosef and Alon 1988; Cauvin 1972; de Contenson 1992; Garstang 1935; Kenyon and Holland 1983; Rollefson 1983, 1986; Rollefson et al. 1992; Simmons et al. 1988, 1990). Paralleling these finds have been equally spectacular remains from the northern Levant and beyond at Çatal Hüyük in south central Anatolia, Çayönü, Hallan Çemi, Nevali Çori, and Göbekli Tepe in southeast Anatolia, including cult rooms, charnel houses, domestic shrines, and, most recently, ritual and funerary sites, together with a rich and distinct iconographic monolithic symbolic statuary (Hauptmann, 1993, 1997; Mellaart 1967; Özdögan and Özdögan 1990; Rosenberg and Davis 1992; Schmidt 1995).

Most ubiquitous among what were clearly deeply rooted belief systems was the practice of postmortem skull removal (Cauvin 1972, 1994), a tradition originating at least two, and more probably four millennia earlier, in the Epi-Paleolithic Natufian of the southern Levant (Belfer-Cohen 1990). There is widespread agreement that this tradition and associated practices, such as the modeling of features on skulls, represent some form of ancestor cult, most probably focusing on veneration of semimythical heroes (Arensburg and Hershkovitz 1989; Bar-Yosef and Alon 1988; Cauvin 1972, 1978; Hershkovitz and Gopher 1990; Kenyon 1979; de Vaux 1966).

Yet detailed archaeological documentation of mortuary practices and the spatial and conceptual relationships between the dead and the living are commonly lacking (see, however, Garfinkel 1994; Kuijt 1995, 1996). It is within this perspective, among others, that an ongoing excavation project at Kfar HaHoresh in the lower Galilee promises to shed light (Goren et al. n.d.; Goring-Morris 1991, 1994b, 1995; Goring-Morris et al. 1994–95, 1995; Hershkovitz et al. 1996).

A brief description of variability in the specifics of mortuary practices already documented at Kfar HaHoresh follows.* This data is complemented by comparisons to other PPNB and PPNA and Epi-Paleolithic Natufian sites in the southern Levant. This demonstrates that, in addition to postmortem skull removal, many other specific aspects of PPNB ideological, ritual, and mortuary practices may be directly traced back some two to four millennia to the Early and Late Natufian. Finally, I attempt to examine the information from a wider developmental perspective, taking into account socioeconomic transformations.

KFAR HAHORESH

Kfar HaHoresh is a 1–2 acre PPNB site located on a secluded north-facing slope in the uppermost reaches of a small, narrow wadi in the lower Galilee Nazareth Hills, north of the Jezreel Valley. Potential arable land is at a premium in the immediate vicinity of the site, as opposed to the settings of most PPNB villages in the region. Though secluded, the opposite hilltop provides a panoramic view from Mt. Carmel and the Mediterranean, across

*This is based on results through the 1996 field season. Subesequent seasons have revealed a much wider range of funerary practices and elucidated stratigraphic and other problems.

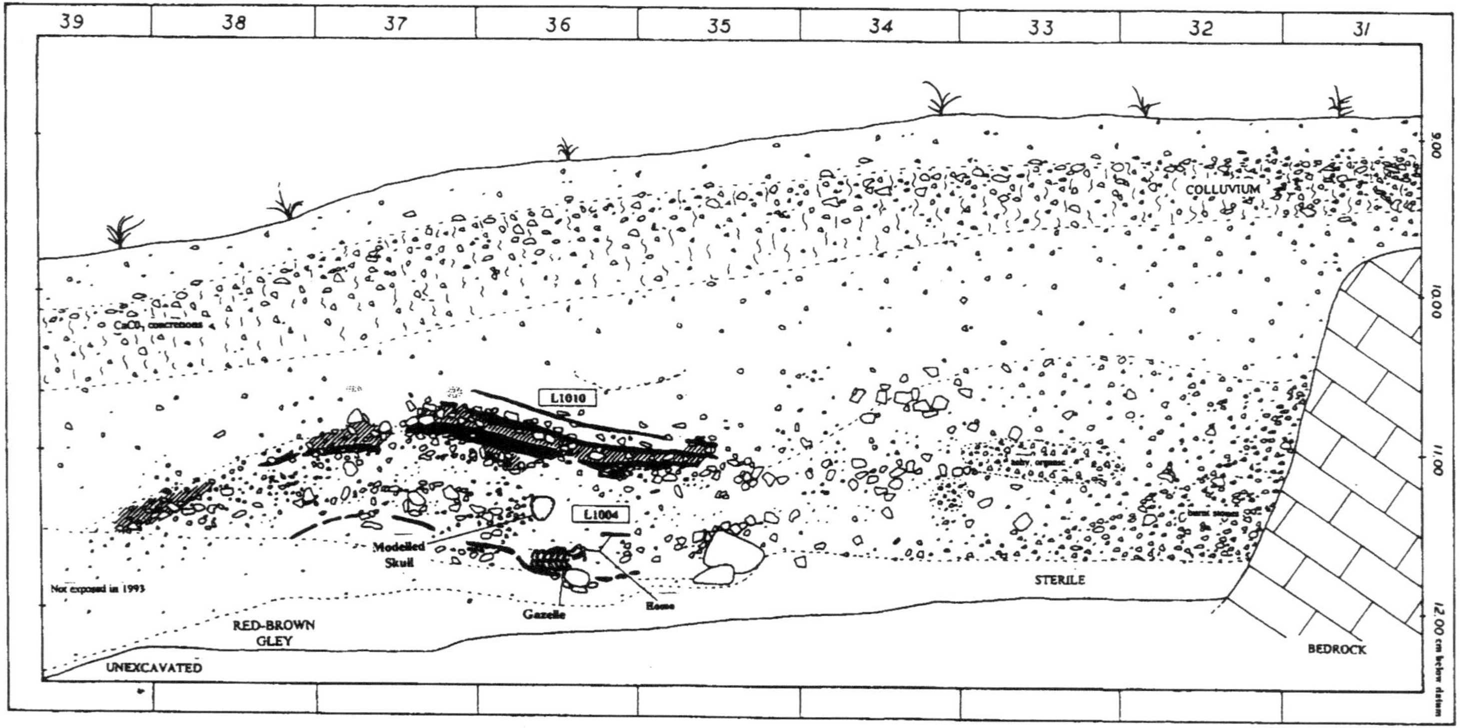

Figure 2. East section of Trench I in upper area. Note location of plastered skull and headless gazelle skeleton.

the Rift to Mt. Hermon and northern Jordan, as well as the Jezreel Valley and Mt. Gilboa. The location and nature of the site are such as to indicate that it may have functioned primarily as a regional funerary center for nearby communities. Technotypological similarities to nearby Yiftahel are used to date it to the first half of the 9th millennium BP, i.e. the Middle PPNB.

The limited excavations to date have revealed a complex stratigraphic series of *ca.* 5–25 m^2, quadrilateral, lime-plaster surfaced structures, terrace or compound walls, and open areas with numerous pits and installations commonly filled with burnt stones, animal bones, and other artifacts (Figure 5). There is also a possible solidly built structure with a red-painted plaster surface containing querns, one of which was filled with lime plaster, though some could also have been used for plant processing and food preparation. Limited probes beneath several of the lime-plastered surfaces have revealed numbers of primary, and especially secondary, human interments, often in direct and hitherto unique coassociations with animal remains, sometimes in partial or complete articulation.

Two plastered skulls have already been recovered from different areas of the site. One, excellently preserved and expertly modeled, was recovered from a lime-plastered pit beneath a lime-plastered surface with a stone-lined and plastered depression containing a single posthole (totem?) lo-

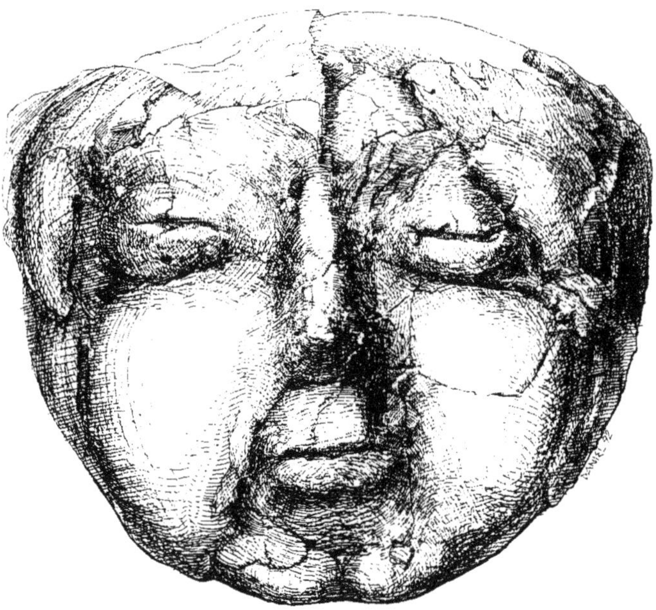

Figure 3. Modeled skull (KHH-H01) from Kfar HaHoresh.

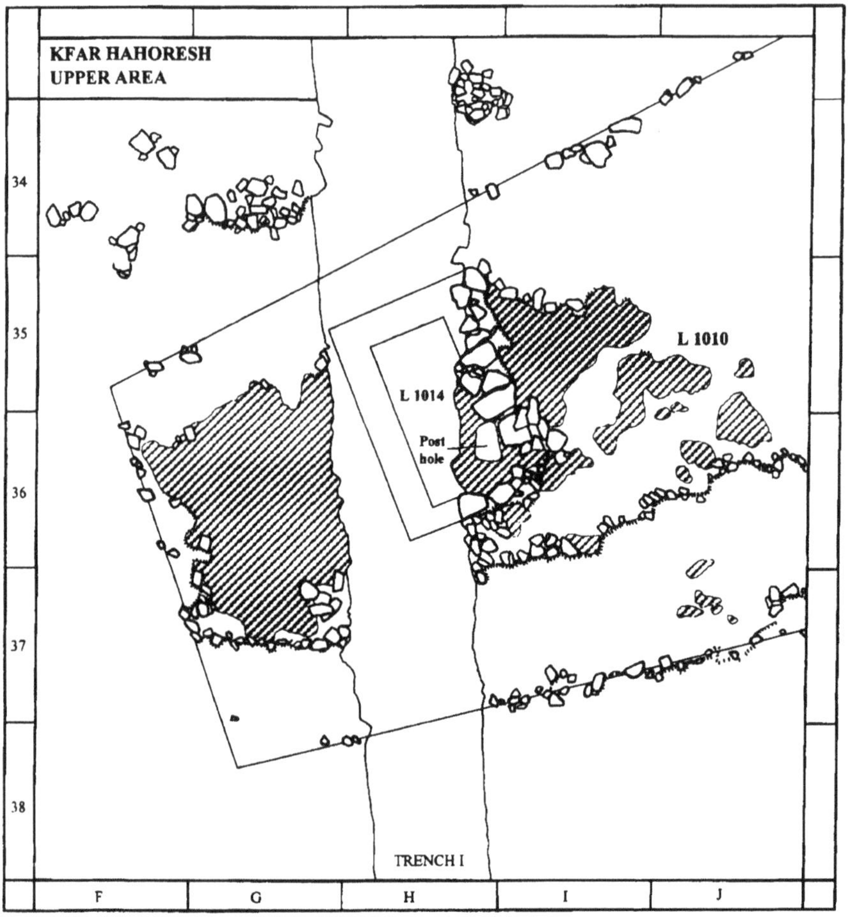

Figure 4. Plan of Locus 1010 in upper area.

cated above the skull (Figures 2–4). It was directly associated with an otherwise complete but headless gazelle carcass. Another fragmentary plastered skull was found in a small oval installation, perhaps also associated with a lime-plastered surface (Figure 5). A probe elsewhere beneath two plastered surfaces revealed parts of an articulated contracted human skeleton lacking the head and mandible (probably removed through a hole in the lower plaster surface) directly overlying a 1-m-diameter and 50-cm-deep pit filled with some 200 partially articulated postcranial aurochs bones, representing at least six adult animals and two immature ones (Figure 6). A limited probe of only 1.5 m^2 beneath yet another plastered surface has already provided numerous isolated human mandibles, other postcranial elements, a few in partial articulation, possibly also in association with articulated gazelle re-

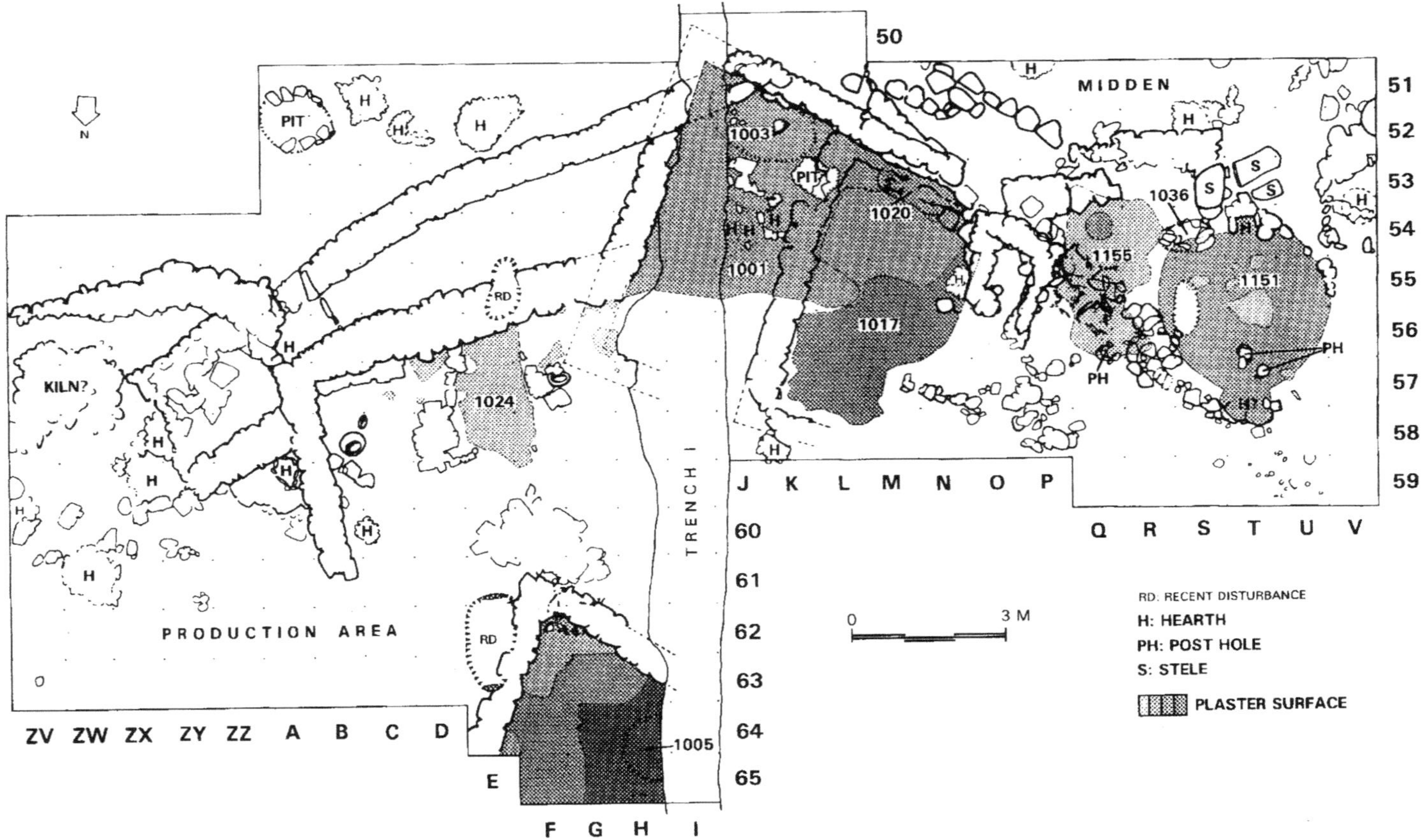

Figure 5. Map of central area at Kfar HaHoresh, 19997.

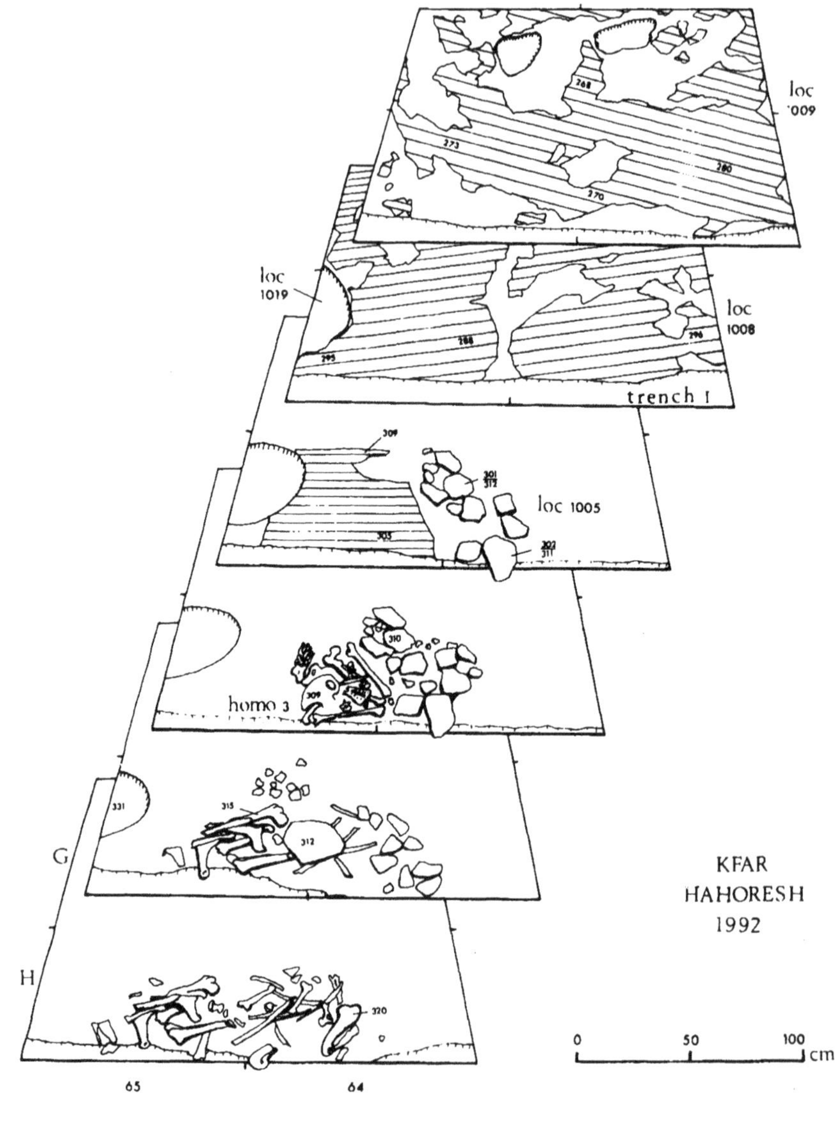

Figure 6. Isometric view of lower area, showing burial pit containing wild aurochs overlain by headless Homo 3, which is covered in turn by lime-plaster floors of Locus 1008 and Locus 1009.

mains. There is also a single primary supine human burial with the cranium removed but mandible present—the only 'typical' south Levantine PPNB burial recovered to date at Kfar HaHoresh. Underlying this, and seemingly associated with an earlier plastered surface, is an apparently untreated skull, though its context remains to be ascertained. Scattered human remains are found almost everywhere in the excavated areas. With the exception of one of the few complete arrowheads recovered on-site, directly associated with the first plastered skull, few other obvious grave goods have been noted, save the animal remains. There is some evidence to indicate monitoring, and even intentional marking of, graves on plaster surfaces in the form of small slabs stuck to the plaster surface or the "totem" posthole noted above.

Faunal evidence indicates that occupants and visitors obtained meat primarily by hunting gazelle, with cattle, boar and deer also represented among the ungulates. The presence of goat in some quantity, however, may indicate that herd animals were in the process of being domesticated (Horwitz 1993). None of the goat remains, however, appear to be associated with obvious grave contexts. No macrobotanical remains have been recovered to date, yet the high frequency of sickle blades seemingly attests to the harvesting of various annuals and/or reeds.

While a number of the 'exotics' at Kfar Hahoresh have not been documented from south Levantine PPNB contexts previously, the overall picture in terms of quantity and variety is not particularly remarkable, though all attest to well-developed regional and long-distance procurement and/or exchange networks. They include obsidian and cinnabar from central Anatolia, asphalt (not from the Dead Sea), and malachite, as well as basalt grinding stones, some of the chipped stone tools on nonlocal flint, and marine mollusks from the Mediterranean and Red Seas. In contrast to Nahal Hemar, the chipped stone industry is not obviously much different from that of "regular" occupation sites: an emphasis on a rough flake technology in the production of ad hoc tools, as well as more refined, specialized technologies for the production of the more standardized tool classes, such as projectile points and sickle blades from naviform cores, and bifacial tools, some, but not all of which, were probably manufactured on site. To the extent that the lithic assemblage differs from published coeval assemblages, it appears to be among the ad hoc element. Intrasite patterning appears quite marked.

The following discussion and reconstruction are necessarily speculative, but are based on the evidence just presented. Though the areas of Kfar HaHoresh excavated to date are limited, the evidence tentatively appears to indicate that the site was primarily used as a secluded regional funerary center by village settlements located in the surrounding lowlands, such as the edges of the Jezreel Valley and the Bet Netofa Valley (e.g. Yiftahel) where arable lands were available in abundance (many other contemporary sites are probably buried beneath subsequent colluvium) (see also Hole, Kuijt, Rollefson and Simmons for further discussions). As such it was probably visited periodically by kin from surrounding settlement(s), though the possibility exists that there may also have been permanent occupation by a few "guardians."

The construction of the various funerary structures, with the production of large quantities of lime plaster, involving the bulk preparation of crushed chalk and limestone and the provision of quantities of fuel, stoking of kilns and groundstone and other tools in its application, would have involved considerable effort. So too, the gearing up for (communal?) hunting parties for feasting and provision of animal offerings, as well as various maintenance activities would all require labor, irrespective of the 'costs' involved in the more obviously ritual and ceremonial activities directly associated with the deceased. Even if conducted only periodically, these could all account for the large quantities of seemingly profane remains and refuse on-site. Of course, it is also quite likely that many such "everyday" items may also have been accorded symbolic and ritual significance.

In addition to the moving of corpses from the death location for primary burial, further funerary treatment(s) and associated rituals would also have necessitated subsequent visits. The open areas, with large quantities of roasting pits containing burnt animal bones, likely reflect more than mundane activities: they could perhaps reflect communal feasting associated with some of these funerary rites. Although both plastered skulls are of young male adults, the other human remains (including the cache of mandibles) are of adults and young children of both sexes. This indicates that the entire population, or an entire segment thereof, was afforded similar basic funerary rites involving primary interment, sometimes together with partially articulated animal carcasses. Subsequent human skull removal, as well as the occasional partial articulation of some limb and other bones in clear secondary contexts, appears to indicate that skull removal, sometimes with complete or partial dismemberment of the rest of the skeleton, was done after most of the soft tissues had disintegrated (a year or so later?). Such generalized treatments seem to indicate ascribed status. In contrast, some individuals (young adult males, at least) were singled out for more elaborate treatment of their skulls (facial modeling), which may reflect attained status. Commonly, mandibles and other postcranial elements seem to have been gathered to be buried and intentionally capped by lime-plastered surfaces, some of which featured low surrounding walls. These surfaces, then, seem to be coverings rather than floors, though a couple of the plastered surfaces at the site may actually represent floors. Plastered skulls may also have been similarly treated, perhaps after completing their reincarnated "life cycle," again sometimes with animal carcasses. It is, of course, interesting to speculate whether the contents of these plastered funerary monuments (at least five have been exposed, though none completely excavated) reflect kin or other social ties, in addition to stratigraphic factors (attempts are being conducted to investigate the matter through DNA studies). Bearing in mind that PPNB burials also occur in occupation sites, it seems that the secluded setting of Kfar HaHoresh, together with the evidence for adults and children of both genders being afforded similar basic, if not full "treatments," may indicate that funerary centers were reserved for segments of the population perhaps related by kinship or other ties. Funerary rituals may have been in part exclusionary, involving only some members or segments of the communities involved.

The hitherto unique coassociation of human and animal remains, the latter commonly in at least partial articulation but also lacking the skulls (symbolic offerings?) may be particularistic to Kfar HaHoresh. Alternatively, it is possible that this was the common practice in mortuary sites, of which Kfar HaHoresh is the only PPN site documented to date in the southern Levant. Whatever the case, the fact that the animals involved are gazelle and aurochs is probably of symbolic significance in the context of incipient animal domestication, since both are wild (notwithstanding the presence elsewhere on-site of possibly domesticated goat). Gazelle was the staple meat resource in the region throughout the Terminal Pleistocene and the beginning of the Holocene, and its symbolic significance is reflected in animal sculptures in Natufian contexts. The symbolic role of the aurochs is already evident in the PPNB (and perhaps even much earlier), e.g. their prominence among the 'Ain Ghazal figurines in comparison with their low representation among the faunal remains. This subsequently develops into the "Bull Cult" of the eastern Mediterranean (see at Çatal Hüyük; also Cauvin 1994). Changes in the nature and division of labor associated with the beginnings of herding involving the reduced role of hunting are likely to have had a considerable psychological impact on those segments of the population most affected (the last of the big game hunters?).

GENERAL OBSERVATIONS ON TERMINAL PLEISTOCENE–EARLY HOLOCENE MORTUARY PRACTICES

Precise documentation of the contexts of burials is vital if we are to be able to at least partially decipher the attitudes of living communities to the dead. For it is clear that the specifics of corpse disposal and subsequent symbolic reintroduction of the deceased into the realm of the living were integral to the fundamental and revolutionary socioeconomic changes in lifeways that occurred during the Terminal Pleistocene and Early Holocene (see also Kuijt 1995, 1996). That these beliefs were deep-rooted and part of a wider cosmology is amply illustrated by the longevity (two to four millennia; see below) of certain funerary rituals and treatments in respect to the deceased by the living communities, even if certain aspects evolved through time, both in specific details as well as probably in their symbolic relevance. Although most discussions focus on postmortem skull removal, there is evidence to indicate other topics were central to mortuary practices, including the locations and contexts of corpse disposal, architectural elements (especially lime-plastered surfaces) associated with funerary practices, and the coassociation of human and animal burials (Table 1) (see also Chapters 6, 7, 10, and 11, this volume).

Ritual Centers, Burial Grounds, and Cemetery Sites

During both the PPN and the Natufian there is some evidence to indicate that, while the deceased were often buried and symbolically "reincarnated" by postmortem head removal in localities whose main functions were as residential occupation sites, other localities functioned primarily as ritual centers and burial grounds. During the PPN throughout the Mediterranean and semiarid regions of the southern Levant, human remains within settlements are generally quite rare, certainly in comparison to their size and duration (e.g., 'Ain Ghazal, Abu Gosh, Beidha, and Nahal Issaron) (Goring-Morris and Gopher 1983; Kirkbride 1967; Lechevallier 1978; Rollefson personal communication). Thus, it seems likely that either most of the deceased were buried elsewhere, beyond the excavated areas, within the confines of settlements, or else, and perhaps more likely, there are specific mortuary sites located some distance away. They may frequently have been overlooked for a combination of reasons: their specific (secluded?) locations, small sizes, and specific attributes. Nevali Çori (and Göbekli Tepe) on the Euphrates in southeastern Turkey also fulfilled roles as major cultic and funerary centers during the PPNB. Interestingly the latter, upon initial discovery, was thought not to date to the Neolithic (Schmidt 1995).

For the PPNA we presently have no data beyond that from within residential sites. The relative abundance of burials in the vicinity of the tower at Jericho may reflect a ritual area within the settlement (see Kuijt 1996). However, during both the Early and Late/Final Natufian certain parallels are available. Thus Early Natufian Erq el-Ahmar and, perhaps, Kebara Cave may be viewed as primarily burial localities rather than occupation sites. At Mallaha the Early Natufian Structure 1, with its enigmatic circular plastered bench, slab outlined perimeter, hearth and underlying cemetery, almost certainly does not represent a simple domestic residence, but more

Table 1. Southern Levantine Cultic Installations and Burial Traditions

	Cultic structures	Headless burials	Skulls	Modeled skulls	Vault treatment	Statues and busts	Masks	Animals
Early Natufian								
Erq el-Ahmar		?	nest	—		—	—	—
Eynan	present	present?	single; scalp	—	dentalia	—	—	dog
Hayonim	in cave?	—	single??	—		—	—	dogs
El Wad	?	present	—	—	dentalia	—	—	—
Ein el-Saratan	—	present?	present	—	ochre	—	—	—
Late/Final Natufian								
Nahal Oren	present	present	?	—	???	—	—	gazelle horns
Eynan	—	present?	nest (6); singles	—	—	—	—	gazelle horns
Hayonim	in cave?	present	singles?	—	—	—	—	dogs
PPNA								
Jericho	present	present	nests[1]	—	—	—	—	—
Netiv Hagdud	—	present[1]	nest[1]+single	—	—	—	—	—
Early/Mid PPNB								
Ain Ghazal		present	nests[1]	nests+single	—	bust	—	—
Jericho	present	present	nests	nests	ochre	nests+single	—	—
Beidha	present	present?	present?	—	—	—	—	—
Mid/Late PPNB								
Ain Ghazal	present	present	nests	nests+single	ochre, bitumen?	nests	—	—
(Nahal Hemar)		—	present[1]	—	"wigs"	fragments	present	—
Ramad	present?	present?	?	nest?	—	present	—	gazelle; auroches; fox; capra?
Beisamoun	—	present	nest[1]	present	—	—	—	—

(continued)

Table 1. *Continued*

	Cultic structures	Headless burials	Skulls	Modeled skulls	Vault treatment	Statues and busts	Masks	Animals
Yiftahel	—	present	—	—	—	—	—	—
Kfar HaHoresh	present	present[1]	nests + single[4]	nests + single	—	present	?	gazelle; aurochs; fox mandibles; capra hors
Abu Gosh	—	nest, single	—	—	—	—	—	—
?Basta	—	present[1]	nests, single	—	—	—	present	
Tell Abu Hureyra	—	present?	—	ochre	—	—	—	—
Final PPNB								
Ain Ghazal "PPNC"	present	absent	single	absent	—	—	—	pig skulls[2]
Athlit Yam	—	absent?	present	—	—	—	—	—
Ramad?	??	?	?	nests?	—	nests	—	—

[1]Including children/infants.
[2]Rollefson and Köhler-Rollefson (1993:38).

probably some form of funerary structure. Indeed, even the more convincing dwellings at Mallaha and Wadi Hammeh 27 display clear evidence as to the incorporation of obvious ritual and ceremonial practices (personal obsservation). As for the Late and Final Natufian, several sites in the Mediterranean zone, such as Shuqba and Nahal Oren V–VI display little, if any, evidence for obvious domestic architectural features and could perhaps be more convincingly interpreted as mortuary centers (notwithstanding problems of early excavation techniques).

Monitoring Burials

From a technical perspective many PPNB burials obviously stratigraphically predate the construction of the overlying architectural features. However, various lines of evidence indicate that the precise location of the burials was monitored by the builders, and that they may sometimes ultimately be viewed as integral to the construction of habitations (see Kuijt 1995, 1996). Skull removal is usually (but not always) prior to construction of plaster floors; that is, there is some lapse in time between primary burial and subsequent construction (see also Chapter 7). Removal of plaster floors, burial, and subsequent replastering appears to be the exception rather than the rule. In at least three instances at Kfar HaHoresh burial pits clearly stratigraphically underlie and are sealed by plaster surfaces. The small slab marker stuck to the plaster floor immediately above the group of mandibles (see above) appears to indicate that it served as a "tombstone" to mark the precise location of the interment. So, too, the "totem" posthole in the rectangular feature, immediately overlying one of the modeled skulls in the upper area, could be interpreted as a marker rather than a purely constructional feature.

The rectangular plastered structure immediately overlying the supine headless burial at Kfar HaHoresh also appears to have been positioned intentionally relative to the skeleton. A somewhat similar situation also pertains at Beisamoun with respect to the burial of two individuals, H201 and H202, in Feature 191, under the lime-plaster floor in the western corner of Habitation 1 (Lechevallier 1978:Figure 47). The same may be true of two other individuals, H207 and H208, as well as the remains of seven other individuals in Locus 188 under the floor of the adjoining antechamber of Habitation 1, Locus 180. All of these interments lacked the cranial elements. So too, at Abu Gosh the headless burial of an individual was recovered from Locus 537 beneath the plastered floor of Locus 507 in the northern corner of Habitation I (Lechevallier 1978:35 and Figures 4 and 9).

Though most of the interments at Yiftahel were located in fills after structures went out of use, the only subfloor burials were one in each of the three units constituting the structure: a neonatal, Homo 5, in the northwest corner of the structure, Locus 700; an infant, Homo 6, in the middle of the central unit, Locus 710; and a young female adult, Homo 3, lacking the cranium, on the east side of the southern unit, Locus 720 (Hershkovitz et al. 1986). At 'Ain Ghazal burials have been reported to be located under floors adjacent to plaster-built hearths as well as close to walls under structures (Rollefson 1983, 1986). In all these cases, the burials and subsequent removal of the crania commonly (though not exclusively) appear to predate the initial construction of the overlying lime-plaster floors. Furthermore, it is of some interest to note that there are also burials actually incorporated within architectural features at 'Ain Ghazal, Jericho, and Basta (Kenyon 1981; Kuijt 1996; Nissen et al. 1991, 1992; Rollefson 1986).

Thus, in some instances at least, architecture was intentionally constructed over what had previously served as cemeteries or burial grounds: the locational data would seem to indicate that the relative position of the burials was monitored (not surprising given the fact that the crania were likely commonly removed at least a year or so later) and taken into account for the subsequent construction. In other cases abandoned structures were subsequently then used as burial grounds, e.g. Yiftahel, Beidha. Kuijt (1996) has also recently emphasized the likely function of some burials (especially children/neonatals) from Kenyon's excavations at PPNA Jericho as foundation deposits, sometimes immediately under a single "post socket" ("totem" hole) or plastered "bin." It is interesting to note possible parallels with the Late/Final Natufian cemetery at Nahal Oren, where there are circular features/postholes ("totem" holes?) as well as "stone pipes" directly associated with some graves (Stekelis and Noy 1963). Indeed Stekelis discussed the possible symbolic significance of the 'stone pipes' in terms of connections between the living and the deceased.

Somewhat similar practices can be documented already from the Early Natufian, such as at Eynan, where Cemetery B predates Shelter 131 (Cemetery A and Shelter 1 being interpreted here as an actual funerary monument), and perhaps at el Wad B2 and even Wadi Hammeh 27 (Edwards 1991; Goring-Morris 1996; Perrot and Ladiray 1988). Though seemingly not absolutely contemporary in stratigraphic terms, I would venture that such coassociations of Early Natufian cemeteries and stratigraphically later structures are unlikely to be fortuitous. The general monitoring and repetitiveness of (symbolic?) activities is also indicated by the successive living floors exposed by Valla (1990) in Shelter 131 at Mallaha, with piles of colored pebbles. At Erq el-Ahmar there is an overlying pavement which seals and/or marks the burials (Neuville 1951).

Though from the Late Natufian onward burials are generally single, the tradition of subsequent construction after removal of the cranium is common. Thus the demonstrated instance of Locus 1005 at Kfar HaHoresh, with stratigraphic evidence for the subsequent opening of the floor to remove the skull, prior to the replastering of the entire floor of the structure, appears to be the exception rather than the rule; interestingly the immediately overlying plaster is notably thin and overlain by yet another plaster surface, which has yet to be excavated.

Human and Animal Burials

At least two, and perhaps four, examples of burials being accompanied by almost complete carcasses of wild animals are documented at Kfar HaHoresh (Figs. 9.5 and 9.7). The fact that they are all wild animals, at a time of the possible introduction of domestic herd animals is likely to be significant. The absence or clear underrepresentation of cranial elements of these animals is also noteworthy and parallels the treatment of humans. KHH-H01 was also directly associated with one of the few intact arrowheads recovered from the site. It seems plausible to suggest that this practice may reflect beliefs associated with the status of hunters in the community at a time of profound economic change. Although these are the first cases documented in the PPNB (or PPNA for that matter), the practice of human burials accompanied by animals is known already from the Early Natufian: in all cases dogs, though domesticated, were not apparently a subsistence source but may have been primarily invested with spiritual significance rather than simply used as a hunting aid (Davis and Valla 1978; Valla et al. 1991; especially Valla 1996). There may also be an aurochs cranium associated with a Khiamian/Sultanian burial at Hatoula (Lechevallier and Ronen 1994). Cauvin (1978) has also described the intentional incorporation of aurochs skulls in architectural features at Mureybet II. As such, they provide further evidence for the longevity and deep-rooted beliefs underlying such traditions.

The somewhat later PPNC levels at 'Ain Ghazal have revealed the presence of several instances of sporadic Sus bones (also wild) accompanying burials (Rollefson 1986). Furthermore, it is of interest to note that at 'Ain Ghazal cattle (again nondomesticated) are disproportionately represented among animal figurines, in contrast to their actual frequency among the faunal remains of the PPNB levels (Rollefson 1986:47), a pattern seemingly present at other contemporary sites. The symbolic significance of cattle is also emphasized by the incorporation of skulls in architectural features at Çatal Hüyük in eighth millennium BP levels (Mellaart 1967). But, so far, no similar evidence for the special positioning or treatment of animal skulls has come to light at Kfar HaHoresh.

Ranking as Indicated by Skull Removal and Further Treatment

One of the hallmarks of the PPNB is postmortem cranial removal following primary interment (Table 2). The tradition is prevalent, though by no means

Table 2. Distribution and Nature of Burial and Ritual Paraphernalia in Pre-Pottery Neolitic Sites in Southern and Central Levant

	Settlement size (dunams)	Residence	Cult structures and installations	Cemetaries	Interments	Postmortem treatment	Iconographic themes
Early & Middle Epi-Palaeolithic	Camps: 0.05–0.15 Aggregation 10–15 (Kharaneh, Jilat)	Nuclear family huts: flimsy, 3–5 m diam. (Ohalo II, Ein Gev I)	None	None In occupation sites	Primary single, under or near huts	None	Ladder (Urqan e-Rubb); Chevron (Jiita)
Early Natufian	Hamlets: <2.0 Basecamps 0.5–1.0 (Eyan, W. Hammeh, U. Bessor 6	Moieties/sodalities: Stone-built lodges: 7–16 m diam. (Eynan, W. Hammeh)	Funerary struct: slablined & lime-plaster bench (Eynan habit 1.) Cave (El-Wad) Feasting (Beidha) Engraved slabs (W. Hammeh) Basins & grooves (el Wad, UB6)	Cemeteries in occupation sites Also cave sites? (Erq el-Ahmar? Kebara? Hayonim?)	Primary, single group Secondary	None Skull removal rare (Eynan, E. el-Ahmar?) pigment (A. el Saratan) Dog (Eynan, (Eynan, Hayonim)	Meander Lozenge Animal figurines Male genitalia (el Wad, Fazael VI) (sickle handles?)
Late & Final Natufian	Hamlets: <2.0 Basecamps 0.5–1.0 Aggregation: <5 (R. Horesha)	Nuclear family huts? Circular stone-built structures: 3–5 m diam. (Eynan, R. Zin, R. Harif)	Cave with circular huts (Hayonim) Kidney-shaped hut of large slabs (R. Horesha) Feasting (R. Horesha) Monolith (R. Zin)	Funerary center (Nahal Oren) Caves? (Shuqba) Also in occupation sites	Primary, secondary single, group	None Skull removal (N. Oren, Eynan, Hayonim)	Meander Chevron Male genitalia (R. Zin?) Zoomorphic figurines

PPNA	Hamlets & villages:<25	Nuclear/extended family Oval houses: 4–7 m diam.	Tower + ? shrine, vats/silos (Jericho)	None?—occupation sites	Primary, single	Skull removal With Aurochs head? (Hatoula)	Meander Female figurines Male genitalia (A. Darat) Zoomorphic figurines
PPNB	Hamlets & villages: <120	Nuclear/extended family: Two-story pier houses with plaster floors	Shrines (Beidha A. Ghazal) Basin & grooves (N. Oven, Beidha) Favissa (N. Hemar, A. Ghazal, Kfaar HaHoresh) Monoliths (A. Ghazal, Beidha, Kfar HaHoresh)	Funerary centers (Kafar HaHoresh) Also in settlements	Primary, single Secondary, multiple	Skull removal Modeled skulls Wigs With headliess aurochs, gazelle (Kfar HaHoresh)	Female figurines Animal figurines especially aurochs (A. Ghazal) (sickle handle?) N. Hemar Male genitalia (Kfar Hafforesh)

ubiquitous, during the PPNA and PPNB. Recently it has been demonstrated that the tradition originated during the Late Natufian (Belfer-Cohen 1990). However, though rarer, there seems to be some evidence to indicate that it may have even begun sporadically during the Early Natufian, based upon Erq el-Ahmar, Mallaha (e.g., the skull of Homo 37 on the floor of Shelter 1), and 'Ain el-Saratan (Garrard 1991; Neuville 1951; Perrot and Ladiray 1988).

Seemingly applied in roughly equal proportions to male and female adults, there is solid evidence to indicate that it was applied to children and even neonatals, albeit less ubiquitous, from at least the PPNA (indicative of ascribed status). More elaborate treatment of a minority of (adult) skulls clearly reached a zenith during the Middle to Late PPNB, paralleling the increased sizes of some communities. This is one of the few obvious expressions of some form of ranking or hierarchy (attained status), interpreted as representing a cult of the "ancestors" or "heroes." Many discussions focus exclusively upon the striking modeled or plastered skulls. While some physical anthropologists have argued that such treatment is exclusively reserved for males, others have suggested that women were also so treated. It has also been suggested that these individuals are more markedly brachycephalic than their contemporary population, and that there is an attempt to emphasize a gerontocracy (Arensburg and Hershkovitz 1989). I would venture that, while young males were certainly so treated, the status of women is open to debate. There is, however, little, if any, solid evidence for an emphasis on elder members of the community. In vivo skull deformation by bandaging or headgear may also account for some of the seeming pathological features, e.g. Bouqras (Meiklejohn et al.1992).

Indeed many studies have tended to overlook the variety of mortuary treatments accorded during the PPNB, which include

1. Primary burial with no skull removal (interestingly absent at Kfar haHoresh).
2. Primary burial with subsequent skull removal.
3. Caches of skulls either singly or in "nests" of variable numbers.
4. The daubing of pigment, whether black or red, upon the vault, as at PPNB 'Ain Ghazal. While no modeled skulls have been forthcoming to date from PPNA contexts, the daubing of pigment already occurs on a male adult in the Early Natufian at Ein el-Saratan (Garrard 1991).
5. The application of "wigs" or headgear of some resinous or bituminous substance, but no treatment of the facial region, as documented to date only from Nahal Hemar (perhaps due to taphonomic processes of preservation).
6. The full modeling of facial features in lime plaster and other materials but with the vault seemingly untreated, as known from Jericho,

Ramad, Beisamoun, 'Ain Ghazal, and, most recently, Kfar HaHoresh. In light of the Nahal Hemar skulls, the question arises as to whether the plastered skulls were originally adorned with some form of wig or cap of perishable material.

7. Secondary burial of disarticulated or partially articulated (usually postcranial) remains, commonly of several individuals. The contents beneath the plastered surface at Kfar HaHoresh and the charnel houses and sanctuaries at Nevali Çori and Çayönü represent obvious examples. However, the practice commonly appears elsewhere in less obvious contexts, e.g. Beisamoun Locus 188. Such treatment also occurs as early as the Late Natufian at Mallaha Graves 9 and 10 (Perrot and Ladiray 1988:85), Hayonim Cave Grave V (where, interestingly, the contents include a cache of mandibles; see Belfer-Cohen 1990), and even at El-Wad and Erq el-Ahmar.

Such varied and differential postmortem treatments during the PPNB would appear to represent some form of social and/or ritual ranking, the only obvious evidence to date for any manner of hierarchical ordering within the community, though to what extent it reflects status of the deceased premortem in the living community is moot.

Favissae as Repositories for Modeled Skulls

The possibility exists that those individuals designated for more elaborate treatment than simple skull removal were subsequently deposited in specially constructed installations. Whether these favissa at Kfar HaHoresh represent the final repositories for the skulls at the end of their "reincarnated life cycle" (see also Garfinkel 1994) or, perhaps more plausibly, whether they acted as temporary storage facilities while the modeled skulls were still 'in circulation' presently remains unclear. If Kfar HaHoresh functioned primarily as a ritual center, then it is likely that ceremonies utilizing the skulls were periodic and/or seasonal.

Although the precise setting of many of the plastered skulls at Jericho, Ramad, and 'Ain Ghazal remain vague (on and/or under floors, in installations, pits, though see Kuijt 1996), the two plastered skulls at Beisamoun were found in a vestibule at the back of the large rectangular structure, where several complete arrowheads were also recovered. In PPNA Netiv Hagdud some skulls were recovered from on the floor of a dwelling. What appears to be a somewhat similar, clay-coated installation to that at Kfar HaHoresh has been briefly described at 'Ain Ghazal (Simmons 1990). To date no conclusive evidence has been forthcoming concerning the placement or caching of detached skulls in Natufian contexts, beyond the aforementioned skull (Homo 37) in front of the hearth of Shelter 1 at Mallaha (Perrot and Ladiray 1988).

Use of Lime Plaster to Integrate Profane and Symbolic Realms.

The massive use of lime plaster for profane construction purposes is characteristic of most permanent PPNB settlements in the south-central Mediterranean Levant (Kingery et al. 1988; Goren and Goldberg 1991). As such it represents the first large-scale pyrotechnical production, probably manufactured on a family or clan basis. It has even been claimed that the technology was largely responsible for the ultimate demise of the PPNB, through major ecological degradation by deforestation for fuel and construction (Garfinkel 1987; Rollefson 1991; Rollefson and Köhler-Rollefson 1989). Though certainly a contributory factor, it is improbable that this was a primary cause. It has also been suggested that it probably reflects some form of social hierarchy (Garfinkel 1988). Its intensive use, involving considerable skill and investment of labor, is unrivaled until the Classical period. However, lime plaster probably also had a major symbolic significance, far beyond the mere utilitarian, as indicated by its elaborate use for modeling facial features on skulls and constructing large sculptures (Goren et al. n.d., 1993). Again, it is of interest to note that the earliest major use of lime plaster for construction purposes was in the Early Natufian, e.g., the circular 'bench' in Shelter 1 at Eynan, on the floor of which a single skull was recovered. Though precise stratigraphic details are unclear this structure also overlays Cemetery A.

As noted, one of the subfloor inhumations at Kfar HaHoresh was sprinkled with lime plaster prior to construction of a surface (Fig. 9.6). CAT scans, thin sections, and other observations of both modeled skulls at Kfar HaHoresh have revealed that they were constructed in several elaborate stages, with at least three or four separate and distinctive mixtures applied to provide durability and to reach the desired final product (Figure 9.3). Both modeled skulls at Kfar HaHoresh appear to be directly associated with architectural plastered surfaces. And the sealing of burials by lime plaster is again common not only in the PPNB but also is documented in the Early Natufian. It seems plausible, then, to suggest that lime-plaster production may also be viewed as an attempt to simultaneously physically and symbolically segregate and integrate the realms of the quick and the dead.

EARLY HOLOCENE RITUAL AND MORTUARY PRACTICES IN LEVANT

Discussions of the nature of Epi-Paleolithic Natufian sites have tended to focus on such questions as settlement patterns and sedentism, commonly within the framework of settlement size and subsistence. Yet perhaps other subdivisions may be no less relevant. Throughout the Natufian in the Mediterranean zone some sites appear to have functioned primarily, if not solely, as cemeteries, e.g. Erq el-Ahmar and Nahal Oren among others. In other (especially Early) Natufian sites in the Mediterranean zone there would appear to be greater physical integration of the cemeteries and habitation areas (e.g., Early Mallaha, El Wad, and perhaps Wadi Hammeh 27) (Goring-Morris 1996). These differences could be related to the degree of mobility of specific Natufian communities in time and space (as indeed the contrast between primary and secondary burials). Indeed the overall parallels to PPNB Kfar HaHoresh and many other contemporary sites are notable in many respects.

Furthermore the longevity (some four millennia or more) of the habit of postmortem skull removal in the Levant crosscuts cultural and economic subsistence modes. Again, beginning in the Early Natufian and accelerating thereafter, the practice was commonly, though not ubiquitously, applied to members of both genders. There also appear to be hints that, from at least the PPNA, as at Jericho and Netiv Hagdud (Belfer-Cohen 1990; Kenyon 1981), if not earlier, children and perhaps even neonatals may have been occasionally treated in the same manner, albeit much less commonly.

Thus, initially seemingly limited in scope, there is evidence to indicate a time-transgressive trend of increasingly sophisticated embellishments in the subsequent treatment of adult skulls (of both sexes?), particularly during the course of the PPNB. This appears to parallel the growth in site (and hence community) sizes (Table 1). This includes the daubing of ochre or other pigments on cranial vaults (again beginning in the Early Natufian, for example at Ein el Saratan; Garrard 1991) , the application of "wigs," and the modeling of facial features in the PPNB. This seems to be accompanied by instances of in vivo skull deformation.

Skull removal was thus deeply rooted in the belief systems and psyches of the Early Holocene communities in the area. It has been widely accepted that the ritual probably reflects some form of "ancestor cult" or "cult of the heroes" (Bar-Yosef and Belfer-Cohen 1989; Bienert 1991; Cauvin 1972; de Vaux 1966). In light of the major shift in the nature and size of domestic residential units in the Early Natufian (Goring-Morris 1996), could it be that initially skull removal reflects a "cult of the founders"? It thus seems likely that skull removal at that time reflected some form of attained status within the community. However, it is also possible that the significance of the practice evolved through time to one of inherited (ascribed) status and the emergence of social/ritual hierarchies, perhaps in part connected with the distinction between community and private power, title, rights, and wealth and served to encourage social cohesion (see also Bar-Yosef and Belfer-Cohen 1989:63).

Codification of Funerary Practices through Time

As such, the tradition probably functioned as a bridge between living communities and their deceased forbears. In contrast to later periods in the Levant (from the eighth millennium bp through protohistoric and even historic periods), when cemeteries were often physically separated from living areas, the earliest Near Eastern permanent settlements appear to initially display a tendency to physically integrate the quick and the dead by the removal of the skull following interment and its reintroduction back into the context of the living community. In contrast, in later periods there is a notable preference for separating the world of the living from the domain of the dead in a tangible manner, in the form of separation of settlements and cemeteries and in the provision of grave goods and furnishings to succor the deceased in the next world. In this respect these would appear to reflect quite separate philosophical approaches to bridging the void separating the quick and the dead.

Furthermore, differences in the specific treatments of the deceased indicate some form of increasingly sophisticated and, probably ultimately, if not initially, inherited, hierarchical system in operation in a time trajectory from the Natufian through PPNB (but also combining elements of attained status). In the latter this treatment ranges from the nature of primary and secondary interment, the removal or nonremoval of skulls, through the variable treatment of skulls, including seemingly unmodified, the application of ochre and other pigments to vaults, the application of headdresses or wigs, and the modeling of facial features in plaster or other substances. Interestingly there are hints that the modified skulls are limited to adults, though the gender remains unclear.

Although the custom of primary interment beneath lime-plaster floors of habitation structures has been widely reported from the central Levant (e.g. Jericho, Abu Gosh, Nahal Oren, Beisamoun, Yiftahel, 'Ain Ghazal), it is commonly not clear as to the precise chronological order of events: were the interments made during, before or after the structures were in use? The evidence from Kfar HaHoresh would appear to indicate that at many PPNB sites burials and subsequent skull removal commonly precede floor plastering. As such this would seem to indicate that they may have functioned as markers of ownership and even, on occasion, as foundation deposits. This interpretation appears to be bolstered by the presence of what appear to be markers on plastered surfaces at Kfar HaHoresh.

There would appear to be a chronological development from the Natufian through the Final PPNB in the specifics of the practice of skull removal, from its seemingly simple innovation in the (Early) Natufian and PPNA through more elaborate and variegated treatment in the Middle and

Late PPNB. It is likely that this reflects the increasingly larger size of communities and the complex socioeconomic social interactions involved with the introduction of domesticated animals and plants, as well as the breakdown of previous egalitarian lifeways associated with mobile foraging societies. Differential accumulations of wealth are likely to have emerged during the course of the PPNA and become even more pronounced in the PPNB in at least the form of fields and herds, not to mention the exchange of goods. Moreover, the direct economic advantages of larger, extended families would potentially have led to stresses in terms of inheritance unless some form of widely accepted and applied guidelines and deep-rooted belief systems were functioning. Under such circumstances high status members of the deceased would have played a crucial role, hence the physical "return" of selected ancestors back into the world of the living in the form of skulls. In light of the above it appears likely that such high status was inherited, at least in part, given the cross-gender and cross-age nature of the phenomenon. Furthermore, the seeming absence of obvious differentiation in the contexts, such as size and nature of domestic architecture with which the various burial and disposal modes are associated, hint that status was primarily within the context of the extended family or clan rather than at the community level. This would accord with the notion of simple village society lacking any externally obvious community-level hierarchical ranking (see also Chapter 6, this volume). Without doubt the symbolic significance of lime-plaster use as well as combined human and animal interments can also be traced back to the Natufian, presumably reflecting the wide range of ritual beliefs involved.

CONCLUDING REMARKS

In addition to the development of more local provinces, there would appear to be a gradual increase in the sophistication and nature of funerary practices through a time trajectory in the south-central Levantine PPNB. They include, in addition to the integration of the quick and the dead within living quarters, the possible emergence of ritual centers on-site (Beidha, Jericho PPNA, and probably also 'Ain Ghazal), as well as localities that perhaps functioned primarily as regional funerary centers (Kfar HaHoresh–Nahal Hemar was less likely a ritual center rather than a repository of symbolic artifacts similar to the later Chalcolithic Nahal Mishmar hoard).

These elaborations of a deep-rooted belief system stretching back to the Early Natufian reach an apogee in the Central/Southern Levant about the mid/late PPNB and parallel the growth and density of site sizes at about the time of the introduction of animal husbandry into the region. As such it may be suggested that they reflect mechanisms attempting to accommodate the social and economic stresses and dislocations concerned with the growth of large-scale, permanently occupied population centers.

In particular the central role of skull removal and the increasingly elaborate specific treatments afforded them reflect concerns with integrating the living communities with their forebears—that is, the veneration of ancestors, as has been widely hypothesized. However, it seems probable that this concerned not only the community "elders" as has been commonly emphasized. Rather it seems more likely that it reflects inherited or otherwise acquired status, as evidenced by its possible cross-gender and cross-age application. On the other hand, the more elaborate treatments to a minority of mostly young (male?) individuals could indicate attained status. The accumulation of power, in the form of prestige, rights, and more tangible wealth (property, fields, and herds), would have necessitated the development of regulatory social mechanisms in their orderly transfer from one generation to the next.

In the Southern and Central Levant it is clear that in the second half of the ninth millennium BP several factors, such as population pressure, local long-term ecological degradation of resources around larger sites, and the general overextension of the system, and possible environmental deterioration with the onset of the Atlantic period and the withdrawal of the monsoonal system, combined to necessitate a shift to more mobile economies and less dense populations. This was perhaps facilitated or even necessitated by the local introduction of herded animals during the mid/late PPNB. Consequently, many of the ritual regulatory mechanisms seemingly became obsolete in the face of the new realities, as evidenced by their decline during the Late/Final PPNB/PPNC and their eventual disappearance by the Yarmukian (Rollefson and Köhler-Rollefson 1993).

Indeed, it is interesting to speculate that if a major role of these mortuary practices and associated belief systems was directed toward enhancing social cohesion and the ideologically egalitarian nature of the living communities, then they may have been initially successful when community sizes were still quite small. However, as some communities expanded during the PPNB, internal social stresses became more acute, as reflected by increasingly more sophisticated embellishments to previous practices. However, in the (ideologically enforced) absence of some form of centralized and powerful decision-making authority or apparatus, such large communities would have been unsustainable over the long run, irrespective of ecological degradation and similar factors (see Kuijt 1995).

In the northern Levant and Anatolia, however, many sites along the major rivers continued to prosper and grow during the eighth millennium bp, with an increasingly sophisticated emphasis on centralized ritual and mortuary behavior, seen at Nevali Çori, Göbekli Tepe, Çayönü, and Çatal Hüyük. The reasons for this dichotomy appear to be a mixture of geographic and climatic factors, involving specific ecological settings (linear settlement patterns along major river systems), resources, and communications and exchange networks. However, it also possible that mechanisms were found to modify the nature and relevance of the belief systems associated with skull removal. Ironically then, such deep-rooted and long-lasting beliefs and the inability to develop strong centralized political/ritual authority thus may be viewed as contributing to the ultimate demise of PPNB village society in the southern Levant, based as it was on much earlier, Epi-Paleolithic egalitarian principles.

ACKNOWLEDGMENTS

Following initial support by the Israel Antiquities Authority, the 1994 season of excavations at Kfar HaHoresh was conducted under the joint auspices of the Israel Antiquities Authority and the Institute of Archaeology, Hebrew University. The 1996 season was conducted under the sole auspices of the Institute of Archaeology, Hebrew University of Jerusalem. Funding has been provided by the Israel Antiquities Authority, the Irene Levi Sala CARE Foundation, the Jerusalem Center for Anthropological Studies, and R. Klein. The author has benefited from stimulating discussions with and is indebted to Israel Hershkovitz, Liora Kolska Horwitz, Patricia Smith, Gary Rollefson, and, especially, Yuval Goren concerning various topics discussed in this chapter. The author, however, takes sole responsibility for the views expressed herein.

REFERENCES

Arensburg, B., and Hershkovitz, I., 1989, Artificial Skull "Treatment" in the PPNB Period: Nahal Hemar. In *People and Culture in Change: Proceedings of the Second Symposium of Upper Paleolothic, Mesolithic, and Neolithic Populations of Europe and the Mediterranean Basin,* edited by I. Hershkovitz, pp. 115–131. British Archaeology Review, International Series 508. Bar, Oxford.

Aurenche, O., and Calley, S., 1988, L'architecture de l'Anatolie du sud-est au Néolithique acéramique. *Anatolica* 15: 1–24.

Baird, D., Garrard, A., Martin, L., and Wright, L., 1992, Prehistoric Environment and Settlement in the Azraq Basin: An Interim Report on the 1989 Excavation Season. *Levant* 24: 1–31.

Banning, E. B., and Byrd, B. F., 1987, Houses and the changing residential unit: Domestic Architecture at PPNB 'Ain Ghazal, Jordan. *Proceedings of the Prehistoric Society* 53: 309–325.

Banning, E. B., Rahimi, D., and Siggers, J., 1994, The Late Neolithic of the Southern Levant: Hiatus, Settlement Shift or Observer Bias? The Perspective from Wadi Ziqlab. *Paléorient* 20(2): 151–164.

Bar-Yosef, O., 1981, Neolithic Sites in Sinai. In *Contributions to the Environmental History of Southwest Asia,* edited by W. Frey and H-P. Uerpmann, pp. 217–235. Beihefte zum Tubinger Atlas des Voderen Orients, Reihe A, Nr. 8, Wiesbaden.

Bar-Yosef, O., 1984, Seasonality among Neolithic Hunter–Gatherers in Southern Sinai. In *Animals and Archaeology: 3. Early Herders and Their Flocks,* edited by J. Clutton-Brock and C. Grigson, pp. 145–159. British Archaeological Review International Series 202. BAR, Oxford.

Bar-Yosef, O., 1985, The Stone Age of the Sinai Peninsula. In *Studi di Paleontologia in Onore di Salvatore Puglisi,* edited by M. Liverani, A. Palmieri, and P. Peroni, pp. 107–122. Universita di Roma "La Sapienza," Roma.

Bar-Yosef, O., and Alon, D., 1988, Nahal Hemar Cave. *Atiqot* 18.

Bar-Yosef, O., and Belfer-Cohen, A., 1989, The Levantine "PPNB" interaction Sphere. In *People and Culture in Change,* edited by I. Hershkovitz, pp. 59–72. British Archaeological Reports International Series 508. BAR, Oxford.

Bar-Yosef, O., and Belfer-Cohen, A., 1992 , From Foraging to Farming in the Mediterranean Levant. In *Transitions to Agriculture in Prehistory,* edited by A. B. Gebauer and T. D. Price, pp. 21–48. Prehistory Press Madison, Wisconsin.

Bar-Yosef, O., and Meadow, R. H., 1995, The Origins of Agriculture in the Near East. In *Last Hunters, First Farmers: New Perspectives on the Transition to Agriculture,* edited by T. D. Price and A. B. Gebauer, pp. 39–94. Schools of American Research Press, Santa Fe, New Mexico.

Belfer-Cohen, A., 1990, The Natufian Graveyard in Hayonim Cave. *Paléorient* 14(2): 297–308.

Belfer-Cohen, A., Arensburg, B.,Bar-Yosef, O., and Gopher, A., 1990, Human remains from Netiv Hagdud—A PPNA site. *Mitekufat Haeven* 23: 79–86.

Betts, A., 1988, Excavations at Dhuweila, Eastern Jordan. A preliminary report. *Levant* 20: 7–21.

Betts, A., 1990, The Pre-Pottery Neolithic B Period in Eastern Jordan. *Paléorient* 15(1): 150–156.

Bienert, H. D., 1991, Skull Cult in the Prehistoric Near East. *Journal of Prehistoric Religion* 5: 9–23.

Byrd, B. F., and Monahan, C. M., 1995, Death, Mortuary Ritual, and Natufian Social Structure. *Journal of Anthropological Archaeology* 14: 251–287.

Cauvin, J., 1972, *Religions Néolithiques de Syro-Palestine.* Publications du Centre de Recherches d'Ecologie et de Préhistoire, Paris.

Cauvin, J., 1978, *Les Premiers Villages Syrie-Palestine du IXe au VIIe Millénaire av. J.C.* CMO 4, Arch. 3, Lyon et Diffusion de Bocard, Paris.

Cauvin, J., 1994, *Naissance des divinités Naissance de l'agriculture.* CNRS, Paris.

Contenson, H. de, 1992, Les Coutumes Funéraires dans le Néolithique Syrien. *Bulletin de la Société Préhistorique Française* 89(6): 184–191.

Davis, S. J. M., and Valla, F. R., 1978, Evidence for Domestication of the Dog 12,000 years ago in the Natufian of Israel. *Nature* 276: 608–610.

Edwards, P. C. , 1991, Wadi Hammeh 27: An Early Natufian site at Pella, Jordan. In *The Natufian Culture in the Levant,* edited by O. Bar-Yosef and F. R. Valla, pp. 123–148. International Monographs in Prehistory, Ann Arbor, Michigan.

Edwards, P. C., and Thorpe, S., 1986, Surface Lithic Finds at Kharaysin, Jordan. *Paléorient* 12(2): 85–87.

Flannery K., 1995, *Prehistoric Social Evolution. In Prehistoric Social Evolution,* edited by C. R. Ember and M. Ember, pp. 1–26. Prentice-Hall, Englewood, New Jersey.

Garrard, A. N., 1991, Natufian Settlement in the Azraq Basin, Eastern Jordan. In *The Natufian Culture in the Levant,* edited by O. Bar-Yosef and F. R. Valla , pp. 235–244. International Monographs in Prehistory, Ann Arbor, Michigan.

Garfinkel, Y., 1987, Yiftahel: A Neolithic Village from Seventh Millennium B.C. in Lower Galilee, Israel. *Journal of Field Archaeology* 14: 199-212.

Garfinkel, Y., 1988, Burnt Lime Products and Social Implications in the Pre-Pottery Neolithic B villages in the Near East. *Paléorient* 13(1): 69–76.

Garfinkel, Y., 1994, Ritual Burial of Cultic Objects: The Earliest Evidence. *Cambridge Archaeology Journal* 4(2): 159–188.

Garstang, G., 1935, Jericho: City and Necropolis, fifth report. *Liverpool Annals of Archaeology and Anthropology* 23(3-4): 143–184.

Gopher, A., 1989, Horvat Galil and Nahal Betzet I: Two Neolithic Sites in the Upper Galilee. *Mitekufat Haeven* 22: 82–92.

Gopher, A., 1990, Neolithic arrowheads of the Levant: Results and Implications of a Seriation Analysis. *Paléorient* 15(1): 43–56.

Gopher, A., 1994, *Arrowheads of the Neolithic Levant.* American School of Oriental Research: Dissertation Series, Volume 10. Winona Lake, Indiana: Eisenbrauns.

Gopher, A., and Gophna, R., 1993, Cultures of the Eighth and Seventh Millennium BP in the Southern Levant: A review for the 1990s. *Journal of World Prehistory* 7(3): 297–353.

Gopher A., Goring-Morris, A. N., and Rosen, S. A., 1995, A Pre-Pottery Neolithic B Occupation near Ein Qadis at the Edge of the Central Negev Highlands. *Atiqot* 27: 15–33.

Goren, Y., and Goldberg, P., 1991, Petrographic Thin Sections and the Development of Neolithic Plaster Production in Northern Israel. *Journal of Field Archaeology* 18: 131–138.

Goren, Y., Goring-Morris, A. N., and Segal, I., n.d. The Technology of Skull Modeling in the Pre-Pottery Neolithic B: Regional Variability, the Relation of Technology and Iconography and their Archaeological Implications.

Goren, Y., Segal, I., and Bar-Yosef , O., 1993, Plaster Artifacts and the Interpretation of Nahal Hemar Cave. *Mitekufat Haeven* 25: 120-131.

Goring-Morris, A. N., 1991, A PPNB Settlement at Kfar Hahoresh in Lower Galilee: A Preliminary Report of the 1991 Season. *Mitekufat Haeven* 24: 77-101.

Goring-Morris, A. N., 1994a, From Foraging to Herding in the Negev and Sinai: The Early to Late Neolithic transition. *Paléorient* 19(1): 63-87.

Goring-Morris, A. N., 1994b Aspects of the PPNB lithic assemblage from Kfar HaHoresh, near Nazareth, Israel. In *Neolithic Chipped Lithic Industries of the Fertile Crescent. Proceedings of the First Workshop on PPN Chipped Lithic Industries,* edited by H-G. Gebel and S. K. Kozlowski, Studies in Early Near Eastern Production, Subsistence, and Environment 1, pp. 427–444. Ex Oriente, Berlin.

Goring-Morris, A. N., 1995, Daily and ritual life at the PPNB Settlement of Kfar HaHoresh, Lower Galilee. *Qadmoniot* XXVIII, No.1 (109): 20–26 (in Hebrew).

Gorng-Morris, A. N., 1996, The Early Natufian Occupation at El Wad, Mt. Carmel Reconsidered. In *Nature et Culture,* edited by M. Otte, Actes du Colloque International de Liège, E.R.A.U.L. No. 68: 417–427.

Goring-Morris, A. N., and Gopher, A., 1983 Nahal Issaron: A Neolithic Settlement in the Southern Negev. *Israel Exploration Journal* 33: 149–162.

Goring-Morris, A. N., Goren, Y., Horwitz, L. K., Hershkovitz, I., Lieberman, R., Sarel, J., and Bar-Yosef, D., 1994-5 The 1992 Season of Excavations at the Pre-Pottery Neolithic B Settlement of Kfar HaHoresh. *Mitekufat Haeven* 26: 74–121.

Goring-Morris, A. N., Goren, Y., Horwitz, L. K., Bar-Yosef , D., and Hershkovitz, I., 1995 Investigations at an Early Neolithic settlement in Lower Galilee: Results of the 1991 Season at Kfar HaHoresh. *Atiqot* 27: 37–62.

Hauptmann, H., 1993, Ein Kultgebaude in Nevali Çori. In *Between the Rivers and over the Mountains: Archaeologica Anatolica et Mesopotamia, Alba Palmieri Dedicata,* edited by M. Frangipane, pp. 37–69. Rome.

Hauptmann, H., 1997 Nevali Çori. In *Archaeology in the Near East,* Volume 4, edited by E. M. Meyers, pp. 131–134. Oxford University Press, Oxford.

Hershkovitz, I., Garfinkel, Y., and Arensberg, B., 1986, Neolithic skeletal remains at Yiftahel, Area C, Israel. *Paléorient* 12(1): 73–81.

Hershkovitz, I., and Gopher, A., 1990, Paleodemography, Burial Customs, and Food-Producing Economy at the Beginning of the Holocene: A Perspective from the Southern Levant. *Mitekufat Haeven,* 23: 9–47.

Hershkovitz, I., Zohar, M., Speirs, M. S., Segal, I., Meirav, O., Sherter, U., Feldman, H., and Goring-Morris, N., 1995, Remedy for an 8,500-Year-Old Plastered Human Skull from Kfar HaHoresh, Israel. *Journal of Archaeological Science* 22: 779-788.

Hershkovitz, I., Zohar, I., Wish-Baratz, S., Goren, Y., Goring-Morris, N., Speirs, M., Segal, I., Meirav, O., Sherter, U., and Feldman, H., 1996 High Resolution Computed Tomography and Micro-focus Radiography on an Eight Thousand Year Old Plastered Skull: How and Why It Was Modeled. In *Culture et Nature,* edited by M.Otte, Actes du Colloque International de Liège, E.R.A.U.L. No. 68: 669–682.

Horwitz, L. K., 1987, The Fauna from the PPNB Site of Yiftahel: New Perspectives on Domestication. *Mitekufat Haeven* 20: 181–182.

Horowitz, L. K., 1993, The development of Ovicaprine Domestication during the PPNB of the Southern Levant. In *Archaeozoology of the Near East,* edited by H. Buitenhuis and A. T. Clason, pp. 27–36. Universal Book Services, Leiden.

Kenyon, K. M., 1979, *The Archaeology of the Holy Land,* 4th edition. Ernest Benn, London.

Kenyon, K. M., 1981, *Excavations at Jericho, Volume III: The Architecture and Stratigraphy of the Tell.* British School of Archaeology in Jerusalem, London.

Kenyon, K. M., and Holland, T. A. (eds.), 1983, *Jericho V.* British School of Archaeology in Jerusalem, London.

Kingery, D. W., Vandiver, P. B., and Pickett, M., 1988 The Beginnings of Pyrotechnology. Part II: Production and Use of Lime and Gypsum Plaster in the Pre-Pottery Neolithic Near East. *Journal of Field Archaeology* 15: 219–244.

Kirkbride, D., 1966, Five seasons at the Prepottery Neolithic Village of Beidha in Jordan. *Palestine Exploration Quarterly* 98: 8–72.

Kirkbride, D., 1967, Beidha 1965: An Interim Report. *Palestine Exploration Quarterly* 99: 5–13.

Kirkbride, D., 1968, Beidha 1967: An Interim Report. *Palestine Exploration Quarterly* 100: 90–96.

Köhler-Rollefson, I., 1989, Changes in Goat Exploitation at 'Ain Ghazal between the Early and Late Neolithic: A Metrical Analysis. *Paléorient* 15: 141–145.

Köhler-Rollefson, I., Gillespie, W., and Metzger, M., 1988, The Fauna from Neolithic 'Ain Ghazal. In *The Prehistory of Jordan,* edited by A. N. Garrard and H-G. Gebel, pp. 423–430. British Archaelogical Review International Series 396. BAR, Oxford.

Kuijt, I., 1995, *New Perspectives on Old Territories: Ritual Practices and the Emergence of Social Complexity in the Levantine Neolithic.* Ph.D. dissertation, Harvard University, University Microfilms, Ann Arbor, Michigan.

Kuijt, I., 1996, Negotiating Equality through Ritual: A Consideration of Late Natufian and Pre-Pottery Neolithic A Period Mortuary Practices. *Journal of Anthropological Archaeology* 15: 313–336.

Lechevallier, M., 1978, *Abou Gosh et Beisamoun: Deux Gisements du VIIe Millénaire avant L'ére Chretienne en Israel.* Mémoires et Travaux du Centre de Recherches Préhistoriques Français de Jerusalem, No. 2.

Lechevallier, M., and Ronen, A., 1994, *Le Gisement de Hatoula en Judée Occidentale, Israël.* Association Paléorient, Paris.

Meiklejohn, C., Agelarakis, A., Akkermans, P. A., Smith, P. E. L., and Solecki, R., 1992 , Artificial Cranial Deformation in the Proto-Neolithic and Neolithic Near East and Its Possible Origin: Evidence from Four Sites. *Paléorient* 18(2): 83–97.
Mellaart, J., 1967, *Çatal Höyük, a Neolithic Town in Anatolia.* New York: McGraw-Hill.
Mellaart, J., 1975, *The Neolithic of the Near East.* Thames and Hudson, London.
Moore, A.,1985, The Development of Neolithic Societies in the Near East. *Advances in World Archaeology,* edited by F. Wendorf and A. E. Close, pp. 1–69. Academic Press, New York.
Neuville, R., 1951, *Le Paléolithique et le Mesolithique de Desert du Judée.* Archives de l'Institut de Paleontologie Humaine, Mémoire 24.
Nishiaki, Y., 1994, The Naviform Method at Douara Cave II, Palmyra, Syria. In *Neolithic Chipped Lithic Industries of the Fertile Crescent. Proceedings of the First Workshop on PPN Chipped Lithic Industries,* edited by H-G. Gebel and S. K. Kozlowski, pp. 363–378. Studies in Early Near Eastern Production, Subsistence, and Environment 1, Ex Oriente, Berlin.
Nissen, H. J., Muheisen, M., Gebel, H-G., Becker, C., Neef, R., Pachur, H. J., Qadi, N., and Schultz, M., 1991, Report on the First Two Seasons of Excavations at Basta (1986–1987). *Annual of the Department of Antiquities of Jordan* 35: 79–119.
Nissen, H. J., Muheisen, M., Gebel, H. G., Becker, C., Hermansen, B. D., Kharasneh, W., Qadi, N., Schultz, M., and Scherer, A., 1992 , Report on the Excavations at Basta 1988. *Annual of the Department of Antiquities of Jordan* 36: 13–40.
Noy, T., Legge, A. J., and Higgs, E. S., 1973, Recent Excavations at Nahal Oren, Israel. *Proceedings of the Prehistoric Society* 39: 75–99.
Özdögan M,. and Özdögan, A., 1990, Çayönü, A Conspectus of Recent Work. *Paléorient* 15(1): 65–74.
Perrot, J., and Ladiray, D., 1988, *Les Hommes de Mallaha (Eynan), Israel.* Mémoires et Travaux du Centre de Recherche Française de Jérusalem, No. 7. Association Paléorient, Paris.
Redman, C., 1978, *The Rise of Civilization.* W. H. Freeman and Company, San Francisco, California.
Rollefson, G. O., 1983, Ritual and Ceremony at Neolithic t'Ain Ghazal (Jordan). *Paléorient* 9(2): 29–38.
Rollefson, G. O., 1986, Neolithic 'Ain Ghazal (Jordan): Ritual and Ceremony, II. Paléorient 12: 45–52.
Rollefson, G. O., 1990a, The Uses of Plaster at Neolithic 'Ain Ghazal, Jordan. *Archeomaterials* 4: 33–54.
Rollefson, G. O., 1990b, The Late Aceramic Neolithic of the Levant. *Paléorient* 15(1): 168–173.
Rollefson, G. O., 1991, The critical role of technological analysis for prehistoric anthropological inference. *Materials Research Society Symposium Proceedings* 185: 365–374.
Rollefson, G. O., and Köhler-Rollefson, I., 1989, The Collapse of Early Neolithic Settlements in the Southern Levant. In *People and Culture in Change: Proceedings of the Secind Symposium on Upper Paelolithic, Mesolithic, and Neolithic Populations of Europe and the Mediterranean Basin,* edited by I. Hershovitz, pp. 59–72. British Archaelogical Reports, International Series 508, Oxford.
Rollefson, G., and Köhler-Rollefson, I., 1993, PPNC Adaptations in the First Half of the 6th Millennium B.C. *Paléorient* 19: 33–42.
Rollefson, G. O., Simmons, A., and Kafafi, Z., 1992, Neolithic Cultures at 'Ain Ghazal, Jordan. *Journal of Field Archaeology* 19: 443–470.
Rosenberg, M., and Davis, M. K., 1992, Hallan Çemi Tepesi, An Early Ceramic Neolithic Site in Eastern Anatolia. *Anatolica* 18: 1–18.
Schmidt, K., 1995, Investigations in the Upper Mesopotamian Early Neolithic: Göbekli Tepe and Gürcütepe. *Neo-Lithics* 2/95: 9–10.

Schmidt, K., and Beile-Bohn, 1996, A LPPNB Variant of Byblos Points from Gürcutepe II-"Palmyra Points?" *Neo-Lithics* 2/96:9–11.
Simmons, A., Rollefson, G., Köhler-Rollefson, I., Mandel, R., and Kafafi, Z., 1988, 'Ain Ghazal: A Major Neolithic Settlement in Northern Jordan. *Science* 240: 35–39.
Simmons, A., Kafafi, Z., Rollefson, G., and Moyer, K., 1989, Test excavations at Wadi Shu'eib, a Major Neolithic Settlement in Jordan. *Annual of the Department of Antiquities of Jordan* 33: 27–42.
Simmons, A., Boulton, A., Butler, C., Kafafi, Z., and Rollefson, G., 1990, A Plastered Skull from Neolithic 'Ain Ghazal, Jordan. *Journal of Field Archaeology* 17(1): 107–110.
Stekelis, M., and Noy, T., 1963, Excavations at Nahal Oren. *Israel Exploration Journal* 13: 1–13.
Valla, F. R., 1990, Aspects du sol de l'abris 131 de Mallaha (Eynan). *Paléorient* 14(2): 183–296.
Valla, F. R., 1996, L'animal "Bon a Penser": La domestication et la Place de l'Homme dans la Nature. In *Culture et Nature,* edited by M. Otte, Actes du Colloque International de Liège, E.R.A.U.L. No. 68: 651–663.
Valla, F. R., Le Mort, F., and Plisson, H., 1991, Les Fouilles en cours sur la Terrasse d'Hayonim. In *The Natufian Culture in the Levant,* edited by O. Bar-Yosef and F. R.Valla, pp. 93–110. International Monographs in Prehistory, Ann Arbor, Michigan.
Vaux, R., de 1966, *Palestine during the Neolithic and Chalcolithic Periods.* Cambridge, Cambridge University Press.

Chapter 6

Keeping the Peace

Ritual, Skull Caching, and Community Integration in the Levantine Neolithic

IAN KUIJT

> Yet, alternatively, if the seeds of inequality are recognized to be present in egalitarian human groupings, then the focus of this central research question must be shifted. Rather than endeavoring simply to account for inequality, the emphasis should be expanded. Attention must be placed both on those mechanisms in nonstratifed societies that have served to level extant inequalities before they come institutionalized, as well as on the internal and/or external conditions that work to negate those leveling strategies and sanctions so that existent inequalities are permitted to become more institutionalized (Feinman 1995:262)

INTRODUCTION

In studying the emergence of social differentiation in the past, researchers have focused considerable attention on the critical role of ritual behavior as a framework in which people and communities define and modify social relationships. From one perspective ritual can serve as a device of powerful

IAN KUIJT • Department of Anthropology, University of Notre Dame, Notre Dame, Indiana 46556.

Life in Neolithic Farming Communities: Social Organization, Identity, and Differentiation, edited by Ian Kuijt. Kluwer Academic/Plenum Publishers, New York, 2000.

social regulation, as a consolidator of economic, political, and social power among select individuals within communities, and as a potential mechanism to challenge egalitarian belief systems (e.g., Earle 1987; Fried 1960, 1967; Johnson 1982; Hayden 1995). An alternative, but by no means mutually exclusive, perspective is offered by a number of anthropological and archaeological studies that explore how at times people can use ritual practices to maintain egalitarian social systems, as social leveling mechanisms within communities, and to maintain or increase solidarity between individuals and households by stressing shared egalitarian themes (Berreman 1981; Boehm 1993; Flanagan 1989; Flanagan and Rayner 1988; Gerlach and Gerlach 1988; McKinnon 1991; Paynter 1989; Rayner 1988). Viewed collectively, these studies have increased our awareness of the dynamic and multidimensional nature of equality and inequality inherent in most social relationships. One of the most visual, complex, and powerful examples of ritual behavior is that of mortuary rituals, for they often serve as both symbolic and physical expressions of the views and beliefs of general communities (Kan 1989; Metcalf and Huntington 1991; Weiner 1976).

The question of how social structure, ideology, and worldviews are expressed through, and mediated by, mortuary practices has been an active focus of anthropological and archaeological debate for some time (e.g., see Binford 1971; Chapman et al. 1981; O'Shea 1984, 1996; Tainter 1978; Weiner 1976). As part of this dialogue, several recent studies (Carr 1995; Hodder 1982; Kan 1989; McGuire 1992; Metcalf and Huntington 1991) have directed new attention to how mortuary practices often idealize and mask daily social relations; additionally researchers have explored the importance of the living in relation to the perceived status of the deceased in structuring mortuary practices to further understand the social impact of specific mortuary ritual upon individuals and communities. (Readers are referred to Carr 1995 for expanded discussion of different approaches to mortuary analysis adopted by archaeologists, as well as for arguments for their historical developments.) Following many of these works, I view mortuary practice as a form of human behavior actively chosen by actors in relation to specific beliefs and a broader worldview and symbolic themes rather than a direct reflection of social organization. Mortuary practices are often a communal event, usually controlled and directed by a limited number of individuals, and enacted for an audience of individuals present at the event. The power of ritual as a cohesive force is based, in part, on the realization that mortuary rituals are a form of public action, a social drama designed and conducted by the living in such a way that the broader social ethos and mortuary practices are interlinked and mutually reinforcing, and is not always, therefore, a direct reflection of the status, authority, and importance of the deceased (Geertz 1973; Hertz 1960; Metcalf and Huntington 1991; van Gennep

1960). Among others, Geertz (1973:131) emphasizes the centrality of the relationship between ritual symbols and the broader social ethos, stating that "The force of a religion in supporting social values rests, then, on the ability of its symbols to formulate a world in which those values, as well as the forces opposing their realization, are fundamental ingredients." Moreover, the standardization of symbols in household ritual or mortuary practices, such as the number of objects and significance, is central to their intended meaning and can be employed to reinforce broader spiritual beliefs and community ethos within and between households (Hodder 1982; McKinnon 1991; Metcalf and Huntington 1991).

It is important to keep in mind that under different conditions specific mortuary practices can have different political and social impacts upon the individual, household, and community (Blanton 1995; Carr 1995; Hodder 1982; Metcalf and Huntington 1991). In many societies ritual action provides the framework for community cohesion, the arena in which links between households are established, supported, and extended by elaborate codes of social reciprocity that ensure participation in collective rituals by individuals from multiple households. Following other researchers (Blanton 1995; Joyce 1993; Lévi-Strauss 1983; McKinnon 1991), I use the term "household" to refer to the cooperative coresidential economic unit exemplified by internal ranking and some centralized decision-making authority. Membership within households would have been through kinship links, but not all members of the household were kin. A household is viewed as a corporate body that perpetuates itself through the exchange of goods, titles, and membership along real or imaginary lines. Within such small-scale social groups, individual and household level relationships are negotiated, based on real or perceived reciprocity, and are frequently reaffirmed through gift exchange and reciprocal participation in household ritual events, such as mortuary rituals.

While addressing the links between mortuary practices, social distinction, and material culture, several researchers have recognized that mortuary practices not only reaffirm the kin and economic links between households, but also that the actual or perceived coparticipation in mortuary practices impacts communities by symbolically and physically linking and defining individuals. For example, in a recent examination of mortuary practices and their determinants, Carr (1995) illustrates that while funeral attendance and the overall energy expended in mortuary rites often reflects the social position of the deceased within communities, it can also be linked to communal ancestor worship, responsibility to the deceased, beliefs about the soul's nature, and the nature of the afterlife (Binford 1971; Hodder 1982; McGuire 1992; Metcalf and Huntington 1991; O'Shea 1984; Radcliffe-Brown 1964). Importantly, individuals may not recognize the sentiments or actions that reiterate group membership, but the very act of coparticipating in such actions will minimally serve to strengthen existing feelings and develop new relationships. From this perspective, then, mortuary practices fulfill an important integrative function within communities by encouraging participation in a powerful communal act that symbolically and physically links community members in a logical and articulate form, leads to the development of new networks or the extension of existing networks, and reaffirms broader beliefs and worldviews (see Fentress and Wickham 1992; McKinnon 1991; Metcalf and Huntington 1991; Radcliffe-Brown 1964 for further discussion of these themes).

RITUAL PRACTICES IN THE MPPNB: SHARED THEMES AND NUMERIC STANDARDIZATION

In exploring changes in social organization during the Aceramic Neolithic I have previously argued that Middle Pre-Pottery Neolithic B period communities from the Mediterranean zone of the south-central Levant, including the modern states of Jordan, Israel, eastern Egypt, and southern Syria, can be envisioned as "House Societies" characterized by Houses that interacted within a social framework of coexisting competition and cooperation, where different aspects of material culture were employed to express degrees of affinity. I utilize Lévi-Strauss' (1983:174) definition of the house as a foundation: "a corporate body holding an estate made up of both material and immaterial wealth, which perpetuates itself through the transmission of its name, its goods and its titles down a real or imaginary line, considered legitimate as long as this continuity can express itself in the language of kinship or of affinity and, most often, of both" (Lévi-Strauss 1983:174). Although often viewed from a material perspective, Lévi-Strauss' definition clearly acknowledges the importance of nonmaterial expressions of social relations within House societies. Moving beyond this core framework, I share with Waterson (1995:49) a more flexible definition emphasizing elements of temporal continuity, the hereditary transfer of valued property and authority, and the strategic exploitation of the language of kinship and affinity. In this interpretive context, then, the 'House' exists simultaneously as a social, ritual, and economic unit, and can be comprised of multiple households dwelling in separate residential structures, and can serve as a physical and symbolic place of origin residence of fictive and real ancestors. From this perspective, then, I have explored Neolithic household social organization as reflecting a series of complex social rules that reaffirmed the egalitarian values and ethos of general society, and at the same time, permitted the development of social differentiation which cross-cut household and kin-group lines (Kuijt 1995, 1996, 2000, in press).

As an alternative to treating egalitarianism as a descriptive category, a number of researchers (e.g., Boehm 1993: Feinman 1995; Hodder 1991; McKinnon 1991; Plog 1995), approach this concept as a form of ideology, a crafted social identity or worldview that is expressed through material culture, carefully maintained by community leaders so as to deliberately affect community behavior and social relations by emphasizing the shared identity and affinity between individuals within and between Houses. In contrast to some research that views egalitarianism and social differentiation in middle-range societies as being mutually exclusive, I believe that egalitarian ideology and hierarchy are fundamentally interrelated and co-exist in many, if not most, social system (see also Berreman 1981; Feinman 1995; Flanagan 1989; Flanagan and Rayner 1988; Gerlach and Gerlach 1988; Kan 1989; McKinnon 1991; 1995; Myers 1986; Plog 1995; Price 1995; Rayner 1988). In many ways, all egalitarian systems are a reaction by community members against formalized social differentiation, establishing a system of highly complex social rules which ultimately ensures relatively equal treatment for individuals. Based on this ethos, attempts to accumulate power and authority by select individuals and Houses are sharply limited and controlled by other Houses and the collective community. Above all else, Houses are usually traditional, conservative, and focused on maintaining existing social arrangements to limit the gains of other Houses (Lévi-Strauss 1983; McKinnon 1991). One possible reason for this is that within many societies, the emergence of formalized, hereditary, social hierarchy is highly fissive to group membership, because differential access to resources and status runs counter to broader social beliefs (Berreman 1981; Myers 1986), and perhaps more importantly, counteracts the success and longevity of competing Houses. Drawing on a number of studies (Boehm 1993; Flanagan 1989; Hodder 1990; Keene 1991; McKinnon 1991, 1995), I believe that competition plays a major role in the maintenance and expansion of egalitarian ideology in House Societies, and by extension, requires us to recognize that the articulation of social arrangements varies in different House Societies through time.

Based upon previous studies of mortuary practice and evidence for long-term architectural continuity I argue that MPPNB communities, such as at Jericho and 'Ain Ghazal, were comprised of multiple, distinct, yet interrelated 'Houses' consisting of multiple households and residing in multiple residential structures. Ritual practices, focused on the House or household as a social unit, and in particular, the mortuary practices and the daily rituals conducted in individual households and House non-residential structures, were employed as physical and symbolic vehicles for negotiating social arrangements. The physical organization of ritual and mortuary practices, their location, and the organization of physical space provided the symbolic and social language to form social connections and alliances that extended beyond the physical boundaries of individual structures to encompass multiple residential structures in a collective economic, social, and ritual social unit of the House. As noted elsewhere (Arensburg and Herskovitz 1988; Bienert 1991; Cauvin 1994; Cornwall 1981; Kuijt 1995; Kurth and Röhrer-Ertl 1981), MPPNB communities practiced multiple kinds of cranial deformation, painting, plastering, and caching of human skulls of highly select deceased individuals along with a range of differential mortuary practices as a material language of affinity to symbolically and physically differentiate some individuals and Houses over others. Paradoxically, it appears that community members, likely drawn from Houses throughout the community, intentionally limited the material and symbolic ways in which social differentiation was expressed within and between Houses, while actively distinguishing themselves and other individuals by their ritual service to the community and by their differential treatment of the dead.

It is not my purpose to review existing ethnographic and anthropological works documenting how ritual can be, and often is, employed as a means by which social differentiation and, by extension, inequalities of power and authority are accrued and maintained in communities. Nor do I disagree in any fundamental way with arguments that mortuary rituals and symbols embody multiple messages within and between communities, and, as such, it is impossible to develop universal explanations for the importance of ritual in different cultures. Rather, I want to outline how south-central Levantine Middle Pre-Pottery Neolithic B period people expressed and maintained their beliefs and values through a series of highly standardized household ritual practices, expressed in multiple media, and understandable by all community members. In brief, these behaviors are reflected in the archaeological record through (1) systems of ritual practices that symbolically and physically link individuals together, (2) the standardization in the location in which ritual practices occur, (3) the standardization and reiteration of certain themes common to all observers, and (4) the reiteration of these themes in multiple media. For ritual practices to effectively express household and community values and beliefs, participants and observers must be aware of the broader meanings and messages presented in the ritual activities. This point is of particular significance to researchers, since systems of significance often require highly structured patterns—patterns that are archaeologically observable and can be monitored through time. In considering the social impact of these ritual practices, I argue that the reiteration of several ritual/spiritual themes served to constantly reemphasize common rules recognizable by both participants and observers, as well as the broader belief systems that they represent. To illustrate these points, I briefly address three interrelated dimensions of MPPNB ritual practices: (1) the social impact of secondary mortuary practice of cranial removal at the household and community level, (2) the physical and symbolic ways in which systematic interments of human skull caches and anthropomorphic figurines were made intelligible to community members, and (3) how people used these practices to differentially select some members of the community over others.

SECONDARY MORTUARY RITUALS AND COMMUNITY IDENTITY

A number of ethnographic and archaeological studies have illustrated how broader beliefs and worldviews are expressed in secondary mortuary practices and often reflect aspects of ancestor worship and responsibility to the deceased (see Crocker 1977; Hertz 1960; Kan 1989; Kuijt 1996; Lopatin 1960; Metcalf and Huntington 1991; Weiner 1976). Within this paper secondary mortuary practice is defined as a social act focused on the regular and socially sanctioned removal of objects, pieces, or entire remains of a deceased individual from some place of temporary storage to a permanent resting place. Physically this is expressed by the intentional removal of skeletal materials from one location to another location and is often, but not always, represented by the recovery of disarticultated and relatively incomplete skeletal remains. It is important to keep in mind, however, that primary and secondary mortuary practices need not be mutually exclusive from a classification standpoint (especially given that they are usually perceived by ethnographic groups as being interlinked as parts of a broader belief system). Although the means and rationale behind secondary mortuary rituals varies considerably between ethnographic groups, secondary mortuary rituals throughout diverse cultures impact individuals within and between households of a community as an extremely powerful means of defining, shaping, and maintaining identities and social relationships. One aspect to this is that, even if they focus on specific individuals from separate households, secondary mortuary practices involve, be it perceived or unperceived, an element of communal ancestor worship as part of a collective social memory and identity. The broader articulation of a shared identity requires the message be conventionalized as well as simplified so as to make it understandable to all. This is partially accomplished by reference to generalized ancestors and the development of highly standardized social rules. Importantly, secondary mortuary practices permit scheduling of funeral events at a pre-arranged time that does not conflict with other tasks, and are at times envisioned as a season of festivities (Hertz 1960; Metcalf and Huntington 1991). This facilitates extensive co-participation in secondary mortuary events from within the community as well as for initial primary mortuary events and by extension, broader recognition of a worldview and beliefs (Figure 1).

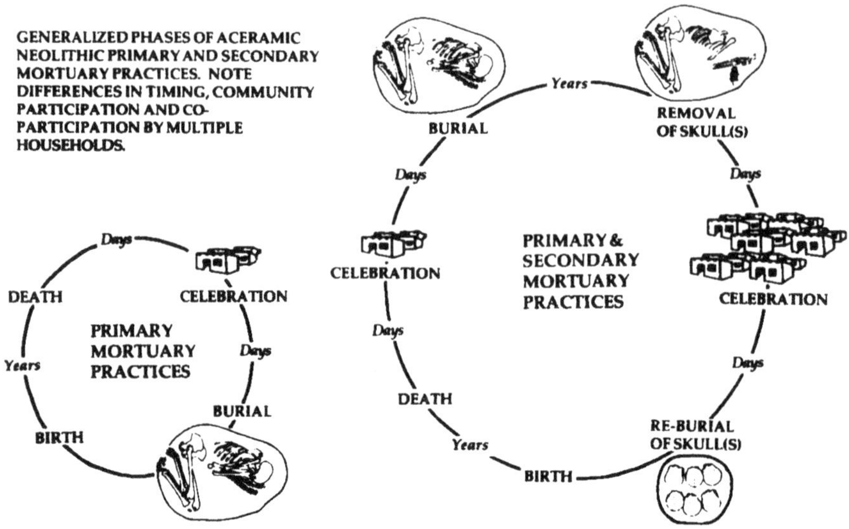

Figure 1. Generalized stages of Pre-Pottery Neolithic primary and secondary mortuary practices. Note differences in timing, community participation, and possibilities of coparticipation by multiple households.

In many societies secondary mortuary practices are organized to facilitate participation in community events that cross-cut kin and household lines (Downs 1956; Hertz 1960; Hudson 1966; Metcalf and Huntington 1991). Among the Ma'anyan of Borneo, for example, corpses from different households are removed from a primary burial context after a number of years and are collectively given funeral rites as part of a week-long community festival (Hudson 1966:361–98). Similarly, Downs (1956:78–91) outlines how community level secondary mortuary practices occur among the Toradja of central Celebes for deceased individuals from multiple households. These studies reiterate that broader beliefs and worldviews fundamentally affect and perpetuate secondary mortuary practices. Ultimately, these ethnographic data illustrate how purposefully ritual practitioners and communities organize secondary mortuary rituals as part of high profile public ceremonies; therefore, we can view these as spiritual and symbolic acts that have social, political, and personal meanings. In contrast to primary, single-stage, mortuary practices, aspects of multi-stage secondary mortuary practices are planned in advance, often held in conjunction by multiple households as part of a community festival, and require extraordinary levels of community involvement. In observed ethnographic societies in which skull caching occurs, the secondary burial of the skull is often viewed as an obligation by the deceased's household and a necessary stage of a multi-year mortuary ceremony, requiring substantial participation from other households in the village and from neighboring communities (McKinnon 1991). Beyond these logistical dimensions, secondary mortuary practices, with the deliberate removal of some or all of the skeleton, such as skull removal, are often linked to broader beliefs in ancestor worship. For all of these reasons, secondary mortuary rituals differ from primary burial of individuals, as these ceremonies often crosscut kin and household lines, thereby emphasizing the community over the individual.

Social Impact of MPPNB Secondary Mortuary Practices

Archaeological research at a number of major MPPNB settlements in the south-central Levant has revealed that mortuary rituals, including skull removal as a secondary mortuary practice, were organized as a series of elaborate mortuary ordering principles based on the age of the deceased. A review of archaeological mortuary data indicates that these practices focused on (1) the primary interment of adults, probably both males and females, in single graves, (2) the secondary removal and caching of many adult crania singularly and in groups, (3) the interment of infants in single graves, usually without cranial removal, and (4) the occasional internment of adults in extramural and intramural locations without secondary cranial removal (see Cornwall 1981; Kuijt 1995; Rollefson Chapter 7 this volume; Rollefson et al. 1992 for more detailed consideration of these themes). Previous studies have outlined that some of these mortuary systems originally emerged in the PPNA and the Late Natufian period (Belfer-Cohen 1991; Hershkovitz and Gopher 1990) (Figure 2). Excavations from the MPPNB occupations at Jericho and 'Ain Ghazal have demonstrated that after primary burial of adults community ritual practitioners often, but not always, later removed the skull. Primary interment of adults was usually associated with architecture, although not always so, and burials occurred without any grave goods (Figures 3 and 4), a pattern that is also seen at the MPPNB settlements of Beidha, Yiftahel, and Kfar HaHoresh (Byrd 1994; Garfinkel 1987, Goring-Morris 1991; Kirkbride 1968).

Interestingly, children were treated differently from adults at time of death in MPPNB communities. Infants were usually buried as individuals; although occasionally buried in intramural areas, they are found in fills and courtyard contexts. Crania were occasionally removed from the skeletons of infants and youths (Cornwall 1981; Kirkbride 1968; Moore 1985; Rollefson et al. 1992). Some evidence from MPPNB 'Ain Ghazal and Jericho suggests a variation in these practices, where at times infant remains were associated with adults with intact skulls. It is not clear, however, if this association was

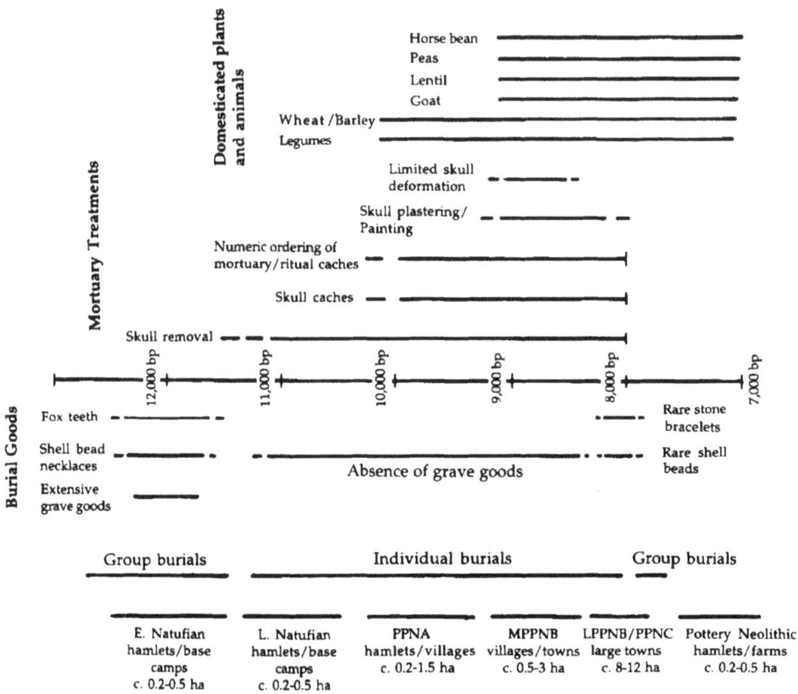

Figure 2. Changing dimensions of south-central Levantine Natufian and Neolithic mortuary practices.

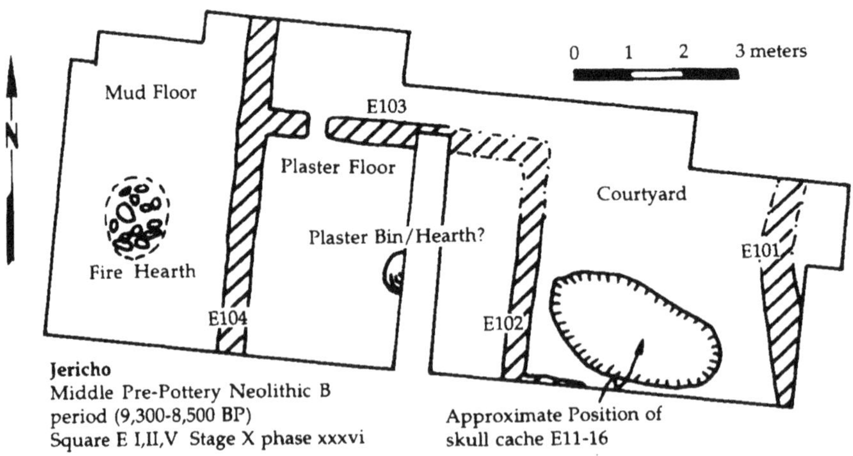

Figure 3. Plan view of MPPNB structure in square M1, stage XV, phase lxxviii.

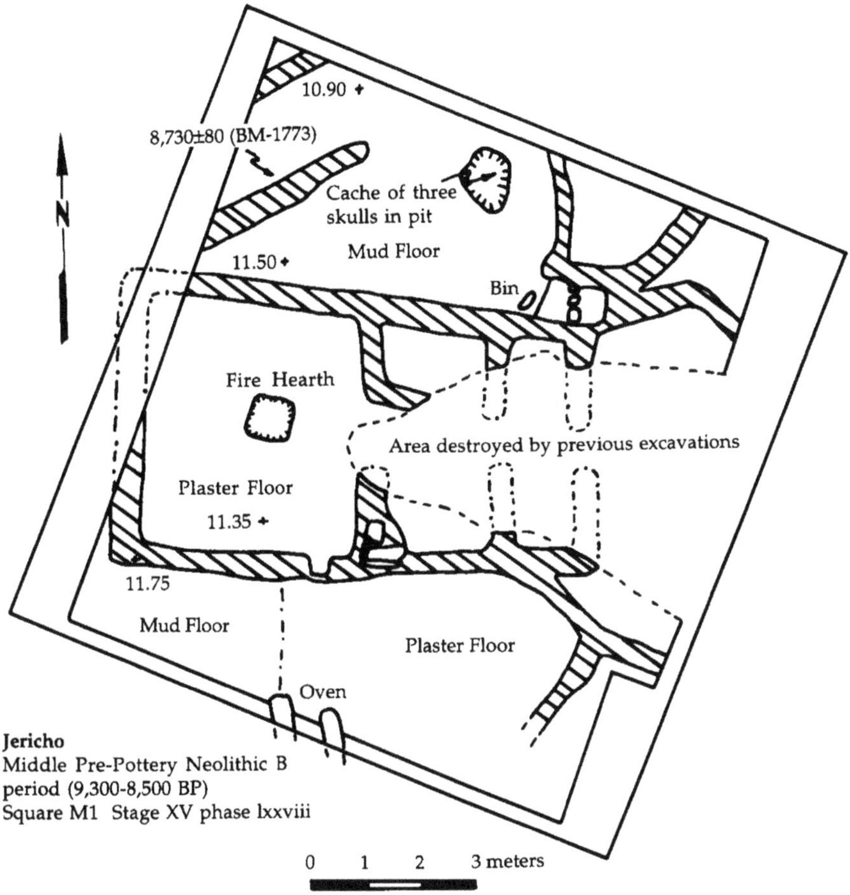

Figure 4. Plan view of MPPNB structure in square EI, II, V, stage X, phase xxxvi. Note the location of human skull cache E11–16 in a courtyard area (see Figure 5).

intentional or a coincidental by-product due to the repeated interment of individuals over time. On other occasions at 'Ain Ghazal and Jericho, infants were clearly interred in a ritual context, such as in subfloor pit features and as dedicatory offerings within the foundation or the walls of a building (Cornwall 1981; Rollefson et al. 1992).

In considering the function and context of, and meanings for, Neolithic caches, archaeologists have adopted a range of interpretive approaches. Describing ritual caching in the Neolithic period, Garfinkel (1994) adopts a functional perspective, believing that ritual caches in the Middle Pre-Pottery Neolithic period represent discarded objects worn out though ritual use.

Undoubtedly Garfinkel is correct in arguing that many cultic objects were used in a ritual context before caching, but at the same time this descriptive work is less than satisfying since it provides no convincing evidence to suggest that all painted and plastered skulls were worn out in ritual use before burial. Most of the skulls recovered at Jericho, for example, were remarkably well preserved and, in many cases, still possessed intact shell inlays for the eyes and clearly visible lines of paint. More importantly, while many previous treatments have focused on the depositional context of materials, they often treat secondary burial systems and skull and anthropomorphic caches as isolated entities and, ultimately, fail to explore the more important past social context reflected by material culture.

As with other researchers (e.g., Bar-Yosef and Belfer-Cohen 1989; Cauvin 1994; Rollefson 1986; Schmandt-Besserat 1998), I argue that it is more profitable to view these cultic objects, skull caches, plastered skulls, and foundation offerings as a series of thematically interrelated aspects of MPPNB ritual beliefs and community ideology. Cultic objects were intentionally manufactured for use in ritual practices by the living, and rather than viewing ritual caching as refuse of worn-out cultic items, we can envision it as a later stage within a series of symbolic and spiritual acts that have social, political, and personal meanings. A wide range of cultic objects was purposefully manufactured by ritual practitioners for interment as part of highly visible public ceremonies focusing on the secondary burial of skulls of deceased community members and, therefore, can be viewed as a means of symbolic expression of a shared system of beliefs and values.

One, but by no means the only, important dimension to secondary mortuary practices in ethnographic and archaeological contexts is that they are often deliberately held in highly visible public contexts to maximize participation in this shared experience in a meaningful way. In the case of Jericho, many of the skull and anthropomorphic caches appear to have been interred in extramural locations, although the excavation methods employed at Jericho make it impossible to confidently reconstruct the location of all skull caches. Despite some ambiguity in the stratigraphic placement of different kinds of caches, current data indicate that many of the larger caches were situated outside of structures (see Chapters 4, 5, and 7, this volume, for further consideration). For reasons to be outlined later, I believe that all of these different dedicatory rituals involving caches were organized and implemented by one or more households and were geographically focused on extramural areas, such as courtyards, which furthered participation in these rituals by members of other households. In light of the physical location of skull caches in public areas and ethnographic accounts outlining the high degree of community participation in secondary mortuary rituals, I argue that these caches represent the physical expression of very important household-level ritual events organized for the veneration or worshipping of ancestors while serving to reaffirm relationships within and between households linked by marriage, political, and economic ties.

Decapitation as a Common Theme: Skull Removal and Anthropomorphic Figurines

Examination of the ways in which MPPNB households and communities physically and symbolically structured mortuary ritual provides us with insights into the high degree of standardization and, by extension, shared meanings within ritual practices through time (Figure 2). For example, in the MPPNB cranium removal was a reoccurring theme within communities that was physically and symbolically expressed in multiple media, including (1) actual skull removal as a secondary mortuary practice (e.g., 'Ain Ghazal, Jericho, Nahal Hemar, Yiftahel, and Çatal Höyük), (2) symbolic pictorial representation of decapitation in wall paintings (e.g., Çatal Höyük), and (3) the actual decapitation of anthropomorphic figurines (e.g., 'Ain Ghazal, Jericho, and Çatal Höyük). MPPNB communities employed secondary mortuary practices that included special treatment of adults in mortuary practices, such as painting and plastering, and caching of human skulls in groups. Archaeological excavations at Jericho, 'Ain Ghazal, Nahal Hemar, and other MPPNB sites illustrate that, while skull removal starts in the Late Natufian/PPNA, systematic collective skull caching in large numbers occurs predominately between *c.* 9,300/200 bp and 8,500 bp, with intermittent continued use through to *c.* 8,000 bp (Byrd and Monahan 1995; Kuijt 1995, 1996). Excavations at 'Ain Ghazal have recovered multiple examples of caches characteristic of MPPNB secondary mortuary practices. One example was recovered from beneath the floor from the southeast corner of the house in Sq 3083 (placed in a row facing away from the center of the room). In the same house, but in a separate room, a single adolescent skull was recovered from beneath the southwest corner of the floor. The rear portion of this cranium was thinly coated with black pigment, possibly bitumen (Rollefson 1986:51). A second cache of skulls from four individuals includes two that were plastered and was found in a burial pit in a courtyard context (Butler 1989; Rollefson 1986). The most extensive evidence for secondary mortuary practices in the MPPNB occurs at Jericho. Excavation of the MPPNB horizon resulted in the recovery of 232 skeletons (compared to 254 skeletons from the PPNA), with 33 skulls individually interred and 52 skulls recovered from 12 caches (Kurth and Röhrer-Ertl 1981). Reconsideration of Kenyon's stratigraphic subdivisions and the typical subfloor pit depositional context of individual caches at other MPPNB sites indicates that all but five cached skulls were associated with the MPPNB occupation at Jericho (see Kuijt 1995).

The reoccurring theme of cranial removal in ritual practices is also seen in the decapitation and/or mutilation of anthropomorphic statues in south-central Levantine MPPNB communities. Excavations of MPPNB and LPPNB period (*c.* 8,500–8,000 bp) sites often recover small clay anthropomorphic figurines. For example, the 'Ain Ghazal researchers have recovered many MPPNB human figurines, upwards of forty from the 1983–1984 season alone. Other than six human figurines, which Rollefson et al. (1992) classify as "fertility figurines," most of the anthropomorphic figures recovered are heads or torsos, with no indication of secondary sexual characteristics. A stylized human figure (recovered from Sq 3282), roughly conical in shape with a mutilated head, typifies this assemblage. Other forms include the upper torso of a human figurine broken in half in antiquity (Rollefson 1986). Although fewer in frequency, human figurines have also been recovered at Beidha, including one female figurine (Kirkbride 1968:272). Similar observations have also been made at Çatal Höyük in Anatolia in which multiple, if not most, figurines were recovered without their heads, a theme reiterated in both the famous wall painting of vultures and decapitated individuals and in the multiple wall sculptures where the heads appear to have been deliberately defaced (Mellaart 1967; Omura 1984; Voigt Chapter 11, this volume).

Examination of damaged anthropomorphic figurines recovered in the excavation of Levantine Neolithic sites has led a number of authors to argue that in many cases these figurines were deliberately mutilated, damaged, or, in some cases, constructed so that there was no head (Bar-Yosef and Belfer-Cohen 1989; de Contenson 1971; Goring 1991; Rollefson 1986, Chapter 7, this volume; Voigt 1983, Chapter 11, this volume). In examining different patterns of damage for various types of cultic objects, Voigt (1983:192) has argued that damage to anthropomorphic figurines is often " . . . due to 'killing' at the time of disposal." Similarly, Goring (1991:52) remarks that "the apparent associations between damage and burial leads one to seriously consider the possibility that this damage was deliberate and applied ritually." These authors offer persuasive arguments for the deliberate decapitation of small anthropomorphic statues. Viewed collectively, these observations reflect the importance of cranial removal as symbolically significant to Neolithic peoples, both with human burials as well as with figurines in Neolithic ritual practices over many thousands of years.

Large Anthropomorphic Statue Caches

Another important aspect to MPPNB ritual practices is related to the construction, dedication, and interment of large anthropomorphic statues in caches and the continued emphasis on real or constructed skulls, in this case expressed through deliberate emphasis of the face and head in statues. Anthropomorphic figurines have been recovered in excavations of Jericho, Ramad, and, most importantly, from 'Ain Ghazal. One of the more exciting results of the excavation at 'Ain Ghazal has been the recovery of multiple plastered human statues from three pit features with highly detailed naturalistic painting and molding of the face and heads (Grissom 1996; Rollefson 1986; Rollefson et al.1992; Schmandt-Besserat 1998). Most of these figurines are half-size replicas of human skeletons or busts of the upper torso. The large human replicas have clearly formed legs and arms, and both busts and replicas were usually painted to draw attention to the elements of the face, even employing shells and bitumen for the eyes. Study of the methods of construction by Tubb and Grissom (1995) indicates that building these statues would have required considerable time investment. In the case of the statue cache from Sq 3282, eleven statues/busts were recovered, four in the lower layers and seven in the upper zone of the cache, from a pit that was cut through the floor of an abandoned house. Although the excavations of the MPPNB deposits at 'Ain Ghazal have not involved extensive horizontal exposure, Rollefson (1986) nevertheless has convincingly provided evidence to support the argument that these caches were from extramural locations. Although poorly preserved and from an unclear context, Garstang's (Garstang et al.1935) excavations at Jericho recovered anthropomorphic statues made of plaster in the context of four statue caches, two with three statues and two with single statues (Garfinkel 1994:164). As at 'Ain Ghazal, it appears that all of the caches from Jericho are from pit contexts. Viewed collectively, I believe that the deliberate focus on the head of these large statues, both in the construction as well as decoration, and the removal of the head of small figurines and the secondary removal of skulls from human skeletons, were all parts of an internally consistent shared system of ritual practices. As will be seen, this argument is based, at least in part, on the observations that many of these objects appear to have been buried in similar public contexts and that they are at times numerically organized along the same lines.

NUMERICAL ORGANIZATION OF MPPNB RITUAL CACHES

As with the earlier consideration of how many secondary rituals appear to have been practiced in highly public contexts, it is important to consider some of the ways in which MPPNB communities standardized mortuary practices and dedications, making them meaningful to other community members. For example, while there are exceptions, it is important to recognize that in many cases ritual practitioners in MPPNB communities organized the caching of cultic objects in groups of three or multiple sets of three (Tables 1 and 2, Figure 5). Kenyon's excavation of Sq DI, stage VIII C, phase xxix–xxx at Jericho recovered nine skulls placed in three sets of three from a single pit feature. All of these were placed in a single row facing the same direction, clearly separated into three groups (Kenyon 1981:52; Kurth and Röhrer-Ertl 1981:436) (Figure 5b). Excavation of a cache from Sq EI, II, V, stage VIII, phase xxxii revealed a total of six crania, three adults and three infants, placed in a circle facing inward (Kenyon 1981:287) (Figure 5a). Similarly, excavations in Sq MI, stage XIII, phase lxxiv uncovered three skulls in a single cache from an extramural location. Finally, a cache from Sq DI, stage XVI–XVII, phase xlii–xliii contained nine plastered skulls. Even

Table 6.1. Number, Location, and Probable Dating of Skull Cache Discovered in Village Sites in the South-Central Levantine MPPNB and LPPNB

Archaeological Site	Period	Number	Location of skulls	Reference
Jericho	MPPNB	6	Sq E I, II, V stage VIII, phase xxxii	Kenyon 1981
Jericho	MPPNB	4	Sq MI, stage VIII, phase xlv b	Kenyon 1981
Jericho	MPPNB	9	DI, stage VIII C, phase xxix–xxx	Kenyon 1981
Jericho	MPPNB	9	DI, stage XVI–XVII, phase xlii–xliii	Kenyon 1981
Jericho	MPPNB	3	Sq M I, stage XIII, phase ixxiv	Kenyon 1981
Jericho	MPPNB	2	Tr. I, stage XV A, phase xx	Kenyon 1981
Jericho	MPPNB	3	SQ EIII, IV, phase NNi	Kenyon 1981
Jericho	MPPNB	2	Sq EIII, IV, phase NNi	Kenyon 1981
'Ain Ghazal	MPPNB	4	Pit feature	Rollefson 1983
'Ain Ghazal	MPPNB	3	Subfloor	Rollefson 1983
'Ain Ghazal	MPPNB	3[1]	Exterior pit	Griffin n.d.
Nahal Hemar	PPNB	6[2]	Sqs E9, D8a, D7	Bar-Yosef & Alon 1988
Beisamoun	LPPNB	2	Loc. 188	Lechevallier 1978
Ramad	LPPNB	6	M4.S.Q. 4.0 m	de Contenson 1966
Ramad	LPPNB	3	M4.N.E. 1.6 m	de Contenson & Van Liere 1966:170
Ramad	LPPNB	12	H 10	de Contenson 1966:20–21
Es-Sifiya	LPPNB	3	Sq. C13, Burial 10	Mahasneh, 1997

[1]Rather than skulls, these objects were clay skull masks that fit over the front of the crania (Schmandt-Besserat 1998).

[2]Three of these were decorated at the back of the crania with bitumen.

Table 6.2. Number, Location, and Probable Dating of Anthromorphic Statutes Discovered in Village Sites in the South-Central Levantine MPPB (Based on Garfinkel 1994; Rollefson 1983, 1986)

Archaeological site	Period	Anthropomorphic statues	Location	Reference
Jericho	MPPNB	3	Locus 195	Garstang *et al.* 1935
Jericho	MPPNB	3	Locus 190	Garstang *et al.* 1935
Jericho	MPPNB	1	Locus 208	Garstang *et al.* 1935
Jericho	MPPNB	1	EI, II, V	Garstang *et al.* 1935
'Ain Ghazal Cache 1	MPPNB	25 (13 full figures & 12 one-headed busts)	Sq 3083 Loc. 20	Rollefson 1983
'Ain Ghazal Cache 2 (possibly incomplete)	MPPNB	7 (2 full figures, 3 two-headed busts and 2 unidentified pieces	Sq 3282 Loc. 049	Rollefson 1986

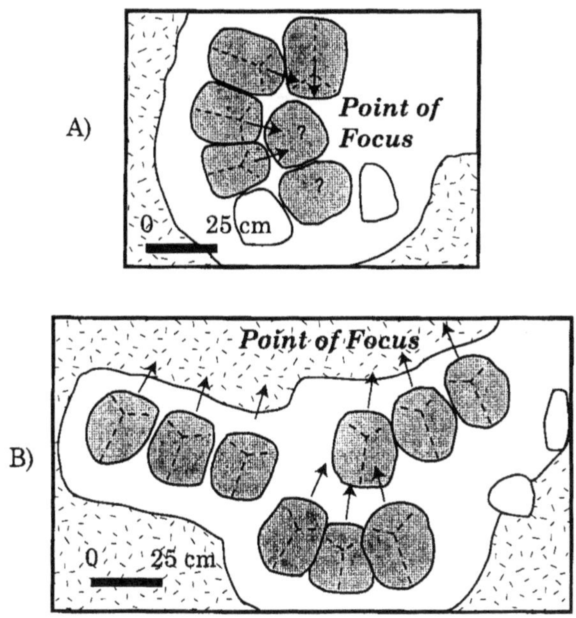

Figure 5. (A) Plan view of Jericho MPPNB skull cache E11–16 (EI, II, V, phase xxxiii). After Kenyon (1981:Plate 155). (B) Plan view of Jericho MPPNB skull cache D35–44 (DI phase xxix–xxx). After Kenyon (1981:Plate 36). Note physical grouping of nine skulls in three groups of three.

in the caches with associated skeletons, such as that of Tr.I, FI, stage xxxi, in which nine mandibles were associated with two skeletons, the number 3 or multiples of 3 was often used to organize secondary mortuary interments.

This same pattern of threes also occurs at Beidha, Nahal Hemar, and 'Ain Ghazal in the MPPNB and at several LPPNB settlements, such as Es-Sifiya and Basta. In the excavations at 'Ain Ghazal, several plastered skulls have been recovered. Most of these are individual skulls or portions of skulls, and one cache of a group of three. Rollefson also reports the recovery of several unplastered caches of skulls, one of which involves a group of three (Rollefson 1986). At Beidha, in an abandoned Level II workshop, Kirkbride reports recovering nine infant skeletons (Kirkbride 1968:272). Similarly at Nahal Hemar, six skulls were uncovered in the excavation of the cave, three of which were decorated at the back with the same net pattern, although it is not entirely clear if these were all part of a single cache (Bar-Yosef and Alon 1988).

Current archaeological evidence indicates that MPPNB ritual practitioners often cached other cultic objects, such as dedicatory offerings, in sets of three (Table 6.2). The excavators at 'Ain Ghazal uncovered from the corner of a room a circular stone storage feature which contained three Bos metacarpals, one of which had three longitudinal incisions. Underneath these three bones, lying directly upon the plaster base of the feature, was a single Bos figurine (Rollefson 1986:47). Similarly, in the excavations undertaken by Garstang (Garstang et al. 1935:166) at Jericho, two sets of three statues were recovered, each of these apparently consisting of a man, a woman, and a child. Needless to say, there are exceptions to the practices of interring skulls and skeletons in groups of threes at the most extensively excavated sites of Jericho and 'Ain Ghazal. These exceptions account for fewer than half of the examples and usually involve sets of two cultic objects, possibly as a symbolic and physical expression of couples. Interestingly this emphasis on twinning is also seen at 'Ain Ghazal in the recovery of three double-headed statues in which two heads were placed upon a single torso (Schmandt-Besserat 1998). Although the specific meanings of such standardization are lost to us, the clear organization of such practices outlines that MPPNB ritual practitioners at Jericho, 'Ain Ghazal, and Beidha intentionally organized ritual practices in a highly standardized way, presumably so as to be meaningful to others observing these practices and in a way that was consistent with community beliefs and ideology.

How then do data from MPPNB secondary mortuary and anthropomorphic figurine caches help us to understand some of the links between household ritual practices and broader social beliefs in early agricultural communities in the Middle East and to understand why the number of cached objects was probably predetermined? While it is not possible for us to directly interpret the specific spiritual beliefs associated with skull caching in predetermined units in the MPPNB, it is possible to reflect upon why such ritual practices were standardized at the household level and to comprehend their impact on relations within early agricultural village communities.

Skull and Statue Caching: Implications for the Scale of Social Action

For a moment let us consider how household and community-level participation in MPPNB skull caching may have been affected by the caching of multiple skeletal and ritual elements within the same context—in some cases up to nine skulls—and the degree to which different households may have been represented. Specifically, were these caches the result of ritual practices by single households (thereby stressing the importance of an individual household), or were the caches composed of human remains from several households (thereby stressing the collective participation by multiple households)? If the skull caches were from single households then it would have been necessary to postpone the secondary interment of skulls until there were enough deceased household members for a collective cache. Depending on the number of human skulls required, the death rates within individual households, and the amount of required for decomposition of the flesh, this could easily require four to six years and possibly as many as twenty years for smaller households. In such a scenario, households would probably hold a primary mortuary celebration for deceased individuals quickly after death, bury the individual with other deceased household members, and then hold a secondary mortuary celebration at some point in the future when there would have been enough deceased individuals for a burial cache. A second scenario is that secondary MPPNB mortuary skull caches represented relationships between multiple households. In this scenario, the skulls from deceased members of various households would have been removed from their primary location and collectively buried by ritual practitioners in a single ceremony at the same time. Depending on the death rate within households, as well as the size of the overall community, such events probably would have occurred at least once a year. Such an annual large community-level ritual event would have required elaborate preparations and organization by the host households and, by linking a number of households within and between villages, would have created a massive community event with widespread participation.

Needless to say, from an archaeological perspective it is very difficult to distinguish between these two scenarios, as both are likely to have very similar material expressions. There is, however, one very important difference between these two scenarios: the first prioritizes the individual household over others, whereas the second one prioritizes the community over individual households. If skull caches had been organized by individual households then one would anticipate a material emphasis of the economic and political status of the host households over others, presumably expressed by an extensive variation in grave goods, residential housing, and material culture within individual structures or household compounds. It is important to note, however, that this is not supported by archaeological evidence (see Byrd 1994; Kuijt 1995, 1996; Rollefson Chapter 7, this volume). Thus, we must ask ourselves if individual members of a single household were being differentiated in skull caching, as it is very tempting to believe; then one must ask why did these people not differentiate themselves from others in more visible ways? Why, for example, did household members not draw attention to their household by including grave goods, building larger and different residential structures, and varying mortuary practices rather than adhere to conservative social codes prohibiting grave goods? Simply put, if individual households wanted to differentiate themselves over others, then why do we not see more evidence for variations in contemporaneous residential architectural practices and the differential treatment of people in life and death? I believe that the most parsimonious explanation for these questions is that the acts involved in, and organization behind, the differential selection of individuals in life and death, such as the painting and plastering of skulls, dedicatory caches, and caching of human skulls and large statuary, focused on emphasizing ties between multiple households within MPPNB communities.

Social Differentiation in the MPPNB

Having argued that MPPNB community members adhered to social codes that limited social differentiation, it is also necessary to note that recent research has drawn increasing attention to archaeological evidence for the emergence of social differentiation in the MPPNB (Bar-Yosef and Belfer-Cohen 1989, 1991; Bar-Yosef and Meadows 1995; Byrd 1994; Kuijt 1995; Rollefson et al. 1992) and possibly as early as the PPNA (Bar-Yosef and Meadows 1995; Kuijt 1994, 1996). While it is not possible to provide a detailed examination of this topic within the scope of this chapter, it is necessary to outline some of this evidence. Briefly, while MPPNB material culture and architecture suggest an emphasis on standardization in size, shape, and internal organization of residential buildings within the greater community (see Byrd 1994; Kuijt 1995; Özdögan and Özdögan 1989), consideration of mortuary data reflect subtle, yet observable, dimensions of competition and tension between individual Houses within the community-level egalitarian worldview. Materially these tensions were expressed through the employment of a series of mortuary practices that differentially selected some individual over others. For example, approximately 70% of adult community members at 'Ain Ghazal and Jericho were selected for secondary mortuary practices with the interment of skulls. Similarly, only 20% of all individuals appear to have been interred in caches, with a much lower percentage of individuals selected at death for skull plastering or painting.

While maintaining the overall emphasis on secondary mortuary practices that first appeared in Late Natufian and PPNA contexts, MPPNB mortuary practices included the intentional caching of crania from select individuals in collective public rituals, with the secondary interment of crania in extra- and intramural locations. Some individuals were also differentially selected through the modification of the crania in the elaborate plastering and painting of skulls with clay, asphalt, and pigments and inlaying shells in the eye sockets. Field work at Jericho and Nahal Hemar documented several kinds of cranial deformation practiced in the MPPNB, and interestingly, many of the individuals identified with deformed crania were from the skull caches (Arensburg and Herskovitz 1988; Cornwall 1981; Kuijt 1995; Kurth and Röhrer-Ertl 1981). At Jericho the physical separation of different kinds of cranial deformation in different areas of the settlement illustrates how differentiation within the community was expressed during life, how highly select individuals were physically and symbolically identified through physical appearance, and how individual Houses may have been differentiated from each other. Thus, certain individuals and groups were selected from the community and treated differently during life with skull deformation and again distinguished in death through skull caching and plastering.

CONCLUSION AND IMPLICATIONS

In light of the growing body of archaeological evidence for social differentiation within MPPNB communities, as well as the symbolic and physical use of material culture to stress real and fictive affinity within and between individuals, households, and communities, I have argued that MPPNB ritual practices reinforced a collective ethos with the continued use of social mechanisms to encourage social cohesion and solidarity. Consideration of the archaeological record in question, with the almost total absence of grave goods with MPPNB primary and secondary interments and the homogenous design of residential architecture, illustrates a pattern that is consistent with communities attempting to emphasize a real or perceived parity between individuals, and the existence of political and economic cooperation and relationships between households. Archaeological studies provide a number of specific material patterns that inform researchers as to how community members may have dealt with new social and organizational pressures associated with increased population aggregation in early agricultural communities. Among the observable patterns from the MPPNB are (1) a significant expansion of secondary mortuary ceremonies in comparison to the PPNA, (2) the caching of cultic objects in extramural locations, and votive offerings in inter-mural contexts, including human figurines and faunal remains, (3) the development of other forms of ritual, probably focused on the household, that involved caching of animal figurines, and (4) evidence for the deliberate standardization in the number of votive offerings in groups of three. Collectively, I believe that consideration of these developments, as well as the limited development of social differentiation in the MPPNB, indicates that future research is facilitated by envisioning MPPNB social systems as organized by a series of complex social rules that reaffirmed the egalitarian values and ethos of general society and at the same time permitted the development of social differentiation that crosscut household and kin-group lines.

This reconstruction of the MPPNB community and its chronological placement raises a number of interesting implications for our anthropological and archaeological understanding of the relationships between emerging hierarchy, community relations, and existence of House Societies many thousands of years ago. First of all, available evidence indicates that the earliest systematic appearance of social differentiation in the Aceramic Neolithic occurred in the MPPNB, between *c.* 9,200–8,500 bp, in the form of cranial deformation, skull plastering and painting, and the select use of secondary cranial removal and caching to differentially identify some community members over others (also see Chapter 4, 5, and 7 in this volume). This realization is important, for if Bar-Yosef and Meadows (1995:88) are correct in arguing that size reduction in goats had already occurred by the MPPNB and that domesticated wheat and barley first appeared in the PPNA communities of Tel Aswad, Jericho, Gilgal, and Netiv Hagdud (Hillman and Davies 1990), then our most convincing evidence for systematic social differentiation in the Levantine Pre-Pottery Neolithic occurs after the domestication of plants and probably after that of goats as well. Such an awareness has profound implications for how archaeologists and anthropologists model the relationship of social differentiation and the origins of food production in the Levantine Neolithic, for many, if not most, models of this transition (e.g. Hayden 1995) either assume that they occurred simultaneously or fail to consistently situate these models within the archaeological record for this period of time. In light of our current knowledge of the Neolithic archaeological record, I argue that detailed consideration of PPNA and MPPNB mortuary and architectural evidence from the south-central Levant do not support arguments that the limited social differentiation seen in the MPPNB led to successful political and economic consolidation among competing individuals or households. Thus, despite the dramatic increase in the scale of PPNA and MPPNB communities (see Bar-Yosef and Belfer-Cohen 1991; Bar-Yosef and Meadows 1995) and competition between different households within MPPNB communities over the control of new resources, I believe that community leaders developed and maintained a series of elaborate social controls, materially expressed through mortuary, ritual, and architectural practices, that emphasized membership and affinity at the household and community level. In short, while social differentiation existed within and between households in communities, widely accepted social codes restricted the consolidation of this into some form of hereditary power, authority, or status during this period.

This realization, while counter to our expectation of food production leading to the emergence of hereditary social inequality, actually makes sense when we envision Pre-Pottery Neolithic communities as organized with competing and cooperating Houses and households, founded on an egalitarian ideology. From this viewpoint, the earlier development and maintenance of an egalitarian ideology in the PPNA and MPPNB may have become one of the major venues (if not the major venue) in which individual and household-based competitions were expressed (see also Boehm 1993; Kan 1989; McKinnon 1991). Similar to the system outlined by McKinnon (1991), I believe that the continual negotiation of existing, and development of new, social rules in the MPPNB and PPNA facilitated the cooperative and competitive relationships between individuals, households, and communities. In this light, community members may have transformed the ideological identity of egalitarianism from the earlier Late Natufian/PPNA into a form of egalitarianism focused on competitive exclusion between MPPNB households, to the eventual usurping of authority and power by ritual practitioners in the name of this same ideology in the LPPNB (*c.* 8,500–8,000 bp). Communities in the MPPNB period, therefore, provide one example of how the conceptual boundaries of, and tensions within, kinship-based systems were constantly negotiated and challenged, as people attempted to reconcile a world still ordered and conceptualized in the conservative kinship terms of their ancestors with the increasingly dynamic and unanticipated economic, social, and political interests of their children in a rapidly changing world.

ACKNOWLEDGMENTS

This study has been directly and indirectly supported by the Department of Anthropology, University of California at Berkeley, the Social Sciences and

Humanities Research Council of Canada (grant #753-91-0218), the Department of Anthropology at Harvard University, the Peabody Museum at Harvard, the Mellon Foundation, the National Science Foundation (grant #91-12725), the Sigma Xi Foundation, the American Schools of Oriental Research, the British Institute in Amman for Archaeology and History, and a Frederick Sheldon Fellowship from Harvard University. This study would not have been possible without the kindness and assistance of B. Byrd, E. B. Banning, H. Mahasneh, Z. Kafafi, A. Belfer-Cohen, N. Goring-Morris, I. Gilead, A. Gopher, and A. Betts. Discussions with and critical editorial comments by B. Byrd, R. Joyce, O. Bar-Yosef, R. Preucel, and, most of all, G. Rollefson and M. Chesson, have been crucial to the development of this chapter. While not agreeing with some of the concepts and interpretations presented in this manuscript, their constructive criticism has immeasurably improved the clarity and organization of it.

REFERENCES

Arensburg, B., and Hershkovitz, I., 1988, Neolithic Remains. In *Nahal Hemar Cave,* edited by O. Bar-Yosef and D. Alon, pp. 50–58. Israel Department of Antiquities and Museums, Jerusalem.

Bar-Yosef, O., and Alon, D., 1988, Excavations in the Nahal Hemar Cave. *Atiqot* 18.

Bar-Yosef, O., and Belfer-Cohen, A., 1989, The Levantine "PPNB" Interaction Sphere. In *People and Culture in Change: Proceedings of the Second Symposium of Upper Paleolithic, Mesolithic, and Neolithic Populations of Europe and the Mediterranean Basin,* edited by I. Hershkovitz, pp. 59–72. British Archaeology Review, International Series 508. BAR, Oxford.

Bar-Yosef, O., and Belfer-Cohen, A., 1991, From Sedentary Hunter-Gatherers to Territorial Farmers in the Levant. In *Between Bands and States,* edited by S. A. Gregg, pp. 181–202. Center for Archaeological Investigations, Carbondale, Illinois.

Bar-Yosef, O., and Meadows, R., 1995, The Origins of Agriculture in the Near East. In *Last Hunters–First Farmers: New Perspectives on the Prehistoric Transition to Agriculture,* edited by Price, T. D. and A. G. Gebauer, pp. 39-94. School for American Research, Santa Fe, New Mexico.

Belfer-Cohen, A., 1991, The Natufian in the Levant. *Annual Review of Anthropology* 20:167–186.

Berreman, G. D., 1981, Social Inequality: A Cross-Cultural Analysis. In *Social Inequality: Comparative and Developmental Approaches,* edited by G. D. Berreman, pp. 3–40. Academic Press, New York.

Bienert, H. D., 1991, Skull Cult in the Prehistoric Near East. *Journal of Prehistoric Religion* 5:9–23.

Binford, L. R., 1971, Mortuary Practices: Their Study and Potential. In *Approaches to the Social Dimensions of Mortuary Practices,* edited by J. A. Brown, pp. 6–21. Memoirs of the Society for American Archaeology, no. 25.

Blanton, R. E., 1995, *The Cultural Foundations of Inequality in Households.* In *Foundations of Social Inequality,* edited by T. D. Price and G. M. Feinman, pp. 105–127. Plenum, New York.

Boehm, C., 1993, Egalitarian Behavior and Reverse Dominance Hierarchy. *Current Anthropology* 34(3):227–254.

Butler, C., 1989, The Plastered Skulls of 'Ain Ghazal: Preliminary Findings. In *People and Culture in Change*. BAR International Series, edited by I. Hershkovitz, pp. 141–145. British Archaeology Review, International Series 508. BAR, Oxford.

Byrd, B. F., 1994, Public and Private, Domestic and Corporate: the Emergence of the Southwest Asian Village. *American Antiquity* 59(4): 639–666.

Carr, C., 1995, Mortuary practices: Their Social, Philosophical-Religious, Circumstantial, and Physical Determinants. *Journal of Archaeological Method and Theory* 2(2):105–200.

Cauvin, J., 1994, *Naissance des divinités, Naissance de l'agriculture*. CNRS, Paris.

Chapman, R., Kinnes, I., and Randsborg, I. (eds.), 1981, *The Archaeology of Death*. Cambridge University Press, Cambridge.

Cornwall, I. W., 1981, Appendix A. The Pre-Pottery Neolithic Burials. In *Excavations at Jericho: The Architecture and Stratigraphy of the Tell*, edited by T. A. Holland, British School of Archaeology in Jerusalem, Jerusalem.

de Contenson, H., 1966, Les Trois Premières Campagnes de Fouilles à Tell Ramad (Syrie). *C. R. Académie des Inscriptions et Belles Lettres*, 531–536.

de Contenson, H., 1971, Tell Ramad, a Village of Syria of the 7th and 6th millennia B.C. *Archaeology* 24:278–283.

de Contenson, H., and van Liere, W. J., 1966, Premier sondage à Bourqras en 1965. Rapport préliminaire. *Annales Archéologiques Arabes de Syrie* 16:181–192.

Crocker, J. C., 1977, The Southern Way of Death. In *The Not-so-Solid South*, edited by J. K. Morland, pp. 114–129. Southern Anthropological Society, Athens, Georgia.

Downs, R. E., 1956, The Religion of the Baré' e-speaking *Toradja of Central Celebes*. Excelsior, The Hague.

Earle, T. K., 1987, Chiefdoms in Archaeological and Ethnohistoric Perspective. *Annual Review of Anthropology* 16:279–308.

Feinman, G. M., 1995, The Emergence of Inequality: A Focus on Strategies and Process. In *Foundations of Social Inequality*, edited by T. D. Price and G. M. Feinman, pp. 255–279. Plenum, New York.

Fentress, J., and Wickham, C., 1992, *Social Memory*. Blackwell, Oxford.

Flanagan, J. C., 1989, Hierarchy in Simple "Egalitarian" Societies. *Annual Review of Anthropology* 18: 245–266.

Flanagan, J. C. and Rayner, S., 1988, *Rules, Decisions, and Inequality in Egalitarian Societies*. Gower, London.

Fried, M., 1960, On the Evolution of Social Stratification and the State. In *Culture and History*, edited by S. Diamond, pp. 713–731. Columbia University Press, New York .

Fried, M., 1967, *The Evolution of Political Society*. Random House, New York.

Garfinkel, Y., 1987, Yiftahel: A Neolithic Village from the Seventh Millennium B.C. in Lower Galilee, Israel. *Journal of Field Archaeology* 14:199–212.

Garfinkel, Y., 1994, Ritual Burial of Cultic Objects: The Earliest Evidence. *Cambridge Archaeology Journal* 4(2):159–88.

Garstang, J., Dropp, J. P., and Crowfoot, J., 1935, Jericho: City and Necropolis. *Liverpool Annals of Archaeology and Anthropology* 22:143–173.

Geertz, C., 1973, *Interpretation of Cultures*. Basic Books, New York.

Gerlach, L. P., and Gerlach, U. M., 1988, Egalitarianism, Collectivism, and Individualism: The Digo of Kenya. In *Rules, Decisions, and Inequality in Egalitarian Societies*, edited by J. G. Flanagan and S. Rayner, pp. 113–144. Gower, London.

Goring, E., 1991, The Anthropomorphic Figures. In *Lemba Archaeological Project II.2: A Ceremonial Area at Kissonerga*, edited by E. Peltenburg, pp. 39–60. Paul Åströms, Gothenburg.

Goring-Morris, A. N., 1991, A PPNB Settlement at Kfar Hahoresh in Lower Galilee: A Preliminary Report of the 1991 Season. *Journal of the Israel Prehistoric Society* 24:77–101.

Grissom, C., 1996, Conservation of Lime Plaster Statues from 'Ain Ghazal. In *Archaeological Conservation and Its Consequences*, edited by R. Ashok and P. Smith, pp. 70–75. International Institute for Conservation of Historic Artistic Works, London.

Hayden, B., 1995, The Origins of Agriculture in the Near East. In *Last Hunters-First Farmers: New Perspectives on the Prehistoric Transition to Agriculture*, edited by T. D. Price and A. G. Gebauer, pp. 273–299. School for American Research, Santa Fe, New Mexico.

Hershkovitz, I., and Gopher, A., 1990 Paleodemography, Burial Customs, and Food-Producing Economy at the Beginning of the Holocene: A Perspective from the Southern Levant. *Mitekufat Haeven,* 23:9–48.

Hillman, G. C., and Davies, M. S., 1990, Measured Domestication Rates in Wild Wheats and Barley under Primitive Cultivation and Their Archaeological Implications. *Journal of World Prehistory* 4(2):157–222.

Hertz, R., 1960, *A Contribution to the Study of the Collective Representation of Death*. Free Press, New York.

Hodder, I., 1988, *Symbolic and Structural Archaeology*. Cambridge University Press, Cambridge.

Hodder, I., 1990, *The Domestication of Europe*. Blackwell, Oxford.

Hudson, A. B., 1966, Death Ceremonies of the Padju Epat Ma'anyan Dayaks. *Sarawak Museum Journal* 13:341–416.

Johnson, G. A., 1982, Organizational Structure and Scalar Stress. In *Theory and Explanation in Archaeology,* edited by A. C. Renfrew, M. J. Rowlands, and B. A. Segraves, pp. 389–421. London: Academic Press.

Joyce, R. A., 1993, Social Dimensions of Preclassic Burials. *Ritual Behavior, Social Identity, and Cosmology in Pre-classic Mesoamerica*. Dumbarton Oaks, Washington.

Kan, S., 1989, *Symbolic Immortality: The Tlingit Potlatch of the Nineteenth Century*. Smithsonian Institution Press, Washington.

Keene, A. S., 1991, Cohesion and Contradiction in the Communal Mode of Production: The Lessons of the Kibbutz. In *Between Bands and States,* edited by S. A. Gregg, pp. 376–394. Southern Illinois University, Carbondale, Illinois.

Kenyon, K. M., 1981, *Excavations at Jericho. Vol. III: The Architecture and Stratigraphy of the Tell, Volume III*. British School of Archaeology in Jerusalem, London.

Kirkbride, D., 1968, Beidha 1967: An Interim Report. *Palestine Exploration Quarterly* 100: 90–96.

Kuijt, I., 1994, Pre-Pottery Neolithic A Settlement Variability: Evidence for Sociopolitical Developments in the Southern Levant. *Journal of Mediterranean Archaeology* 7:165–192.

Kuijt, I., 1995, *New Perspectives on Old Territories: Ritual Practices and the Emergence of Social Complexity in the Levantine Neolithic*. Ph.D. dissertation, Harvard University. University Microfilms, Ann Arbor, Michigan.

Kuijt, I., 1996, Negotiating Equality through Ritual: A Consideration of Late Natufian and Prepottery Neolithic A Period Mortuary Practices. *Journal of Anthropological Archaeology* 15(4):313–336.

Kujit, I., 2000, People and Space in Early Agricultural Villages: Exploring Daily Lives, Community Size, and Architecture in the Late Pre-Pottery Neolithic. *Journal of Anthropological Archaeology* 19.

Kujit, I., in press, Meaningful Masks: Death, Authority, and the Transmission of Social Memory in Early Agricultural Villages of the Near Eastern Pre-Pottery Neolithic. In *Social Memory, Identity, and Death: Intradisciplinary Perspectives on Mortuary Rituals*. M. S. Chesson (ed.), American Anthropological Association, Archaeology Division, Vol. 10. American Anthropological Association, Washington, D.C.

Kurth, G., and Röhrer-Ertl, G., 1981, On the Anthropology of Mesolithic to Chalcolithic Human Remains from the Tell es-Sultan to Jericho, Jordan. In *Excavation at Jericho*, edited by K. M. Kenyon, British School of Archaeology, Jerusalem.

Lechevalier, M., 1978, *Abou Ghosh et Beisamoun, deux gisements de VII millenaire avant l'ere chretienne en Israel*. Memoires et Travaux de Centre de Recherches Prehistoriques Francais de Jerusalem, Volume 2, Paléorient, Paris.

Lévi-Strauss, C., 1983, *The Way of the Masks*. Translated by S. Modelski. Jonathan Cape, London.

Lopatin, I., 1960, *The Cult of the Dead among the Natives of the Amur Basin*. Mouton, The Hague.

Mahasneh, M. M., 1997, The 1995 Season at the neolithic Site of es-Sifiya, Wadi Mujib, Jordan. In *The Prehistory of Jordan II. Perspectives from 1997*, edited by H. G. K. Gebel, Z. Kafafi, and G. O. Rollefson, pp. 203–214. Studies in Early Near Eastern Production, Subsistence, and Environment 4. Ex Oriente, Berlin.

McGuire, R. H., 1992, *Death, Society, and Ideology in a Hohokam Community*. Westview Press, Boulder, Colorado.

McKinnon, S. 1991, *From a Shattered Sun: Hierarchy, Gender, and Alliance in the Tanimbar Islands*, University of Wisconsin Press, Madison, Wisconsin.

Mellaart, J., 1967, *Çatal Höyük, a Neolithic Town in Anatolia*. McGraw-Hill, New York.

Metcalf, P., and Huntington, R., 1991, *Celebrations of Death*. Cambridge University Press, Cambridge.

Moore, A., 1985, The Development in Neolithic Societies in the Near East. In *Advances in World Archaeology*, edited by F. Wendorf and A. E. Close, pp. 1–69. Academic Press, New York.

Myers, F. R., 1986, Reflections on a Meeting: Structure, Language, and the Polity in a Small Scale Society. *American Ethnology* 13: 430–447.

O'Shea, J. M., 1984, Mortuary Variability: An Archaeological Investigation. Academic Press, New York.

O'Shea, J. M., 1996, *Villagers of the Maros: A Portrait of an Early Bronze Age Society*. Plenum, New York.

Omura, M., 1984, A Reinterpretation of Figurines from Çatal Höyük. *Orient* 20:129–155.

Özdögan, M., and Özdögan, A., 1989, Çayönü. *Paléorient* 15:65–74.

Paynter, R., 1989, The Archaeology of Equality and Inequality. *Annual Review of Anthropology* 18:369–399.

Plog, S., 1995, Equality and Hierarchy: Holistic Approaches to Understanding Social Dynamics in the Pueblo Southwest. In *Foundations of Social Inequality*, edited by T. D. Price and G. M. Feinman, pp. 189–206. Plenum, New York.

Price, T. D., 1995, Social Inequality at the Origins of Agriculture. In *Foundations of Social Inequality*, edited by T. D. Price and G. M. Feinman, pp. 129–151. Plenum, New York.

Radcliffe-Brown, A. R., 1964, *The Andaman Islanders*. Free Press, New York.

Rayner, S., 1988, The Rules That Keep Us Equal: Complexity and Costs of Egalitarian Organization. In *Rules, Decisions, and Inequality in Egalitarian Societies*, edited by J. G. Flannagan and S. Rayner, pp. 20–42. Gower, London.

Rollefson, G. O., 1983, Ritual and Ceremony at Neolithic 'Ain Ghazal, Jordan. *Paléorient* 9(2):29–38.

Rollefson, G. O., 1986, Neolithic 'Ain Ghazal (Jordan): Ritual and Ceremony II. *Paléorient* 12:45–52.

Rollefson, G. O., 1987, Local and Regional Relations in the Levantine PPN Period: 'Ain Ghazal as a Regional Center. In *Studies in the History and Archaeology of Jordan*, edited by A. Hadidi, pp. 2–32. Amman: Department of Antiquities of Jordan.

Rollefson, G. O., Simmons, A. H., and Kafafi, Z., 1992 Neolithic Cultures at 'Ain Ghazal, Jordan. *Journal of Field Archaeology* 19:443–470.

Schmandt-Besserat, D., 1998, 'Ain Ghazal "Monumental" Figures. *Bulletin of the American Schools of Oriental Research* 310:1–17.

Tainter, J. A., 1978, Mortuary Practices and the Study of Prehistoric Social Systems. *Advances in Archaeological Theory* 1: 106–143.

Tubb, K. W., and Grissom, C. A., 1995, 'Ayn Ghazal: A Comparative Study of the 1983 and 1985 Statuary Caches. In *Studies in the History and Archaeology of Jordan V: Art and Technology throughout the Ages*, edited by K. 'Amr, F. Zayadine, and M. Zaghloul, pp. 437–447. Jordan Free Press, Amman.

van Gennep, A., 1960, *Rites of Passage*. University of Chicago Press, Chicago.

Voigt, M. M., 1983, *Hajji Firuz Tepe, Iran: The Neolithic Settlement*. The University Museum, University of Pennsylvania, Philadelphia.

Waterson, R., 1995, Houses and Hierarchies in Island Southeast Asia. In J. Carsten and S. Hugh-Jones (eds.), *About the House: Lévi-Strauss and Beyond,* pp. 47–68. Cambridge University Press, Cambridge.

Weiner, A. B., 1976, Women of Value, Men of Renown. University of Texas Press, Austin.

Chapter 7

Ritual and Social Structure at Neolithic 'Ain Ghazal

GARY O. ROLLEFSON

A fascination with the spirit world has drawn the attention of scholars and laymen alike for centuries, if not millennia. The draw is especially strong with prehistoric societies, since the absence of written records leaves the field of interpretation temptingly wide open. In the decades after the exciting finds from Jericho in the 1930s (e.g., Garstang and Garstang 1948) and 1950s (Kenyon 1970) and the dramatic evidence from Çatal Hüyük in the 1960s (Mellaart, 1967), a coherent picture of the realm of ritual and ceremony emerged that has found widespread acceptance (e.g., Cauvin 1972, 1994; cf. Rollefson 1983, 1986). But the recent accelerated pace of field research in the Levant, particularly in the Early Neolithic period, has produced considerable new material that requires us to reconsider the earlier views of what rituals occurred and what functions they served in the local and regional social network. Much of the latest information comes from 'Ain Ghazal, and the evidence from this site will be examined in this chapter.

Ten excavation seasons since 1982 at the Neolithic settlement of 'Ain Ghazal (Figure 1) have produced an enormous body of information concerning changing patterns of life for inhabitants of one of the largest known

GARY O. ROLLEFSON • `Ain Ghazal Research Institute, 64372 Ober-Ramstadt, Germany.

Life in Neolithic Farming Communities: Social Organization, Identity, and Differentiation, edited by Ian Kuijt. Kluwer Academic/Plenum Publishers, New York, 2000.

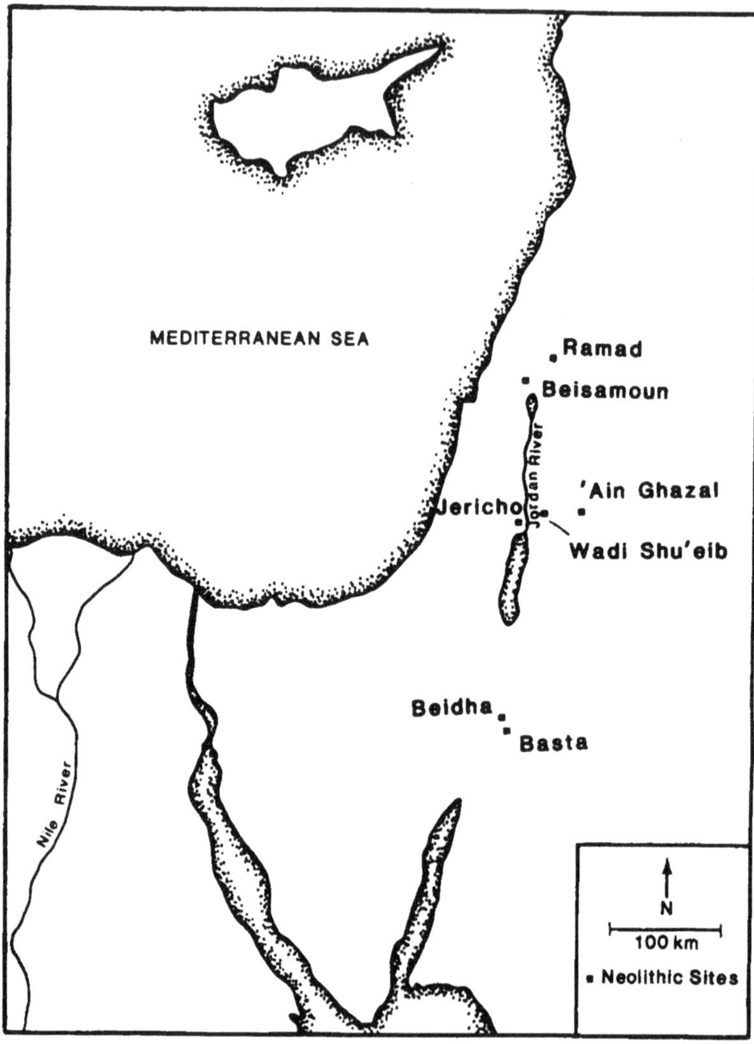

Figure 1. Location of 'Ain Ghazal in relation to other Neolithic sites in the southern Levant.

farming settlements in the Levant. Occupied continuously for more than 2,000 years, the settlement witnessed four major phases of development, including the Middle PPNB (MPPNB, 9,200–8,500 bp), the Late PPNB (LPPNB, 8,500–8,000 bp), the PPNC (8,000–7,500 bp), and the Yarmukian Pottery Neolithic (7,500–?6,500 bp). The general history of the settlement at 'Ain Ghazal has been treated elsewhere (e.g., Rollefson and Köhler-Rollefson 1989; Rollefson et al. 1992; Simmons et al., 1988) and will not be repeated here. Past publications have also described some of the ritual material from the site (e.g., Rollefson 1983, 1986; Simmons et al. 1990), but little attention has been paid so far to the implications such material has on 'Ain Ghazal's social organization and how it changed over the two millennia of the town's growth and decline.

MIDDLE PRE-POTTERY NEOLITHIC B PERIOD (MPPNB)

MPPNB deposits were sampled over approximately 200 m^2 during the 1982–1985 seasons. In addition to the more mundane archaeological remains were numerous clay human and animal figurines, subfloor and courtyard burials, caches of skulls (untreated, painted, or plastered), and ceremonial "burials" of large human statuary made of lime plaster.

Animal and Human Figurines

To date more than 150 clay animal figurines have been recovered from MPPNB deposits, 60% of which were unidentifiable. Of the remainder, more than 90% (fifty-six of sixty-one) were cattle (including a cache of twenty-three in a trash deposit); several were anatomically identifiable as bulls (Rollefson et al. 1985: 87). Notably, cattle bones from the MPPNB show no clear morphological indication of domestication, although pathologies on some *Bos* phalanges suggest some calves may have been tamed and reared in captivity (cf. Rollefson et al. 1984). How many of these animal figurines may have been "toys" cannot be determined, although the posture of the dog (Rollefson 1983:Pl. III-7) and perhaps both goat specimens probably fit this usage. Several clay cattle figurines bore thread impressions around the neck, which can be taken to support the taming hypothesis of these animals. Two bovine figurines received special treatment: after each was ritually "killed" with flint bladelets while the clay was still wet, both were fired and then placed side by side in a pit beneath the floor of a house (Figure 2). Certainly these pieces were not toys, but instead they were used in some solemn ritual, perhaps for luck or magic in the hunt.

MPPNB layers produced forty small clay human figurines that appear to fall into two general categories: "fertility" figurines (Figure 3) and "other." For the former group, it is unlikely that the promotion of fertility was a major concern in a society that had a history of sedentary agricultural abundance for more than a thousand years. But when one considers that pregnancies and the birth events themselves are the most dangerous times of a woman's life, the so-called fertility figurines may have played a more important role as symbolic guardians of the health of the mother as she approached

Figure 2. MPPNB clay cattle figurines. Note lower object was pierced three times with flint bladelets before it was fired; the head of the upper figurine was stabbed twice, although the head was lost before both were placed in a pit cut through a lime-plaster floor. (Photo: C. Blair)

full term and finally gave birth to a child. Although the sample is too small to be statistically valid, among the 'Ain Ghazal mortuary population there is a perceptible and unexpected increase in the death rate for females at *ca.* age 14–15, possibly associated with first birthing experiences. If a child is stillborn or dies shortly after birth, it is a sad event, but another child can be produced within the next year. If the mother dies during birth, this is a major loss of social investment in future generations.

The role of the latter group is unclear, but they might also be interpreted as talismans to protect the owner. The absence of sex-specific features suggests that they could have been used by either men or women. With only one exception, all human figurines were found in a broken state as heads or as bodies. Since this parallels the postmortem treatment of so many of the MPPNB skeletons, this might be taken to reflect ritual death at the decease of the figurine's owner. The exception, represented by an unbroken piece depicted in a reclining position with one arm over a flat abdomen and the other hand wrapped across the face, is possibly an anecdotal expression of grief at the stillbirth of a child (Rollefson 1986:47, Pl. II-3).

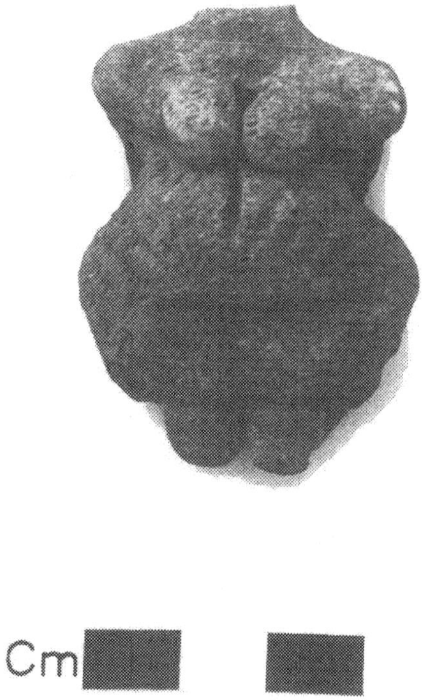

Figure 3. Headless female figurine (MPPNB) with pendulous breasts and extended abdomen. (Photo: C. Blair)

Human Burials and Skull Caching

A total of 81 human interments came from MPPNB contexts. As was the case for Beidha (Byrd 1994:657; Kirkbride 1967:9), the number of burials at 'Ain Ghazal is much too low to represent the general pattern of postmortem treatment. For example, the eight people (or twelve, if the two groups of skulls are included) beneath one house represent a span of approximately 400 years (the use-life of the structure), or one burial every thirty-three to fifty years. Even if all of the burials are considered (eighty-one over the MPPNB span of 750 years), the apparent "death rate" is one burial every nine years. The great majority of the dead were obviously disposed of elsewhere, and most of the bodies recovered in the excavations represent particular individuals selected for special treatment.

Three burial "styles" occurred during the MPPNB: (1) subfloor and courtyard, decapitated; (2) courtyard, skull intact; and (3) infant (Rollefson 1986:50). The differences, aside from the presence or absence of the skull, between the first two styles indicate a clear differentiation in status. In the first case, burials placed beneath house floors (Figure 4) entailed an expense in terms of labor to resurface the floors with lime plaster (cf. Rollefson 1990a) after the removal of the skull. The individuals were placed in the pit in a loosely flexed position, and the fill of the pit was invariably free of any trash; grave goods were very rare. In contrast, individuals in the second "style" appear to have simply been discarded into holes dug into trash deposits, and some people were literally stuffed into small, shallow pits and covered with the debris-laden backdirt. These "trash burials" account for about one-quarter of the noninfant mortuary population.

Infant mortality was high at *ca.* 30% (Rollefson et al., 1985). All infants younger than ca. twelve to fifteen months retained their skulls after burial, although children older than *ca.* fifteen months were routinely accorded the decapitation ritual associated with adults. Except for stillborns accompanying mothers, infants with skulls intact with the postcranial skeleton were often used as house foundation offerings and, in one case, as ritual offerings above a cache of plastered skulls. That children from the age of *ca.* fifteen to eighteen months underwent postmortem decapitation might

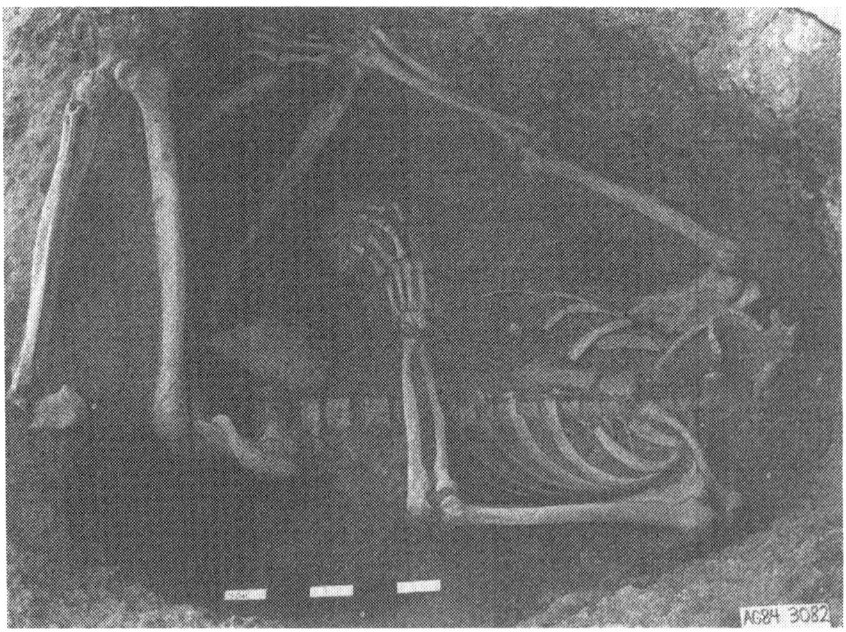

Figure 4. "Typical" MPPNB burial, placed beneath the lime-plaster floor of a house. After some time had passed to allow for decay of the flesh, the burial pit was reopened and the skull removed, after which a new floor was laid. (Photo: C. Blair)

be taken to reflect an ascribed status enjoyed by the older subfloor individuals.

The disposition of the twelve MPPNB skulls removed from subfloor burials varied in several ways. In one case, a cache of three skulls was placed in a separate subfloor pit beneath the same room as six decapitated burials; all three skulls faced the same direction (east), but otherwise there was no evidence of any additional special treatment (Rollefson 1986:50, Pl. II-6). In an adjacent room of the same house, a reflooring episode had removed all but the rear areas of an apparently forgotten skull buried beneath the floor. While nothing can be said of the face, the back of the skull was covered with a thick coat of a black substance, possibly bitumen.

Three temporal and parietal fragments of a shattered skull were found on the floor of another house that appears to have been destroyed by fire. All pieces bore clear evidence of red pigmentation, but since so little of the skull was found nothing can be said of any facial treatment. Its context suggests that the skull had been "on exhibit" before it fell to the floor.

A cache of four skulls was buried in a courtyard accompanied by two infant burials placed above them. Two of the skulls retained small patches of plaster and bitumen "eyeliner" decoration (Rollefson 1983:35, Pl. IV-1, 2). Since the missing plaster was not found in the pit, it is clear that the fragmentation had taken place elsewhere and was not repaired; instead, the group was disposed of ceremonially, possibly replaced by more recent plastered skulls. Another plastered skull, much better preserved and bearing no evidence of facial "cosmetics," comes from an unclear context, although the absence of any nearby architecture indicates it was buried in a courtyard pit (Simmons et al. 1990).

Finally, a courtyard pit dug into sterile clay contained the broken (but essentially complete) remains of three plastered "skulls" (Griffin et al., 1998; cf. Rollefson 1986:45–46, Pl. I-1, mistakenly identifying the material as a single statue head). The skulls themselves were not present, but bone impressions on the interior surface of the fragments prove that the plaster originally coated crania that were clearly disposed of elsewhere (Figure 5). Dated to older than 9,100 bp, these pieces are the oldest known evidence of plastered skulls.

Lime Plaster Statuary

A ceremonial burial of at least twenty-five human statues and busts made of lime plaster was excavated in 1983 (Rollefson 1983; Tubb 1985: Tubb and Grissom 1995), and another badly damaged cache of at least seven more pieces of statuary was recovered in 1985 (Rollefson 1986; Grissom n.d.). Conservation work on the 1983 cache, dated to *ca.* 8,700 bp, has shown

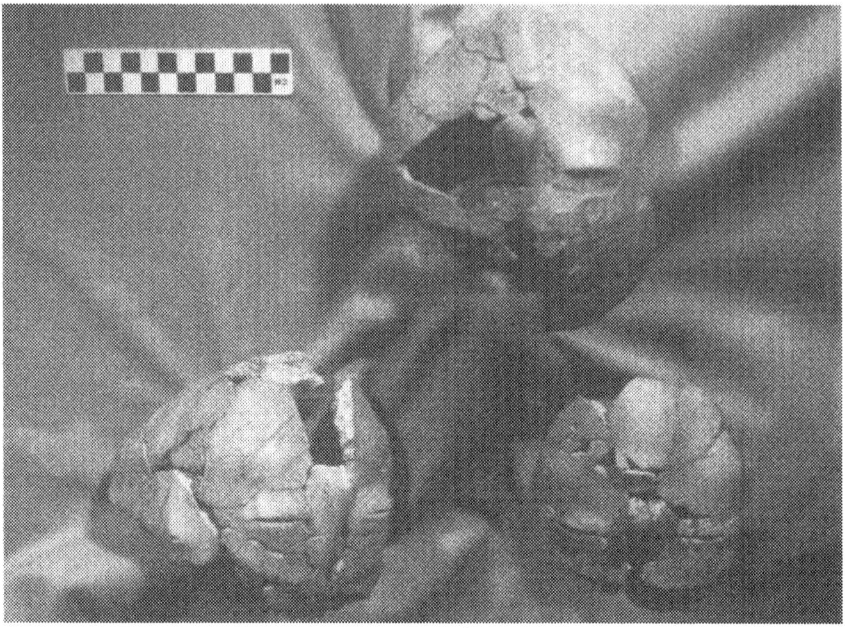

Figure 5. Three MPPNB plaster "faces." Note: impressions on the interior surfaces show these once-covered human skulls. (Photo: C. Grissom)

that both males and females are represented, and it is likely that children are depicted; at least two statues can be interpreted as being associated with birth or fertility. The smaller busts (35–45 cm) share facial details with the statues, including bitumen eyeliners dusted with a green crystal powder. The two sizes (Figure 6) suggest that at least a two-tiered hierarchy is represented. Features of the feet of the statues clearly show that they were anchored through a floor at one time, and it can be assumed that the statuary was on display, at least on special occasions.

The cache excavated in 1985 is younger (*ca.* 8,500 bp), perhaps dating to the transition between the MPPNB and LPPNB. All of the material has been conserved and restored at the Smithsonian (Grissom personal communication), and it is clear that at least three of the statues possessed two heads facing in the same direction (Figure 7) (Grissom n.d.).

LATE PRE-POTTERY NEOLITHIC B PERIOD (LPPNB)

Much less can be said for the LPPNB period at 'Ain Ghazal due to the restricted exposures of intact LPPNB deposits (*ca.* 460 m^2 overall, but much badly damaged or virtually destroyed by later inhabitants). Subfloor excava-

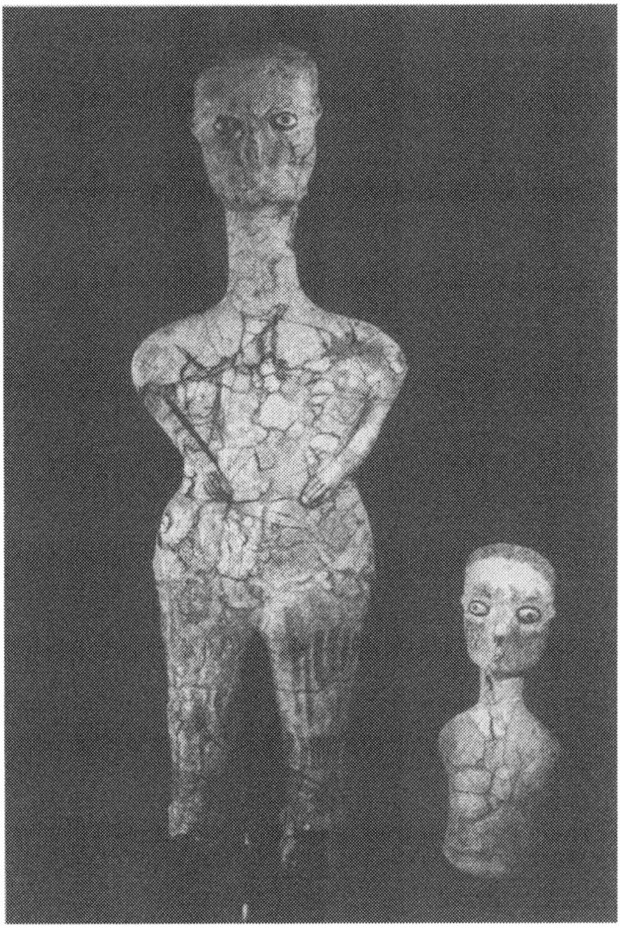

Figure 6. Two of the twenty-six MPPNB lime-plaster statues excavated in 1983. The taller statue is approximately 90 cm high. (Photo: P. Dorrell and S. Laidlaw)

tions have been rare so far, and as a consequence human burials number only seven; in all but one case they resemble the MPPNB styles. The exception to the burial mode is the first example at 'Ain Ghazal of a secondary burial: a fifteen-year old young woman whose bones were stacked alongside her skull. Of the two LPPNB "trash burials," one shows evidence of a violent cause of death: a flint bladelet (snapped at both ends) penetrated the left side of the skull with such force that a 3-cm-diameter piece of the inner wall of the cranium was jammed into the brain. Many of the bones of this individual were also broken, but it has not been determined if this was due to postdepositional disturbance of the burial pit.

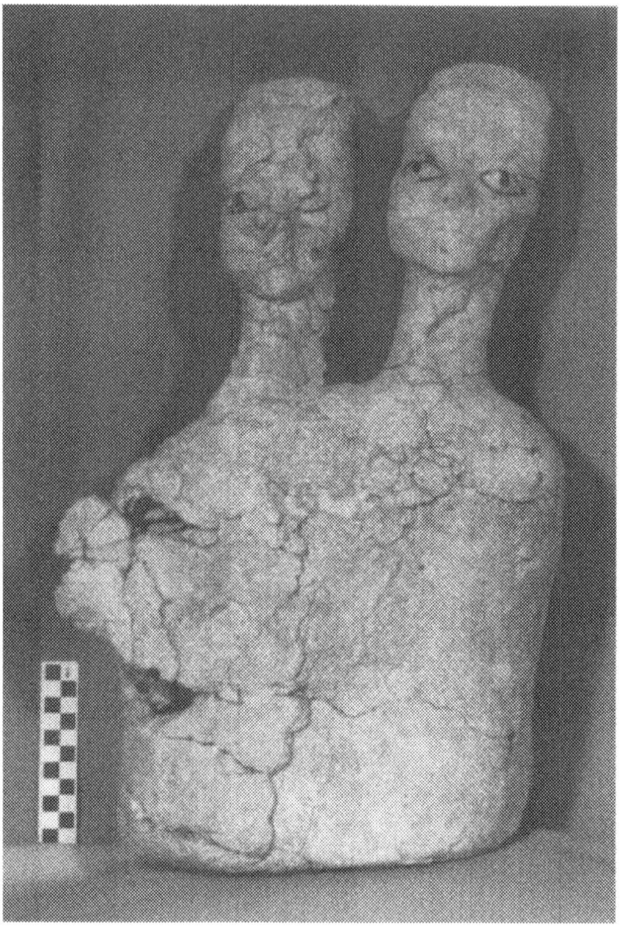

Figure 7. One of three two-headed statues from the 1985 cache at 'Ain Ghazel. (Photo: C. Grissom)

Human and animal figurines have been relatively rare, and there is little reason to comment on them. Plastered skulls have not been found so far, nor has any plaster statuary been recovered (although the 1985 cache may be from the earliest part of the LPPNB). It is possible that sampling error may be responsible in these cases.

LPPNB Ritual Structures

The identification and interpretation of structures associated with ritual practices is not a straightforward task (Renfrew 1985:1). Nevertheless, when floor plans, structural features, and decorative elements are taken into account, some buildings at 'Ain Ghazal are clearly nondomestic and probably served some ritual function. Three kinds of ritual structures have been identified so far at 'Ain Ghazal.

The first kind is an apsidal building, and so far three (and perhaps four) of these constructions have been found over the 500-year LPPNB period at the site: one in the Central Field (once considered to be a Yarmukian "public building") (Rollefson et al. 1990:110–111), one and possibly two in the North Field, and another in the East Field (Kafafi and Rollefson 1994; Rollefson and Kafafi 1994, 1996). Their small size (*ca.* 10 m^2 floor area) and geometric singularity contrast sharply with the contemporaneous domestic architecture (cf. Rollefson 1997), and the overall similarity with the shrine found by Kenyon at Jericho (Kenyon 1981:307) supports a ritual use at 'Ain Ghazal.

The second kind of LPPNB ritual structure may simply be an evolutionary development from the apsidal form. Arguments for this interpretation come from a building that underwent four construction phases: the first two included apses at one end (at least), the third phase closed off a collapsing apse, and the fourth phase witnessed the construction of a circular room with a small antechamber to the east (Kafafi and Rollefson 1994; Rollefson and Kafafi 1994, 1996). The small circular room (2.5 m diameter), with a relatively large hearth/altar installation inside it (Figure 8), was evidently the locus of cult activities. There were eight flooring episodes directly atop each other, each painted red; this cycle of renewing the room's floor would have been very expensive in terms of labor and time (cf. Rollefson 1990a) and suggests that the resurfacing coincided with some special events.

The circular shrine or cult building appears to have been badly damaged by subsidence: three immense cracks in the floor might reflect structural damage that caused its abandonment. Just 4 m to the south and at the same level, an identical circular building was constructed, evidently hastily, to replace the unusable shrine to the north. The walls of the replacement structure were made of small globular stones as opposed to more substantial flint and limestone blocks in the original circular building, and the two flooring episodes of the new building, although painted red, were thin and fragile and placed directly atop the underlying dirt instead of upon the normal foundation layer of a mixture of gravel and lime plaster. Clearly, this replacement building was not intended to last for an appreciable period of time (Rollefson and Kafafi 1997).

The third kind of LPPNB ritual structure can be confidently described as a temple or sanctuary based on the unique "furniture" found inside the edifice (Figure 9). The walls of this building, found high up the steep slope of the East Field across from the main settlement, enclose a space approximately 4 m NS by at least 5 m EW (the western, downslope end was badly

Figure 8. The circular LPPNB room is the last of four stages of alterations to a building at 'Ain Ghazal. The room is about 2 m in diameter and the central hole is 60 cm wide. (Photo: Y. Zoubi)

eroded). The northern, eastern, and southern walls are preserved to a height of about 75 cm, and it is possible that they were never higher than this (see below). The interior arrangement of features suggests at least two phases of use and reorganization.

The earlier phase included three orthostats or "standing stones" about 70 cm high arranged symmetrically between the northern and southern walls, forming a straight NS line around 2 m from the eastern wall. Between the southernmost standing stone and the southern wall was a roughly rectangular arrangement of limestone blocks that surrounded a bed of clay that

Figure 9. View to the west of the eastern room of the upper LPPNB temple. The floor altar is at upper left, the three standing stones upper center, the low platform at upper right, and the red-painted, stone-encircled lime plaster hearth at right center. (Photo: B. Degedeh)

had been burned so intensely and repeatedly that the upper 3–4 cm had been transformed into a ceramic slab; this may reflect an altar-like use. Between the central standing stone (which had fallen toward the downhill side at abandonment) and the eastern wall was a floor hearth of red-painted lime plaster surrounded by seven flat limestone slabs. In the eastern wall was a doorway roughly a meter wide that led uphill. The floor of the structure was unlike any other LPPNB building excavated at 'Ain Ghazal: it was made of plain dirt, with no use of lime plaster at all.

In a later phase, the doorway in the eastern wall was closed with the use of dressed limestone blocks and, in particular, an orthostat of dazzling white limestone shaped to an oval cross section. The top of the orthostat was unfortunately encrusted with a layer of calcretion, so it could not be determined if the central hump was the result of intentional shaping. Even so, the orthostat has an anthropomorphic shape, albeit highly stylized. There is no evidence of the use of pigments anywhere on this feature. The top of the orthostat rises about 10 cm above the walls, which suggests that the enclosure may have been open to the sky, at least during this phase.

Another change in the internal arrangement of the LPPNB temple/sanctuary consists of a low platform built inside a thin wall between the northernmost standing stone and the north wall. This alteration, in addition to the presence of the floor "altar" along the southern wall, effectively created two rooms in the temple, with the only passage from the western room to the eastern through the narrow spaces between the standing stones. It is not clear if this modification of the building coincided with the blocking of the eastern doorway.

PRE-POTTERY NEOLITHIC C PERIOD (PPNC)

Although more than 800 m^2 of PPNC deposits have been exposed at 'Ain Ghazal, they have been sampled deeply only for *ca.* 220 m^2 in the South and North Fields. Despite the small sample for the LPPNB, there are clear differences in ritual activity in the first half of the eighth millennium bp.

Only a handful of figurines have been recovered from PPNC contexts, including at least eight humans, four identifiable animals (one each of cattle, sheep, goat, and equid; all but the last were domesticated species by the PPNC period), and about a dozen unidentifiable fragments. This paltry number contrasts sharply with the MPPNB, suggesting either that figurines played a different role or that they may have been disposed of differently.

One of the human figurines is a fertility statuette elegantly carved in pink limestone. The head is missing, as is the case with most PPNB specimens, and the preserved length is 13.5 cm (Figure 10). Whether "decapitated" fertility figurines are characteristic for the PPNC cannot be determined on this singular basis (no other fertility figurines are known for this phase), and it is possible that this particular piece may originally have derived from an LPPNB layer disturbed by PPNC inhabitants. The context of this figurine is intriguing: it lay facedown atop a small (*ca.* 20 × 20 cm) stone "platform" that was uphill of a line of smaller stone slabs that climbed stepwise uphill over a distance of more than a meter. The ensemble would appear to represent a miniature shrine, although in this case the "steps," "altar," and figurine assemblage could be the creation of children mimicking a scene borrowed from their elders.

Human Burials

A total of thirty-four burials has been excavated from PPNC deposits, although two individuals are represented by crania only. The burial patterns reflect one of the strongest contrasts between PPNC and MPPNB/LPPNB ritual: the ninth millennium practice of skull removal was no longer followed in the PPNC (Figure 11). Another important aspect of PPNC burial practices concerns the relative proportions of primary and secondary buri-

Figure 10. PPNC limestonne "fertility" figurine found in what may have been a miniature, perhaps child's, reconstruction of a shrine. (Photo: Y. Zoubii)

als. Although fifteen of the burials were so badly disturbed by later Yarmukian activity that the contexts are unknown, for the remaining nineteen burials there is rough parity between primary (eleven) and secondary (eight) interments. Furthermore, except for the earliest five PPNC examples, bodies were no longer placed beneath house floors but in courtyards. Finally, many of the burial pits (particularly in the South Field) included pig bones (domesticated by this time?), although additional analysis of the Central Field burial pit contents is necessary to determine if pig bones are "offerings" to be expected in all PPNC inhumations.

The PPNC Temple or Sanctuary*

Excavations in the East Field in 1995 and 1996, where PPNC domestic structures were not found, encountered a sizable (*ca.* 5 · 6 m) structure that appears to have been a temple or a walled sanctuary. The structure has two

*A recent radiocarbon date from the floor of the temple is 8080 ± 60, which argues for a LPPNB age for this structure.

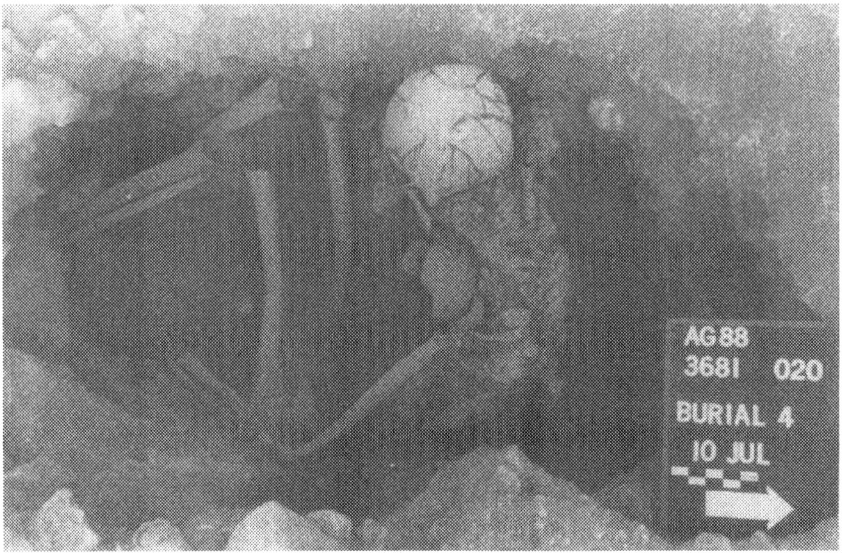

Figure 11. PPNC burials. These normally occur in courtyards with skulls still intact with the postcranial skeleton. (Photo: H. Wada)

primary rooms divided by a north–south wall. The western (downhill) room was badly damaged in antiquity by erosion and recently by bulldozer work.

In the center of the eastern room, which uncharacteristically has a floor made of yellow clay and not common dirt or huwwar plaster, a small, unpainted lime-plaster hearth surrounded by seven flat limestone slabs lay just west of the central cluster of three pairs of standing stones of unequal sizes (from *ca.* 40–70 cm) oriented on a rough N–S axis. The standing stones supported two broad, thick limestone slabs that formed a raised altar in the middle of the eastern wall. At the center of the northern wall a small stone cubicle was built on the floor (Figure 9). The building was protected against erosional damage on the steep slope (35%) by a large retaining wall more than 20 m long (NS) and 2.5 m high that had cut through earlier LPPNB layers. At the base of the retaining wall, an exterior storage (?) chamber 1.5 m EW by 3 m NS was excavated into underlying sterile clay, and presumably the excavated material was used to floor the eastern room of the temple.

At one time there was an entrance into the eastern room via a doorway in the southern wall, but this was later blocked. Another doorway existed in the wall that separated the eastern and western rooms, and the control of movement between the two parts of the temple is particularly interesting. Leading straight west of the doorway into the western room for 60 cm was a thin "screen wall" that abruptly turned to the north, blocking any view from the western room into the eastern altar area. In effect, this thin but impenetrable wall restricted both physical and visual access to the rites being conducted in the eastern room, creating our earliest example of a "holy of holies" in ritual architecture.

The eastern wall of the LPPNB temple was relatively high (preserved to 1.8 m) for its thickness (*ca.* 45 cm), and this structural weakness was susceptible to either subsidence or earthquake. For whatever reason, the eastern wall partially collapsed and led to the temple's abandonment, and as the temple had "died" it was literally "buried" beneath two massive retaining walls, one of which was placed directly over the altar and the other over the exterior "storage" feature. The "burial" of buildings, particularly "special buildings," has been noted at Çayönü, including the Skull Building, Flagstone Building, and the plaza that contained standing stones (Özdogan 1995:84–87; Özdogan and Özdogan 1989: 74).

YARMUKIAN

Yarmukian deposits have not been encountered in either the North or East Field excavations, but *ca.* 1500 m^2 have been investigated in the Central and South Fields. Despite the extensive excavation, only a few figurines have been recovered from 'Ain Ghazal, including seven human and five identifiable animal specimens (one each goat, sheep, and cattle, and two birds) plus a dozen or more unidentifiable fragments. One of the human forms is a typical "coffee bean" fertility specimen (Garfinkel 1995: Plate 10; cf. Perrot 1966:Pl. VI-13, 16–17), and another is a highly stylized, scratched limestone pebble (cf. Stekelis 1950–51:Pl. 2a-b). Superficially, the small number of figurines suggests a minor ritual role continued from the PPNC phase, although it should be stressed that by the beginning of the mid-eighth millennium, the population of Yarmukian 'Ain Ghazal had fallen to its lowest level since the founding of the village some 2,000 years earlier. It is possible that the number of figurines per household was actually not very different from the MPPNB.

The contrasts of burial patterns in the earlier periods with the Yarmukian could not be stronger: not a single Yarmukian burial has been found. Presumably a Yarmukian cemetery has gone undetected by our excavations and test trenches on and off the site.

The "Public Building"

Normal houses for permanent residents of Yarmukian 'Ain Ghazal were mud-floored, rectangular, multiroomed, relatively large (*ca.* 9 × 5 m), and erected on the surface. One structure stands out from this standard. Yarmukian excavators encountered a thin LPPNB lime-plaster floor, which they used for their own needs. Measuring *ca.* 6 × 3.75 m (exterior, NS by EW), the southern end was built with a shallow apse incorporating an orthostat (and an adjacent "pseudo-orthostat") in the center of the arc. The two rooms were of unequal size: a smaller antechamber, with an entry at the NW corner of the structure, measured 1.5 × 2.5 m (interior, NS by EW); the other was 3 × 2.5 m in interior dimensions (Figures 12 and 13).

Beyond the form, size, and reuse of a lime-plaster floor, other differences indicate that this building had a special character. Unlike the common domestic dwellings, only sherds of decorated Yarmukian "fine ware" cups and small jars were found inside the rooms and immediately outside the building. No specific ritual paraphernalia were found, although this negative evidence can be misleading. On the other hand, it might be that this structure was reserved for public functions, such as a meeting hall for a community "council" to reach decisions regarding the profane aspects of 'Ain Ghazal's inhabitants. Or, perhaps it was a building used for both religious and civic purposes at different times of the annual round.

Figure 12. The LPPNB temple or sanctuary at 'Ain Ghazal. A massive retaining wall (top of photo) protects a two-roomed structure below. The raised alter is visible in the center of the east wall of the temple, and the right-angle screen wall is just visible above the meter scale. The building in the foreground is an earlier LPPNB domestic structure. (Photo: Y. Zoubi).

Figure 13. View toward the south of the Yarmukian "public building," which was originally LPPNB in age. Three similar LPPNB apsidal buildings have been found elsewhere at 'Ain Ghazal but not at other LPPNB sites in the southern Levant. (Photo: H. Wada)

MPPNB AND LPPNB RITUAL AND SOCIAL STRUCTURE

The wealth of MPPNB material that was associated with ritual can be seen as a maintenance of earlier traditions in the Levant as well as an ongoing elaboration associated with a rapidly growing population during the late ninth and early eighth millennia. The figurines, burials, treated skulls, and plaster statuary, taken altogether, argue for a social order more complex than a simple egalitarian system of farmers. I propose that ritual behavior during this time occurred on at least three (possibly four) hierarchical levels in the settlement (and perhaps the greater region) and that these levels reflect to some degree the organization of social behavior on a day-to-day basis.

The production of clay animal and human figurines is an activity that requires little talent; a visit to almost any farming village in India will show that children or parents are often engaged in making figurines as toys (Köhler-Rollefson, personal communication). But some figurine manufacture in those same villages is highly restricted in terms of who can make them and how they can be used—social license is strictly governed for some forms and purposes. Certainly some of the MPPNB animal figurines fit into a "toy" or analogous category, but others (particularly human and cattle figurines) appear in contexts that indicate controlled ritual usage (and perhaps production by certain shamans or other ritual practitioners). The dispersal of figurines among the houses and trash dumps at 'Ain Ghazal suggests that figurines were available to everyone for personal protection or prosperity. In this regard, this was the "lowest level" in a hierarchy of ritual activity that involved personal, individual interaction with the magic or luck that is imbued in the talisman he or she possessed.

Human burials involved an intermediate level of the hierarchy. Certainly not every family member was entitled to a subfloor or courtyard burial, complete with ritual decapitation. The size and arrangement of MPPNB architecture suggest that each household was a nuclear family and essentially an independent production and consumption unit. On this "household level" certain individuals of either sex were selected for special postmortem treatment: beneath every MPPNB floor excavated at 'Ain Ghazal there was at least one burial, and decapitated burials included both males and females of all ages above *ca.* fifteen to eighteen months. How this selection process operated is not known. The "trash burials" are difficult to interpret since there are few parallels in the Levant, but the evident lack of respect at the death of these people suggests that they probably enjoyed little respect while they were alive; it is possible that a patron–client relationship was in effect and that the trash burials were essentially of disenfranchised people, with few if any social privileges and claims.

Associated with this household level, and perhaps operating on its own merits, the additional selection of some skulls for particular treatment indicates a higher level of complexity. The skulls of household heads (no pun intended) may simply have been buried elsewhere, untreated except for the fact of decapitation and separate relocation. Some skulls that were painted red or black may have been, for example, particularly respected ritual practitioners. But the elaboration afforded those individuals with a remodeling of the facial features in plaster suggests a higher plateau of respect (and responsibility)—revered ancestors, as Kenyon (1970:54) described them, yes, but perhaps at the level of lineage or even clan leadership.

At the top of the hierarchy come the plaster statues and busts. The smaller, less-impressive busts might be taken to represent mythological lineage or clan founders whose names were passed down the generations through the powerful memory of oral tradition. The larger and more imposing statues tower above them all, and while it is unclear if they were "gods" they may have been the mythical founders of the community of clans (i.e., of "the people").

The twin-headed statues from the 1985 cache introduce an intriguing element in social complexity. Several coincidental aspects provide several possible interpretations. First, the mid-ninth millennium was a time of severe upheaval in terms of the abandonment of MPPNB farming villages, particularly in the Jordan Valley and areas to the west (see Rollefson 1987; Simmons Chapter 9 this volume). Second, it is also about this time that 'Ain Ghazal appears to have undergone a "sudden" increase in size with the establishment of the eastern enclave across the Zarqa River from the main settlement; normal population growth is an unlikely explanation for the rapid growth of 'Ain Ghazal, and it is not improbable that much of the expansion was the result of in-migration of some of families from the deserted settlements. And finally, recent faunal analysis of the ovicaprid remains at 'Ain Ghazal indicate that a rapid shift occurred from a dominance of goats to a strong majority of sheep during the latter half of the ninth millennium (Wasse 1994), a transformation of the subsistence economy that led to the beginnings of pastoral nomadism and increasing contact with steppe and desert populations to the east (Perrot 1993; Rollefson and Köhler-Rollefson1993). The double-headed statues, then, might reflect a symbolic consolidation in the 'Ain Ghazal settlement of two or more related lineages or clan populations formerly spatially segregated, either as separate farming communities or as farmers and steppe/desert-dwelling hunters, or both.

The probability that the statuary was on display at least at certain times of the year logically demands that there was a special building to house them, although we have found no such evidence for a "shrine" or "temple" in our sampling of MPPNB deposits. The burials of the statue caches not far from the spring at 'Ain Ghazal suggest that such a community focus of ritual behavior was located near the center of the settlement, possibly destroyed by highway construction in the 1970s.

Despite the uneven evidence for the LPPNB, there is little to indicate major changes in ritual activity with the exception of the first appearance of secondary burials. With the changeover from goat- to sheep-based animal husbandry, the return of incomplete skeletons from afar for burial in the "home" territory is perhaps reflected here. The apparent replacement of small independent households by extended family economic units is suggested by much larger LPPNB buildings at 'Ain Ghazal and at Basta (cf.

Nissen et al. 1991: Figure 1), for example (Kafafi and Rollefson 1995). Some of the household members may have traveled with the flocks for part of the seasonal round, returning to ‘Ain Ghazal periodically. Current research is addressing this topic, as well as the relationship of architecture and social structure (Rollefson 1997), such as understanding the implications of the violent death (and possible postmortem mutilation) of one community member recovered from a trash burial.

The LPPNB cult buildings/shrines and the temple/sanctuary add new dimensions of ritual practice not previously seen in the MPPNB in the southern Levant (but cf. Hauptmann 1991–92:26–32; Schmidt 1995; Bienert 1995:317–320 for eastern Anatolia). The small size of the circular cult buildings indicates that, although they probably served some ritual purpose, they were not for the use of all the people of 'Ain Ghazal. Instead, they may have served the needs of a part of the population on a lineage or clan level, overseen by a full-time shaman or priest. The temple/sanctuary is also small, but the unique internal features and its location high on the slope of the East Field suggest that it likely served a large part (perhaps all) of 'Ain Ghazal's population. In any event, the two nondomestic structural forms share similarities with the aceramic Neolithic ritual buildings at Jericho (Kenyon 1981), Beidha (Kirkbride 1967), Nevali Çori (Hauptmann 1993) and Çayönü (Özdogan and Özdogan 1989), which argues that the religious sphere of this enormous area of the Near East shared a number of aspects of public expression of ritual meaning.

PPNC AND YARMOUKIAN RITUAL AND SOCIAL STRUCTURE

The PPNC period at ‘Ain Ghazal introduces several major departures from PPNB ritual traditions, among several other sociocultural aspects (Rollefson 1990b; Rollefson and Köhler-Rollefson 1993). The most obvious of these is the suspension of decapitation of burials and its implications for public and private expression of ancestor veneration. Two isolated skulls have been found in PPNC contexts (one context should be described as transitional between the LPPNB and PPNC), although no decapitated skeletons have been found; for the moment, these skulls are being viewed as secondary burials. Second, although subfloor burials occurred in the earliest part of the PPNC, later the common location was in the courtyard, even though the reflooring of a house no longer included the expensive production of lime plaster. Third, the proportion of secondary burials in the PPNC indicates that a larger part of the population consisted of part-time residents of the village compared, at least, to the MPPNB period. This interpretation is supported by the spatially segregated dichotomy of PPNC architecture at ‘Ain

Ghazal (Rollefson and Köhler-Rollefson 1993; Kafafi and Rollefson 1995). It also is evident that the population of 'Ain Ghazal had begun to decline markedly by the onset of the PPNC (Rollefson 1997). Fourth, the apparent decrease in the manufacture and use of figurines indicates that ritual involving the individual level of interaction with the spirit world decreased markedly. Shamans were probably still an important factor in the day-to-day life of PPNC residents, and perhaps they had taken over earlier personal accession to the spirit world. Finally, the apparently intentional inclusion of pig bones in the burial pits (presently they appear to be principally associated with secondary burials) may be indicative of a special relationship of incipient pastoralists with a symbol of nonpastoralist animals; that is, the burials include a symbol of solidarity between the ovicaprid pastoralists and the full-time resident farmers who husbanded pigs at the permanent settlement at 'Ain Ghazal.

The only remaining tie with PPNB traditions concerns the number of burials found beneath or near PPNC houses. Notably, there is no significant change in the sex of either subfloor or courtyard burials for the PPNC. Clearly, these individuals were still special members of the household, and a "common" cemetery (or other means of disposal) has not yet been found by our excavations. The PPNC temple/sanctuary is also relatively small, but in view of the large amount of labor spent in the construction of it and the massive terrace wall most likely it served the community at large.*

With the transition to the Yarmukian Pottery Neolithic, the burial "pattern," in its absence, might indicate that egalitarian treatment of the dead was commonplace. The unfound common cemetery (if it existed) may have included both primary and secondary burials, but such speculation is unproductive. Figurines from the Yarmukian period indicate that shamans were still plying their trade in terms of fertility or "mother protection," but none of the animal figurines indicate anything more special than possible toys or artistic exuberance. With the exception of the "public building," one could get the impression that ritual in any form played a minor role in the daily life of the Yarmukian residents of 'Ain Ghazal, but the unique preservation of ritual objects made of organic material at Nahal Hemar (Bar-Yosef and Alon 1988) is a powerful caution against such an assumption.

CONCLUDING REMARKS

The settlement at 'Ain Ghazal underwent an unprecedented history of uninterrupted habitation in the Levant for more than 2,000 years. During these

*See note, page 179.

two millennia, pressures on the once lucrative environment by a rapidly growing population of farmers and herders resulted in a series of adaptations in the overall cultural management of the landscape and the demands of the resident population. Ecological stresses within and beyond the 'Ain Ghazal territory necessarily involved accommodations in social organization. Ritual behavior is just one of the areas of social interaction that was affected, but like the architectural norms and subsistence economy at 'Ain Ghazal it underwent considerable change, to the point that by the end of the seventh millennium, a time traveler from the end of the ninth would have recognized little (if any) of the ritual forms.

REFERENCES

Bar-Yosef, O., and Alon, D., 1988, Nahal Hemar Cave. *`Atiqot* 18: 1–81.

Byrd, B., 1994, Public and Private, Domestic and Corporate: The Emergence of the Southwest Asian Village. *American Antiquity* 59(4): 639–666.

Bienert, H.-D., 1995, *Kult und Religion in Prähistorischer Zeit: Eine Studie anhand von Fundmaterial epipaläolitischer und frühneolithischer Gesellschaften/Kutlturen Südwestasiens* (12.-6. Jt. v. u. Z.). Ph.D. dissertation, University of Tübingen.

Cauvin, J., 1972, *Religions Néolithiques de Syro-Palestine*. Publications du Centre de Recherches d'Ecologie et de Préhistoire, St.-Andre-de-Cruziérs.

Cauvin, J., 1994, *Nassiance de Divinités, Naissance de l'Agriculture*. CRNS, Paris.

Garfinkel, Y., 1995, *Human and Animal Figurines of Munhata (Israel)*. Les Cahiers des Missions Archéologiques Françises en Israel. Paléorient, Paris.

Garstang, J., and Garstang, J. B., 1948, *The Story of Jericho*. Marshall, Morgan and Scott, Ltd, London.

Griffin, P. S., Grissom, C. A.., and Rollefson, G. O., 1998, Three Late Eighth Millennium Plastered Faces from 'Ain Ghazal, Jordan. *Paléorient* 24(1):59–70.

Grissom, C., n.d., *Conservation of Neolithic Lime Plaster Statues from 'Ain Ghazal*. Unpublished progress report (1995) on file, Conservation Analytical Laboratory, Smithsonian Institution.

Hauptmann, H., 1991-92, Nevali Çori: Eine Siedlung des akeramischen Neolithikums am mittleren Euphrat. *Nürnberger Blätter zur Archäologie* 8: 15–33.

Hauptmann, H., 1993, Ein Kultgebäude in Nevali Çori. In *Between the Rivers and over the Mountains*, edited by Frangiapane et al., pp. 37–69. Università di Roma "La Sapienza," Roma.

Kafafi, Z., and Rollefson, G., 1994, 'Ain Ghazal 1993–1994. *Biblical Archaeologist* 57(4): 239–241.

Kafafi, Z., and Rollefson, G., 1995, The 1994 Season at 'Ayn Ghazal: Preliminary Report. *Annual of the Department of Antiquities of Jordan* 39:13–29.

Kenyon, K., 1970, *Archaeology in the Holy Land*. Ernest Benn, Ltd., London.

Kenyon, K., 1981, *Excavations at Jericho, Volume III*. British School of Archaeology in Jerusalem, London.

Kirkbride, D., 1967, Beidha 1965: An Interim Report. *Palestine Exploration Quarterly* 5–14.

Mellaart, J., 1967, *Çatal Hüyük: A Neolithic Town in Anatolia*. McGraw-Hill, New York.

Nissen, H., Muheisen, M., and Gebel, H. G., 1991, Report on the Excavations at Basta 1988. *Annual of the Department of Antiquities of Jordan* 35:13–40.

Özdogan, A., 1995, Life at Çayönü during the Pre-Pottery Neolithic Period. In *Readings in Prehistory: Studies Presented to Halet Çambel*, pp. 79–100. Graphis, Istanbul.

Özdogan, M., and Özdogan, A., 1989, Çayönü: A Conspectus of Recent Work. *Paléorient* 15(1): 65–74.

Perrot, J., 1966, La Troisième Campagne de Fouilles à Munhata (1964). *Syria* 43:49–63.

Perrot, J., 1993, Remarques Introductives. *Paléorient* 19(1):9–21.

Renfrew, C., 1985, *The Archaeology of Cult. The Sanctuary at Phylakopi*. Thames and Hudson, London.

Rollefson, G., 1983, Ritual and Ceremony at Neolithic 'Ain Ghazal (Jordan). *Paléorient* 9(2):29–38.

Rollefson, G., 1986, Neolithic 'Ain Ghazal (Jordan): Ritual and Ceremony II. *Paléorient* 12(1):45–52.

Rollefson, G., 1987, Local and Regional Relationships in the Levantine PPN Period: 'Ain Ghazal as a Regional Center. In *Studies in the History and Archaeology of Jordan*, edited by A. Hadidi, pp. 29–32. Department of Antiquities of Jordan, Amman, Jordan.

Rollefson, G., 1990a, The Uses of Plaster at Neolithic 'Ain Ghazal, Jordan. *Archaeomaterials* 40:33–54.

Rollefson, G., 1990b, Neolithic Chipped Stone Tool Technology at 'Ain Ghazal: The Status of the PPNC. *Paléorient* 16:119–124.

Rollefson, G., 1997, Changes in Architecture and Social Organization at Neolithic 'Ain Ghazal. In *Prehistory of Jordan II*, edited by H. Gebel, Z. Kafafi, and G. Rollefson, pp. 287–307. Ex Oriente, Berlin.

Rollefson, G., Banning, E. B., Byrd, B., Kafafi, Z., Köhler-Rollefson, I., Petocz, D., Rolston, S., and Villiers, L., 1984, Excavations at the PPNB Village of 'Ain Ghazal (Jordan) 1982. *Mitteilungen der Deutschen Orient Gesellschaft* 16: 139–183.

Rollefson, G., and Kafafi, Z., 1994, The 1993 Season at 'Ain Ghazal: Preliminary Report. *Annual of the Department of Antiquities of Jordan* 38: 11–32.

Rollefson, G., and Kafafi, Z., 1996, The 1995 Season at 'Ayn Ghazal: Preliminary Report. *Annual of the Department of Antiquities of Jordan* 40: 11–28.

Rollefson, G., and Kafafi, Z., 1997, The 1996 Season at 'Ayn Ghazal: Preliminary Report. *Annual of the Department of Antiquities of Jordan* 41:27–48.

Rollefson, G., and Köhler-Rollefson, I., 1989, The Collapse of Early Neolithic Settlements in the Southern Levant. In *People and Culture Change: Proceedings of the Second Symposium on Upper Paleolithic, Mesolithic and Neolithic Populations of Europe and the Mediterranean Basin*, edited by I. Hershkovitz, pp. 59–72. British Archaeological Reports, International Series 508. BAR, Oxford.

Rollefson, G., and Köhler-Rollefson, I., 1993, PPNC Adaptations in the First Half of the 6th Millennium B.C. *Paléorient* 19(1):33–42.

Rollefson, G., Simmons, A., and Kafafi, Z., 1990, The Neolithic Village of 'Ain Ghazal, Jordan: Preliminary Report on the 1988 Season. *Bulletin of the American Schools of Oriental Research Supplement* 27:97–118.

Rollefson, G., Simmons, A., Donaldson, M., Gillespie, W., Kafafi, Z., Köhler-Rollefson, I., McAdam, E., and Rolston, S., 1985, Excavations at the Pre-Pottery Neolithic B (PPNB) Village of 'Ain Ghazal (Jordan), 1983. *Mitteilungen der Deutschen Orient Gesellschaft* 117:69–116.

Rollefson, G., Simmons, A., and Kafafi, Z., 1992, Neolithic Cultures at 'Ain Ghazal, Jordan. *Journal of Field Archaeology* 19(4):443–470.

Schmidt, K., 1995, Investigation in the Upper Mesopotamian Early Neolithic: Göbekli Tepe and Gürcütepe. *Neo-Lithics* 2: 9–10.

Simmons, A., Boulton, A., Butler, C., Kafafi, Z., and Rollefson, G., 1990, A Plastered Human Skull from Neolithic 'Ain Ghazal, Jordan. *Journal of Field Archaeology* 17(1):107–110.

Simmons, A., Köhler-Rollefson, I., Rollefson, G., Mandel, R., and Kafafi, Z., 1988, 'Ain Ghazal: A Major Neolithic Settlement in Central Jordan. *Science* 240:35–39.
Stekelis, M., 1950, A New Neolithic Industry: The Yarmukian of Palestine. *Israel Exploration Journal* 1(1):1–19.
Tubb, K., 1985, Preliminary Report on the 'Ain Ghazal Statues. *Mitteilungen der Deutschen Orient Gesellschaft* 117:117–134.
Tubb, K., and Grissom, C., 1995, 'Ain Ghazal: A Comparative Study of the 1983 and 1985 Statuary Caches. In *Studies in the History and Archaeology of Jordan*, V, edited by K. 'Amr, F. Zayadine, and M. Zaghloul, pp. 437–447. Jordan Press Foundation, Amman.
Wasse, A., 1994, Pastoral Adaptations in the Central and Southern Levant during the Neolithic: The Sheep and Goat Bones from 'Ain Ghazal, Jordan. B.A. honors thesis, Institute of Archaeology, University College, London.

Chapter 8

Is Size Important?

Function and Hierarchy in Neolithic Settlements

FRANK HOLE

INTRODUCTION

The Neolithic witnessed some of the most profound transformations in human history: the inception and spread of agriculture and animal husbandry, permanent settlement in solid houses, growing populations around fixed locales, long-distance trade in raw material such as obsidian, craft specialization in lithic production, the beginnings of simple metallurgy, the use of pyrotechnology in the making of plaster and the production of ceramics, simple methods of marking ownership of objects and goods, and the construction of structures for sacred rituals. To be sure these—and no doubt many more invisible but equally significant developments—took place over a span of several thousand years, with variable emphases from region to region. Nevertheless, it is not too much to say that the foundations of what we consider to be basic attributes of our civilization were established during the Neolithic. This first touch of familiarity encourages us to seek analogs in prehistory for some of the intangibles of human life. "Social configurations" is one of these. What can we say about the ways Neolithic societies

FRANK HOLE • Department of Anthropology, Yale University, New Haven, Connecticut 06520-8277.

Life in Neolithic Farming Communities: Social Organization, Identity, and Differentiation, edited by Ian Kuijt. Kluwer Academic/Plenum Publishers, New York, 2000.

were organized? Can we apply uniformitarian ideas such that the past merges seamlessly with the present? Few would affirm such a notion, yet implicitly we all base our interpretations on combinations of our experiences, ethnographic examples, and more abstract theories about the ways people and societies behave that are derived from the modern world.

Although there is grave danger in imposing our realities on the past, as a start we may consider inequalities or, as expressed more mundanely, differences. The mother's milk of archaeology is difference. Change and variability are operational aspects of difference, and these, rather than unvarying similarity, excite our imagination. In this chapter I question the relationship between two kinds of differences, size of site and special architectural units. This inquiry questions whether large sites are tangible expression of regional importance—that such sites are "centers" affecting social relations within their sphere of influence (Rollefson 1987; Kuijt 1994). The corollary is that small settlements are relatively insignificant and unlikely to have been the residence of regional leaders or the locale of important activities. One need only look at the modern world to find countless examples where relative size is a reliable indication of relative importance: hence, the uniformitarian approach makes eminent sense. However, since the obvious sometimes proves to be wrong, I merely ask whether there are clear indications that the largest sites differ in substantial ways from the smallest settlements during the PPNB.

A fair amount of theory implies that to buffer inherent conflict when people live together in large numbers, there must be organizing principles and customs, usually considered to be manifest in principals and institutions (Wright 1984; Flannery 1995). What problems might arise? In any community, bickering, bullying, thieving and a nearly unending host of similar issues might arise. More serious from a structural point of view would be attempts by individuals or groups to accumulate power and authority and become unequal. Whereas in small, ephemeral communities the aggrieved often merely remove themselves and migrate to a friendlier community, this is not as tenable when one has agricultural fields and stores. In such cases, the first resort is usually to the family and its extensions, and the aggrieved may or may not find an effective solution in these parleys, particularly if the offenses take place solely within the familial setting. Heads of households may try to resolve disputes between their groups, but this relies on personal persuasion, which lacks effective long-term sanctions. Feuds that may last generations keep tension alive. In most preurban (even prestate) communities there is unlikely to have been any formally constituted "police" to keep order and punish transgressors; rather action normally would be taken by *ad hoc* groups of affected relatives.

Even in communities where individuals and families have essential autonomy, there are group norms and customs that regulate behavior. These are ordinarily as unquestioned as the local dialect is to its speakers. Such customs may include reciprocal obligations, burial practices, and rites of passage that the family and community go through. Kuijt has described how an egalitarian ethos might have been promulgated and maintained through group burial rituals (Kuijt 1996). Other customs that might serve to solidify the community might relate to planting ceremonies, occasions for sacrifice, harvest festivals, etc. Insofar as these are archaeologically visible through ritual paraphernalia or other tangible remains, they would tend to tie sites within regions together in a network of "similar cultural practices."

For most of daily living it is unlikely that there arose a need for "regulation" or "control" to be exercised outside the immediate community. It is hard to imagine how any community in the PPNB could exert "control" over another because resources, despite their potential depletion, were still widely available, given the size of local populations. On the other hand, control of a resource, such as high-quality flint for making naviform cores, might engender trade possibilities. We should imagine that people living within walking proximity had regular intercourse through marriage and exchange of locally available goods. While the agricultural fields immediately adjacent to the villages were probably held exclusively, pasture lands were no doubt part of the great "commons," available to all. Hence, herds and herders would come in frequent contact. Autonomy in subsistence among scattered villages fostered equality and did not necessarily lead to social exclusion. One might imagine occasions when perhaps thousands of people gathered for important ceremonies or events such as summer solstice. If they did, there is no apparent archaeological evidence. On the other hand, there is abundant evidence that people did gather in small groups in nondomestic structures. Just what rites they carried out or what politics they engineered in them cannot be known, yet their ubiquity suggests that they played a central role in the workings of PPNB society. The fact that a single such structure seems to have been present at any given time in many sites suggests that only some members of each community actually participated in the rites or discussions.

I consider several such structures as well as the sites in which they occur, asking the general questions: Does size of site bear any relationship to the kind or size of structure? Do these structures have to do with "centrality" and "hierarchy," or are they components of all communities that share customs and rites?

WHAT IS LARGE?

One striking fact of the archaeology of the Early Neolithic in the Near East (known as the PPNB) is that there are a number of very large sites—sites that are anomalously large in the context of what came before and after, and in the broader contemporary picture. Most archaeologists would argue that size and specific function go naturally hand in hand, and a whole literature on the implications of size hierarchies follows from this assumption. For example, Ian Kuijt has made a case for a settlement hierarchy in the PPNA (10,300–9,600 bp) of the southern Levant where site sizes range from 0.1 to 2.5 ha, implying, perhaps, a range from a family camp to a settlement of a few hundred people (Kuijt 1994). Jericho is regarded as the largest such site, but its actual extent in the PPNA cannot be determined and no other site is larger than 1.5 ha. One of these, Netiv Hagdud is estimated to have had some 200 small round structures occupied simultaneously (Bar-Yosef and Belfer-Cohen 1991). Based on such evidence Kuijt suggests that "ritual and community leaders" who coordinated activities among settlements from their bases in the large sites had already emerged during the PPNA.

Using similar reasoning, Gary Rollefson sees as many as four "centers" in the PPNB, namely the largest sites in each region that "served as focal points for the development of local innovations and as centers for the diffusion of new concepts and techniques" (Rollefson 1987:31). Rollefson bases his ideas of centrality more on sheer size of site than on other criteria, although some of these sites have yielded portrait skulls and abundant exotic materials. He also points to subtle differences in local styles as perhaps indicative of regional groups, with these large sites as their foci, but the evidence is far from compelling when we consider the confounding problems of determining contemporaneity and site size and that only small samples of each site have been recovered.

In later times a community of 2.5 ha would qualify merely as a small village, but these things are relative. An incautious application of the site size approach would lead one to expect that the smaller settlements would lack the special features of the large ones, despite abundant ethnographic evidence that even very small communities may hold the kinds of special structures that commonly imply specialization and social complexity (Kramer 1994). Archaeology of the PPNB seems to support ethnography, for there are some very large sites that have yielded no evidence for special activities, as well as very small sites that have.

There are many reasons why a site may have been large or may appear to have been large. It may have been (a) a political, economic, or religious center or (b) in a locally rich environment. In either case the site actually may have had a large population. However, a site may merely appear to have been large because (a) of sequential use so that small settlements accumulating horizontally appear to have the same "archaeological" age; (b) dwellings may be widely spaced, but artifacts cover the entire area, thus giving a false impression of size; (c) structures that may have been used for other than domestic purposes take up a large amount of the space; (d) a large site may represent merely a seasonal agglomeration of many normally smaller units. Clearly each of these situations might lead to different correlates in the sphere of social relations.

On a broad historical scale, Ronald Fletcher focuses on "three great transformations in settlement form and organization . . . the development of sedentary communities, the initial formation of urban settlements and the Industrial Revolution" (Fletcher 1987:65). At each new stage the largest settlements increased by a factor of 100. We are not concerned here with *that* level of growth, but we can identify another stage within that sequence—the 3,000 years from the time of initial sedentary communities to the later PPNB—that is, just prior to the use of ceramics. During this time, nearly all settlements remain a hectare or less, but the largest ones are 8–12 ha, and there are few to no sites of intermediate size. Thus, the largest sites in the Near East are some ten times the size of the typical small site.

ARCHAEOLOGICAL DATA SETS

Let us now turn to a sample of PPNB sites that I have chosen to explore the relationship between site size and the presence or absence of monumental, public, or cultic structures, or evidence of elite status. There are hundreds of known sites from 0.1 to 11–12 ha, from southeastern Anatolia to the southern Levant, that share architectural style, arrowheads, and burial practices from which such a sample could be drawn, although as yet most of these have not been excavated or reported fully.

Bouqras

The first site we shall consider is Bouqras, situated on a terrace overlooking the Euphrates River, in an environmental zone—the arid steppe—that today has some 150 mm of precipitation, well below the needs of agriculture. Despite this limitation, the site is estimated to cover some 2.75 ha and have held as many as 850 persons during the span (8,400–7,900 bp) making it a small village by PPNB standards (Boerma 1989–90).

A careful study of the local environment makes it clear how much more favorable conditions at the site were than in the immediately surrounding areas (Boerma 1983). The site is at the terminus of a wadi whose bed can be farmed and that traditionally served as a route westward toward the oasis of Palmyra (Akkermans et al. 1983:336). Bouqras also lies opposite the mouth of the Khabur River, one of the principal routes to the obsidian sources in Anatolia. Thus, one reason for the agglomeration at this site may have been its location; if people had wanted to settle in this region, Bouqras was the best place.

The village plan was revealed partly by scraping the surface of the site where house walls were exposed and partly through excavation. Houses throughout the site were essentially similar: rectangular structures, ranging in area from 50–140 m^2, with a quarter or more of that area taken up by open courtyards. The remainder of the spaces comprised small rooms or cubicles, some of which were for storage. The buildings have a tri- or quadripartite layout but considerable variability in the way spaces were divided (Figure 1i,j). It appears that the buildings were laid out to full size from the beginning, and any subsequent changes were simply modifications to the interior; there are no auxiliary rooms tacked on to the outsides. Most of the buildings are separated from their neighbors by only a narrow corridor and are entered from one side only. Larger open areas or streets are devoid of structures. The overall impression of the site plan, based on exposure of perhaps fifty buildings, is one of monotony for none of the structures stands out as being unusual or greatly different from its neighbors except in overall dimensions. Large or small, each building seems to have had the same basic structure and function as a domestic and storage unit.

Bouqras was composed of individual households, each of which seems to have held its property separately in one of the house compounds. The excavators remark that later structures were circumscribed by extant buildings so that continuity in size, form, and location was imposed by the history of each building (Akkermans et al., 1983:343). These facts argue for the simultaneous rather than sequential use of the buildings and hence for a larger, rather than smaller, total population. It is not clear how one should estimate the number of people resident, but inasmuch as much of the site has not been revealed there may have been two or three times as many houses as presently known. A population of 850, assuming five persons per house, is not unrealistic, and a considerably greater number of residents seems quite possible. With this population size it would not be unexpected to find signs of social differentiation. However, there is no hint of ceremonial or other integrative structures or evident signs of differential wealth or access to goods. In short, this large community appears to have been essentially the sum of its interchangeable parts, each of which held its property within the walls of its compound. It would be interesting to discover whether similar compound structures would be found in isolation or in small hamlets, or whether this household form is a function of an irrigation economy as Bernbeck has asserted for Tell es-Sawwan (Bernbeck 1995).

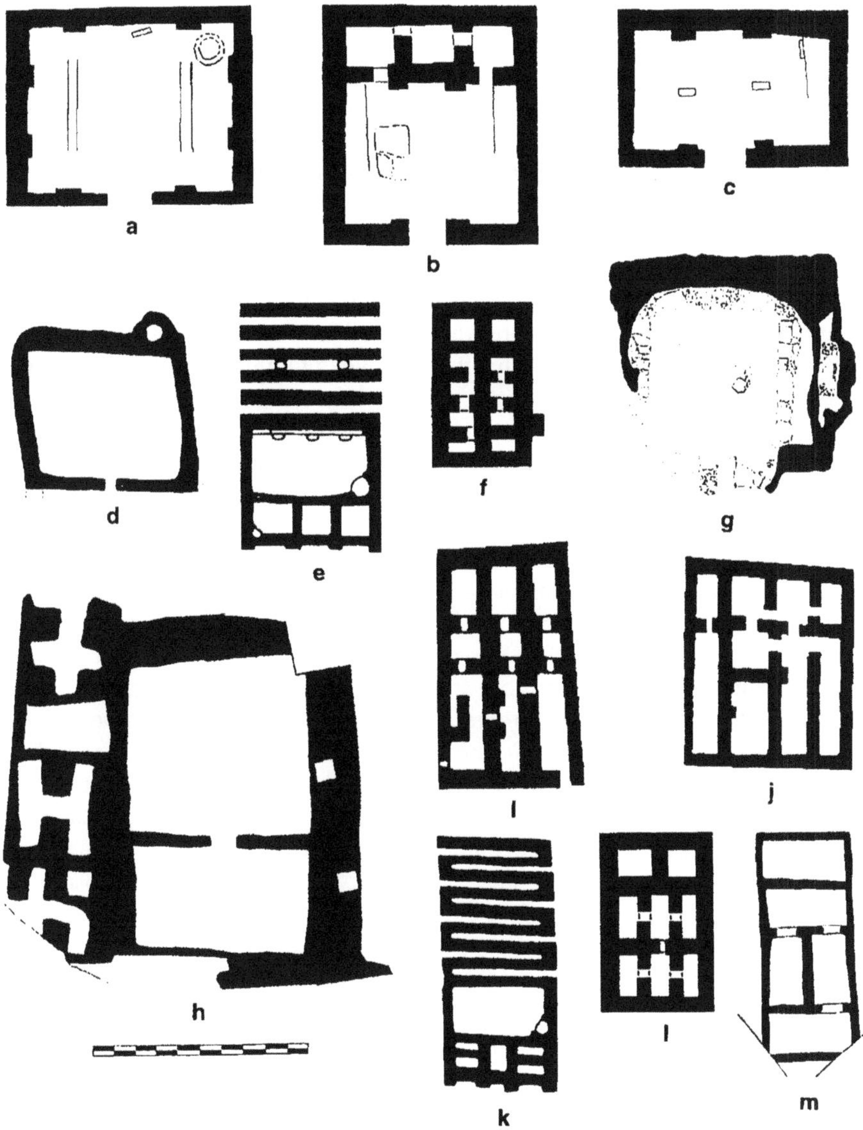

Figure 1. Nondomestic and domestic architecture at several Near Eastern sites: (Çayönü Tepe): (a) Terrazzo building, (b) Skull building, (c) Flagstone building, (e, f, k, l) domestic grill and cell plan buildings (based on Schirmer 1990); Beidha: (d) Building C1, h, Building C8, with attached basements of corridor buildings (based on Byrd 1994); (based on Schirmer 1990); Nevali Çori: (g) Building III, showing stepped entryway, interior bench, and stela; Bouqras: (i, j) domestic buildings; Abu Hureyra: (m) Neolithic house (based on Moore 1985).

Abu Hureyra

Upstream some 200 km from Bouqras, was Abu Hureyra. At 11.5 ha, this was the largest pre-pottery settlement known from Syria during the span 9,600–9,200 bp. Andrew Moore reports finding mud-brick houses of similar style across the entire mound. He describes them as having "several rooms and were separated from each other, even if only by narrow passages and courts. The rooms were always small, no more than from 1.4 to 2 m. wide and 3 to 4 m. long." Some houses may have had lofts supported by poles socketed into the tops of the mud-brick walls (Moore 1975:60). He notes that the settlement was densely crowded and may have held many thousands of inhabitants, with a consequent need for some form of "community government to regulate not only affairs between families but also to oversee the well-being of the community as a whole, to apportion rights of land around the settlement and even to arrange simple irrigation works" (Moore 1981:450). Other contemporary sites along the Euphrates were much smaller, and Moore raises the interesting possibility that Abu Hureyra may have been a "regional center" (Moore 1975:69).

The preliminary reports on this site do not differentiate between buildings that date to the various periods, but only illustrate what we must assume are typical houses of the "Neolithic 2" or latest PPNB (a span he estimates to cover 9,600–8,000 bp). Some of these houses are smaller and have a simpler layout than those from Bouqras. The only complete house plan illustrated from Abu Hureyra shows a structure with five "rooms" arranged in a long rectangle (Figure 1m) (Moore 1981: Figure 2c). Another structure has a sleeping platform and oven in one of its rooms (Moore 1981:Plates 1, 2), but most buildings appear to have lacked such features. On the other hand, small segments of architecture retrieved from other narrow trenches suggest the kind of room arrangements typical of Bouqras (Moore 1981:Figure 2a, b). None of the 162 burials examined displayed unusual signs of status, and some, in keeping with PPNB practices, lacked skulls (Moore 1975:61). In sum, the evidence presented reflects a stable community founded on agriculture (or intensive collecting of cereals, legumes, and pulses) and hunting that later incorporated domestic herds of sheep. As with Bouqras, the family residence was the module, although perhaps it housed fewer people, *if* the single house illustrated is typical. At this stage of our understanding there is no obvious indication of communal facilities, cultic activities, or status differences. We must bear in mind, however, that very little of Abu Hureyra was excavated.

The situation at Abu Hureyra may have been similar to that at Bouqras—a locally propitious place for a settlement at the edge of the steppe and the alluvial river plain. Abu Hureyra receives 50 mm more precipitation than

Bouqras, although such a total does not support predictable agriculture today. The bottom of a nearby wadi may have facilitated either agriculture or harvesting of wild crops and the site was in the migratory range of gazelles whose abundance sustained the inhabitants for well over 1,000 years. These two sites on the Euphrates would seem to argue for the importance of ecological opportunity in the establishment of large settlements during the Early Neolithic. Other known sites in the same general region are all very much smaller, and it remains to be determined whether environmental factors could account for the size differences.

Çayönü

Çayönü, in southeastern Anatolia, covers an area of 2–3 ha, about the same as Bouqras and much smaller than Abu Hureyra. The site was occupied for more than 1,500 years (*ca.* 10,300–8700 bp) during which some twenty-five building levels have been discerned; there is no implication that the entire site was ever occupied simultaneously. In no sense, then, can the site be said to be "large" in comparison with a number of other roughly contemporary sites. Nevertheless, Çayönü—unlike either Bouqras or Abu Hureyra—has a variety of architecture that surely indicates activities that are not simply domestic, as well as some differentiation among the domestic structures. Among the special buildings are a large rectangular building with a terrazzo floor; a "skull building" in which the remains of at least ninety individuals were discovered; and a building with a buttressed wall, paved with flagstones in which two stelae were set. These three structures differ in nearly all respects from the abundant and well-preserved houses (Schirmer 1988, 1990).

Let us first consider the houses (Figure 1e, f, k, l). These have been interpreted to consist of a series of changing styles that share approximately the same sizes and shapes and are built on foundations of stones (Özdögan and Özdögan 1989). Although the superstructures are missing, the plans suggest single-family dwellings that ranged in size from 20 to 60 m^2, not counting exterior spaces assigned to the house. These houses are roughly aligned in the site and stand less than 5 m apart. Although the floor areas are not unusually large by contemporary standards, one sign of social differentiation is seen during the cell phase (*ca.* 9,600–9,200 bp). At this time contemporary houses in the eastern part of the site, surrounding an open plaza of about 1,000 m^2 where the special buildings stood, are some 60% larger and better made than those on the western end of the site (Davis 1991). Only these houses in the plaza area possess skirting stone podiums, each is equipped with a porch, and some have stone pavements on exterior surfaces (Özdögan and Özdögan 1989:74). Michael Davis notes that only in these larger houses are certain "special" artifacts found, including chess-pawn-like stone objects, long obsidian blades, and two life-sized stone sculptures of featureless human heads. Further, all artifacts made of exotic materials are confined to this area of the site, including a huge obsidian nodule and large cores. Finally, in the cell phase, where more than 100 burials are known, three graves are exceptionally rich, containing large obsidian tools. In sum, the distribution, sizes, quality, and contexts of the houses on the east side all suggest greater wealth or prosperity.

The eastern part of the site also contains the special buildings (Özdögan and Özdögan 1989). The oldest of these, the "flagstone building," has a breadth of 10.7 m and is approximately 5.5 m deep (Figure 1c) (Schirmer, 1990:378). The north wall has two buttresses, and standing directly in front of these at a distance of 2 m are two upright stone slabs, perhaps roof supports. A third such slab stood in front of a "bench" on the east side. The floor of the room was paved with large and carefully fitted slabs of limestone. The room has approximately the same area as a house, but it has different proportions and lacks the underlying footings. No other structural or contextual information informs on the uses of this building.

The second structure, belonging to the "intermediate period," is called the "skull building" because the remains of as many as 400 individuals have been discovered there, including some ninety skulls (Figure 1b) (Schirmer 1990:378–382). Apparently this building underwent at least two reconstructions. The first is a crudely built, double-walled apsidal structure, with skulls on the floor and a deep pit in which there were many human skeletons as well as Bos bones. A second skull building was constructed next to the remains of the apsidal structure by first digging a deep trench into which the "cells" were placed. The cells were covered with flagstone lids and two upright slabs were placed opposite them. A pit dug into the sunken floor contained Bos horns and bones. Completing the picture is an interior bench surrounding the extant east and west sides of the room. The final skull building saw the construction of chambers above the cell pits, with buttresses replacing the now-covered upright stelae. The floor now had a coating of lime plaster and a carefully polished large limestone slab was placed on the floor. As before, the chambers held skulls and bones.

One is not engaging in unwarranted speculation to assert that the skull building played an important role in the funerary rites of the community. A clue as to the nature of these rites has emerged from analyses of blood residues taken from the stone slab that lay on the floor. This slab has traces of blood from three species: humans, cattle, and sheep. The evidence is unequivocal. Further possible evidence is a flint dagger found lying in this room, which also had blood traces of Bos and human blood (Loy and Wood 1989). It is probable that sacrifices or operations were performed either as part of the funerary ritual or on other occasions. Since skulls were routinely detached from postcranial elements, it may be that the human blood on the slab resulted from postmortem decapitation, and it seems that on some occasions Bos heads also were severed. As we shall see, there is a similar slab feature at Beidha in southern Jordan.

The final special-purpose building, also in the same area of the site, is the terrazzo building, known from its extraordinary floor surface that has remained largely in tact for some 9000 years (Figure 1a). This structure measures 11.75 × 9 m and is thus the largest of the special buildings. Thin buttresses decorate the interior walls, and the floor is carefully laid red terrazzo with two sets of parallel lines composed of white limestone chips all set into lime plaster. After its laying the floor was carefully polished. In one corner is a round depression that was not covered with terrazzo on the stone rim of which have been found traces of human blood. Blood also occurred on a limestone slab on which a human face is carved in relief.

The similarity among these buildings, as well as their differences from houses, and their location in one part of the site suggest that they had similar functions having to do with community-wide rituals or other activities. That these rituals may have been directed by residents in the larger houses is a reasonable inference.

Nevali Çori

Nevali Çori, a PPNB site in Anatolia, is another example of a small site with unusual architecture. Situated on a spur of land flanked by two wadis and a flowing tributary of the Euphrates, Nevali Çori occupies only about 0.7 ha, yet it contains some of the most interesting structures of the PPNB. Most of the site is composed of rectangular houses set on stone footings similar in style to the intermediate forms at Çayönü. The houses are roughly 12 × 6 m, closely spaced, and oriented in the same way. As at other contemporary sites, detached skulls were found. In the absence of a published plan, it is not clear how much open area there may have been, but our interest today is with the special nondomestic buildings.

Building II is 13.9 × 13.5 m, measuring from the outside of the walls, but with a trapezoidal-shaped interior space measuring 9.15 × 8.40 × 9.20. A bench built of stones and covered with large flat slabs lined the three walls facing the central entryway. This paved entry, which descended two steps to the floor, was flanked by massive stone slabs. Some thirteen stone pillars and sculptures once stood on the encircling bank, some set into sockets, possibly to support the roof. These pillars were 2.3–2.4 m tall and 40 × 50 cm in cross section. A niche, 1.85 m wide × 2.5 m deep, is centered in the southeast wall. A pedestal may have occupied the rear portion of this niche.

Another niche/podium, flanked by T-shaped stelae was set into the corner to the right of the entry. The inner walls have traces of red and black paint and the floor is terrazzo.

Building III, built above Building II, is nearly identical in size (Figure 1g). The previous large niche was covered so that the room was square with benches again on three sides. Again the roof was supported by pillars, some of which were T-shaped, spaced about 2.5 m apart. Two such pillars flanked the entryway, which stepped down 18 cm. A small niche was directly opposite the entrance and set into the wall behind the bench. Two broken sculptures, one of a human head with a snake in high relief on the back (Hauptmann, 1993:Plates 14, 19, 20), and a fragmentary torso (Hauptmann 1993:Plate 22) came from this area. Centered in the room were two additional stelae, again possibly serving as roof supports, one of which had a stylized human form in relief (Hauptmann, 1993:Plate 16). Unfortunately the tops of these stelae were eroded and broken, probably from postabandonment exposure. A "mixed" human/bird sculpture also came from the fill of this building (Hauptmann, 1993:Plate 25). Other bits of sculpture in the site may have been dispersed from earlier buildings during later construction, after abandonment, or possibly deliberately mutilated and buried (Garfinkel 1994).

Further evidence of spectacular remains at small sites occurs at several southern Anatolia sites, such as Göbekli Tepe, Gürcütepe, and Asikli Tepe. Common features are large rooms with paved, sometimes painted and polished plaster floors, orthostats, and realistic sculptures, all of which suggest cultic activities (Esin 1993; Schmidt 1995). These sites make clear that special functions and "monumental" constructions occur even in small sites. The labor involved in moving stone slabs that weighed up to one ton (Carlson 1985) or the burning of enough lime to cement a 30-cm-thick terrazzo floor (Kingery et al. 1988) and its subsequent polishing go well beyond the ability of the individual family. Similarly, the labor and skills required to manufacture stone bowls and carve stelae imply that specialists were at work (Yoffee 1993). Thus, despite their small size and the prevailing similarity of the *domestic* architecture in a number of these Anatolian sites, they also show evidence of social differentiation, a focus on some kind of "cult of the dead," and nonresidential buildings.

'Ain Ghazal

'Ain Ghazal, at 12–13 ha, is one of the largest spreads of PPNB in the Levant. Situated just outside Amman, Jordan, the site has been bifurcated by a major highway and irreparably damaged by agriculture, housing developments, and a sewage treatment plant. Working around these intrusions, Rollefson has demonstrated that the site was occupied and continued growing in size throughout the PPNB and into the PPNC (*ca.* 9,200–8,000 bp) (Rollefson 1992; Rollefson et al. 1992). These excavations have revealed large numbers of burials, many decapitated, but so far no structures like those in Anatolia have been recovered. However, in the Late PPNB (8,400–8,000 bp), two nearly identical, successively built "shrines" have been recovered (Kafafi and Rollefson 1995; Rollefson and Kafafi 1996). One is a circular structure, 2.5 m in diameter, with eight successive red-painted lime-plaster floors. In the center of this floor is a circular hole from which two pairs of subfloor channels radiate. The second example is some 4 × 5 m with oblique-angle room corners. In the center are three stelae-like stones standing in a line, behind which is a possibly anthropomorphic orthostat. Also in the room are a stone bench and a circular hearth of red-painted plaster surrounded by seven flat stones. Clearly "ceremonial" in nature, these structures may parallel the off-site building at Beidha (see below), but their size would seem to preclude community-wide public ceremonies. Again, it may be just a matter of luck that larger, more evidently public facilities have not been discovered in this, the largest of the Levantine sites. Rollefson sees these shrines as centers of "a lineage or clan cult with periodic religions rites." No clue to the nature of either the rites or these structures is given by the most spectacular of remains from the site, the plastered statues. Unfortunately these were buried in caches under the floors of abandoned houses, as were similar statues at contemporary Nahal Hemar and Jericho, thus depriving us of potentially fruitful insight into PPNB ritual practices.

Beidha

A final example comes from near the southern limits of the PPNB, the tiny site of Beidha, south of the Dead Sea in Jordan. Its architectural development has been conveniently and incisively summarized recently by Byrd, who focuses on what he sees as a shift from communal to individual family sharing and the development of mechanisms for integrating the community—in other words, mechanisms to bind potentially fissionable elements (Byrd 1994, n.d.). His thesis is that "architecture organizes, regulates, and delimits contact between individuals and households, especially through the ability to create public and private space" (Byrd, 1994:643). Specifically he relates restriction of physical and visual access to households as indicative of their exclusionary tendencies, whereas the size, plan, quality of construction, and location within the community reveal their public functions.

The site itself is only about 0.1 ha, by any standards a small community, yet there were twenty-four buildings in the uppermost phase, apparently the entire settlement. Only a third as much area was excavated in the oldest phase, yet nineteen buildings were recovered, so the sample from this site is unusually good by typical archaeological standards. Byrd distinguished three classes of buildings—domestic, storage, and nondomestic—and tracked their changes through the three building phases. In essence, the community went from an agglomeration of small huts with easy access, many open spaces, and a small nondomestic structure in phase A, to a densely packed settlement of two-story houses with greatly restricted access and a large public building in phase C (Figure 1h). (Kirkbride shows this as a house surrounded by storage/workshop units (Kirkbride 1960:Figure 2).) The seven public buildings were apparently occupied successively and were all situated in the center of the settlement, opening onto a courtyard (Figure 1h). The public buildings were always at least twice as large as any domestic structure, had very large central hearths, and lacked domestic artifacts (Figure 1d). The floor of the final such building was plastered five times, requiring a metric ton of quick lime each time.

Some 40 m off-site was another series of nondomestic buildings, floored with stone slabs, cobbles, or gravel. One was walled with large stone slabs and had a "massive stone-slab basin, approximately 3.0 × 2.2 m in size." One building also had "a very large, raised stone-slab platform and a large rectangular stone monolith" (Byrd 1994:657). It is likely, as Byrd surmises, that these structures are related to burial ceremonies, analogous to the skull building at Çayönü or the large structures with stelae at Nevali Çori.

CONCLUSIONS

The general question is whether sites with unusually large spatial extent, and presumably large populations, hold "public" facilities that are similarly scaled. I have investigated this question with reference to the few sites whose excavations have revealed a class of nondomestic architecture ordinarily interpreted as "shrine," "temple," or, more ambiguously, "public." Although the sample of sites is not large, it encompasses the full range of site sites and is sufficient to answer my initial question. Nevertheless, a number of caveats must be raised. In the first place, most excavations have not been extensive enough to reveal whether sites have special structures. Perhaps it was only chance that none were revealed at Bouqras or Abu Hureyra. Moreover, it is also possible that really large, central facilities at some sites have been missed entirely, or that they are off-site where large groups could gather. One should also reiterate that the size of most sites at any moment cannot be determined with accuracy and, to a great extent, the absolute size of a site in the PPNB must have more to do with availability of local resources than with regional political or economic interactions.

The evidence presented suggests that PPNB sites typically possess one or more nondomestic structures that could have functioned as places for ritual or cultic activities or just social events. The southwestern kiva, a structure that serves as mens' meeting house, storytelling lodge, repository of sacred material, and place where spirits emerge from the earth, models one possible analog. The ubiquitous sheik's house in the Near East today models a somewhat different set of functions, primarily secular: community reception hall, residence of the leader, and so on. Both of these alternatives are plausible for the PPNB, although the structures seem to lack any domestic function.

The buildings at Nevali Çori, at approximately 9 m on a side, with perimeter benches could have seated thirty-five to forty people at a time. In other words, a large proportion of the community, if not everyone, could have been accommodated. The structures at Beidha and Çayönü may also have served as community halls for they are about twice the size of domestic buildings and were well paved. In contrast, the structures of 'Ain Ghazal, less than half as large as at Nevali Çori, but similar in size to hundreds of southwestern kivas, seem more appropriately sized to accommodate individual households or at least small groups.

Whereas the function of most of the buildings cannot be determined easily, the skull building at Çayönü as well as the many decapitated and plastered skulls found in PPNB sites confirm that the so-called burial cult was pervasive. At Çayönü a large proportion of the population was placed within the burial vaults, and blood residues on the floor of the room suggest that it was used for the preparation of bodies for burial and beheading. Inasmuch as detachment of skulls was a widespread, if not universal, practice during the PPNB, we should expect that buildings or off-site charnel houses at other sites functioned in a similar way, even if bones were not stored in vaults. Thus, despite the common practice of burial beneath houses, where the skull cult was practiced, it seems likely that it was carried out in one of the special buildings. It remains to be determined, perhaps through residue analysis for blood, whether all the buildings that I have reviewed served in the same way.

Domestic structures offer few clues as to the organization of PPNB society. There is no appreciable differentiation among residential structures other than overall size, but there is a tendency for the larger sites and larger houses to hold more evidence of exotic goods than the smaller ones, as at Çayönü. There seems little doubt that the PPNB sites were organized around the household, as a working, economic, and social unit. At some sites, such as Bouqras, the houses were more complex than at sites like Abu Hureyra. This may imply that larger groups composed the household at Bouqras, but in each site all houses look structurally like all others. This leads to the conclusion in all cases that households held property separately from the community, and such integrative activities as took place involved burial rites and perhaps social events, not community storage.

It is hard to find convincing evidence for leadership in any of the villages despite some differences in the sizes of houses. Clearly extraordinary effort was put into the construction of some of the special structures, as manifest in the terrazzo, plaster, and carved stelae at Çayönü, Nevali Çori, 'Ain Ghazal, and other sites, but local artisans had already developed the requisite skills during the building of their residences. I see nothing in the PPNB that would have required a large or highly skilled labor crew.

What then can we say about the overall organization of the PPNB world? To date, despite gross differences in size and even in apparent quality of constructions, there is no convincing evidence that any site served the function of a political or economic center. Rather, each site looks pretty much like all others, including having special structures. It is hard to imagine what a "center" would have done, but one should not dismiss the effect of sheer size on other contemporary settlements. At the least the largest sites must have had a substantial buffer zone to allow for their residents to extract resources without competition from outsiders. Any site, large or small, might have controlled access to a scarce commodity such as high-quality flint for the making of naviform cores. 'Ain Ghazal may have controlled such a source, although it would be hard to find evidence that it did in fact limit access (Quintero 1997).

In conclusion, the concept of site "hierarchy" may not be applicable to PPNB sites, despite manifest differences in size. At least in the PPNB, size of site does not seem to have been a signal of complexity or, in itself, a determining factor of specialization or complexity. In one of this characteristically broadbrush essays, Robert McC. Adams considered that archaeologists' tendencies to construct normative classifications of sites belies the variability that is inherent, particularly as such communities are perceived archaeologically. As he put it, "any social reality involves variable rather than standardized units" (Adams 1988:5). He refers to ways that communities may become differentiated and stresses that when advantage accrues to one community, it alters its relationships with other, "generating conflict, exploitation and enhanced asymmetries of power as well as indeterminancies of outcome" (Adams 1988:6). Nevertheless, asymmetries are transitory for any number of social-economic-political reasons, yet the physical communities may remain largely intact. An older village may be large but have lost its "power," whereas a smaller village with a rising leader may have become the center that in time would possibly outstrip others in size if the many internal and external factors permitted continued growth.

A further consideration regarding indeterminancies is the possibility that there was significant regional and temporal variation in the architectural expression of ceremonial and social activities. The sites in Anatolia have the best preserved and most exceptional features, whereas the sites on the Euphrates have so far yielded none. The sites in the southern Levant show more modest examples in the Anatolian pattern. The reported structures are Middle to Late PPNB. Does this imply that such facilities arose to meet changing needs as the PPNB progressed, perhaps, as Rollefson has argued, in the face of environmental depletion (Rollefson and Köhler-Rollefson 1989, 1992)? Until there has been fuller excavation and reporting, the possible implications of these sources of variability remain obscure.

In the meantime, while there is clear evidence for nondomestic activities that appear to be strongly ritualistic at many sites in the PPNB, apart from the mortuary cult, we are unable to specify what these rituals may have involved or who participated in them. Equally clear is that such rituals were normal components of PPNB society and they may have been universal in settlements. At this time there is no reason to think that large sites were organized differently from small settlements: the module throughout may have been the household whose activities were coordinated at the settlement level by their heads, meeting in "shrine-like" buildings.

REFERENCES

Adams, R. M., 1988, Introductory Remarks: Spatial and Temporal Contexts. In *Conceptual Issues in Environmental Archaeology,* edited by J. L. Bintliff, D. A. Davidson and E. G. Grant, pp. 1–15. University Press, Edinburgh.

Akkermans, P. A., Boerma, J., A., K., Clason, A. T., Hill, S. G., Lohof, E., Meiklejohn, C., le Miere, M., Molgat, G. M. F., Roodenberg, J. J., Waterbolk van Rooyen, W., and van Zeist, W., 1983, Bouqras Revisited: Preliminary Report on a Project in Eastern Syria. *Proceedings of the Prehistoric Society* 49: 335–372.

Bar-Yosef, O., and Belfer-Cohen A., 1991, From Sedentary Hunter-Gatherers to Territorial Farmers in the Levant, In *Between Bands and States,* edited by S. A. Gregg, pp. 181–202. Southern Illinois University, Carbondale, Illinois.

Bernbeck, R., 1995, Lasting Alliances and Emerging Competition: Economic Developments in Early Mesopotamia. *Journal of Anthropological Archaeology* 14:1–25.

Boerma, J. A. K., 1983, Environmental Conditions. *Proceedings of the Prehistoric Society* 49: 362–365.

Boerma, J. A. K. 1989–90, Palaeoenvironment and Palaeo Land–Evaluation Based on Actual Environment Conditions of Tell Bouqras, East Syria. *Anatolica* 16:215–249.

Byrd, B. F., 1994, Public and Private, Domestic and Corporate: The Emergence of the Southwest Asian Village. *American Antiquity* 58(4):639–666.

Byrd, B. F. n.d., The Neolithic Village of Beidha: Architecture, Occupation History and Spatial Organization. In *The Excavations of Mrs. Diana Kirkbride-Helbaek,* by British Academy Monographs in Archaeology. Oxford University Press, Oxford.

Carlson, B., 1985, The World's First Formal Architecture? *University of Chicago Magazine* (fall):8–11.

Davis, M., 1991, *Social Differentiation at the Early Village of Çayönü, Turkey.* Paper presented at the Annual Meeting of the American Anthropological Association, Chicago, Illinois.

Esin, U., 1993, Zum Ursprung der Kastenbauweise in Anatolien. Instanbuler *Mitteilungen* 43:123–128.

Flannery, K. V., 1995, Prehistoric Social Evolution. In *Research Frontiers in Anthropology,* edited by C. R. Ember and M. Ember, pp. 1–26. Prentice-Hall, Englewood Cliffs, New Jersey.

Fletcher, R., 1987, Settlement Archaeology: World Wide Comparisons. *World Archaeology* 18:59–83.

Garfinkel, Y., 1994, Ritual Burial of Cultic Objects: The Earliest Evidence. *Cambridge Archaeological Journal* 4: 159–188.

Hauptmann, H., 1993, Ein Kultgebäude in Nevali Çori. In *Between the Rivers and over the Mountains,* edited by M. Frangipane, H. Hauptmann, M. Liverani, P. Matthiae, and M. Mellink, pp. 37–69. Dipartimento di Scienze Storiche Archeologiche e Anthropologiche dell'Antichità, Università di Roma "La Sapienza," Rome.

Kafafi, Z., and Rollefson G. O., 1995, The 1994 Excavations at 'Ayn Ghazal: Preliminary Report. Annual of the Department of Antiquities of Jordan. 39:13–29.

Kingery, D., Vandiver, P., and Rickett, M., 1988, The Beginnings of Pyrotechnology, Part II: Production and Use of Lime and Gypsum Plaster in the Pre-Pottery Neolithic Near East. *Journal of Field Archaeology* 15:219–244.

Kirkbride, D., 1960, The Excavation of a Neolithic Village at Seyl Aqlat, Beidha, near Petra. *Palestine Exploration Quarterly* 92:136–145.

Kramer, C., 1994, Scale, Organization, and Function in Village and Town. In *Archaeological Views from the Countryside: Village Societies in Early Complex Societies,* edited by G. M. Schwartz and S. E. Falconer, pp. 207–212. Smithsonian Institution Press, Washington.

Kuijt, I., 1994, Pre-Pottery Neolithic A Settlement Variability: Evidence for Sociopolitical Developments in the Southern Levant. *Journal of Mediterranean Archaeology* 7:165–92.

Kuijt, I., 1996, Negotiating Equality through Ritual: A Consideration of Late Natufian and Pre-Pottery Neolithic A Period Mortuary Practices. *Journal of Anthropological Archaeology* 15(4):313–336.

Loy, T. H., and Wood, A. R., 1989, Blood Residue Analysis at Çayönü Tepesi, Turkey. *Journal of Field Archaeology* 16:451–460.

Moore, A. M. T., 1975, The Excavation of Tell Abu Hureyra in Syria: A Preliminary Report. *Proceedings of the Prehistoric Society* 41:50–77.

Moore, A. M. T., 1981, North Syria in Neolithic 2. In *Préhistoire du Levant,* edited by J. Cauvin and P. Sanlaville, pp. 445–456. Colloques Internationaux du Centre National de la Recherche Scientifique, No. 598. Editions du Centre National de la Recherche Scientifique, Paris.

Moore, A. M. T., 1985, The Development of Neolithic Societies in the Near East. *Advances in World Archaeology* 4:1–69.

Özdögan, M., and Özdögan, A., 1989, Çayönü, a Conspectus of Recent Work. *Paléorient* 15:65–74.

Quintero, L., 1997, *Evolution of lLthic Economics in the Levantine Neolithic: Development and Demise of Naviform Core Technology.* Ph.D. dissertation, University of California, Riverside. University Microfilms, Ann Arbor, Michigan.

Rollefson, G. O., 1987, Local and Regional Relationships in the Levantine PPN Period: 'Ain Ghazal as a Regional Center. In *Studies in the History and Archaeology of Jordan,* edited by A. Hadidi, pp. 29–32. Department of Antiquities of Jordan, Amman.

Rollefson, G. O., 1992, Neolithic Settlement Patterns in Northern Jordan and Palestine. *Stud-*

ies in the History and Archaeology of Jordan IV, edited by S. Tell, pp. 123–127. Department of Antiquities, Amman.

Rollefson, G. O., and Kafafi, Z., 1996, 'Ain Ghazal. Ten Seasons of Discovery. *ACOR Newsletter* 8:1.

Rollefson, G., and Köhler-Rollefson, I., 1989, The Collapse of Early Neolithic Settlements in the Southern Levant. In *People and Culture Change: Proceedings of the Second Symposium on Upper Paleolithic, Mesolithic and Neolithic Populations of Europe and the Mediterranean Basin,* edited by I. Hershkovitz, pp. 59–72. British Archaeological Reports, International Series 508. BAR, Oxford.

Rollefson, G., and Köhler-Rollefson, I., 1992, Early Neolithic Exploitation Patterns in the Levant: Cultural Impact on the Environment. *Population and Environment* 13:243–254.

Rollefson, G. O., Simmons, A. H., and Kafafi, Z., 1992, Neolithic Cultures at 'Ain Ghazal, Jordan. *Journal of Field Archaeology* 19:443–470.

Schirmer, W., 1988, Zu den Bauten des Çayönü Tepesi. *Anatolica* 15:139–160.

Schirmer, W., 1990, Some Aspects of Building at the "Aceramic-Neolithic" Settlement of Çayönü Tepesi. *World Archaeology* 21:363–387.

Schmidt, K., 1995, Investigations in the Upper Mesopotamian Early Neolithic: Göbekli Tepe and Gürcütepe. *Neo-Lithics* 2/95:9–10.

Wright, H. T., ed., 1984, *On the Evolution of Complex Societies: Essays in Honor of Harry Joijer.* Undeena Publications, Malibu.

Yoffee, N., 1993, Mesopotamian Interaction Spheres. In *Early Stages in the Evolution of Mesopotamian Civilization: Soviet Excavations in Northern Iraq,* edited by N. Yoffee and J. J. Clark, pp. 257–269. University of Arizona Press, Tucson, Arizona.

Chapter 9

Villages on the Edge

Regional Settlement Change and the End of the Levantine Pre-Pottery Neolithic

ALAN H. SIMMONS

INTRODUCTION

Archaeologists have studied the origins of food production in the Near East for over a hundred years. Since the inception of true interdisciplinary projects such as the Braidwood's pioneering Iraqi research (Braidwood and Howe 1960; Braidwood et al. 1983), our comprehension of the complex processes involved in this monumental adaptive shift have increased substantially. What is now emerging in some innovative contemporary research is systematic investigation not on the origins of domestic economies but on their consequences. In many ways, this is a more elusive topic than origin studies, since the archaeological signature of agricultural and herding impacts can be difficult to precisely define. Some researchers, however, have recently examined certain aspects of the profound social consequences of food production on Late Neolithic societies in the Levant, particularly at large sites such as 'Ain Ghazal in west-central Jordan. These investigations have shown that developments in Jordan differed considerably from those

ALAN H. SIMMONS • Department of Anthropology and Ethnic Studies, University of Nevada, Las Vegas, Nevada 89154-5012.

Life in Neolithic Farming Communities: Social Organization, Identity, and Differentiation, edited by Ian Kuijt. Kluwer Academic/Plenum Publishers, New York, 2000.

closer to the Mediterranean (Rollefson and Köhler-Rollefson 1992; Simmons 1995a; Simmons et al. 1988).

This research has demonstrated that the Late Neolithic was a turbulent time that witnessed major changes in many aspects of culture. While most studies have focused on the ecological and economic consequences of domestic subsistence strategies, it seems clear that social organization also must have been substantially affected. This, unfortunately, is difficult to directly demonstrate in the archaeological record. In this chapter I present some admittedly speculative scenarios that may have operated during this time, and discuss social as well as ecological and subsistence implications of these events.

THE NEOLITHIC IN THE LEVANT

Chronology and Phasing

Many of the Near East's best known Neolithic sites are located in the Levantine Near East, and there is a substantial radiocarbon-supported chronology for the region (Kuijt and Bar-Yosef 1994). Although there is disagreement on details of this sequence, most researchers concur on the general outline of the Pre-Pottery Neolithic (PPN) and the Pottery Neolithic (PN). As with the PPN, the PN consists of several phases (Garfinkel 1993; Gopher and Gophna 1993; Kafafi 1982; Stekelis 1973). In the past, many researchers (e.g., de Vaux 1966; Kenyon 1960; Perrot 1968) felt that the PN was temporally separated from the PPN by a gap of up to a millennium, a proposition supported at major sites. Furthermore, they believed the PN represented new and intrusive populations, filling a gap that developed with the cessation of the "PPNB Interaction Sphere" (Bar-Yosef and Belfer-Cohen 1989). Increased aridity at the end of the PPN often was cited as a major stimulus for this abandonment. Recently, however, this position has been challenged, due to better dating techniques resulting in a shrinkage of the presumed gap, more sophisticated understanding of paleoclimates, and research at newly discovered sites. Many researchers now believe that the PN was characterized by shifts in site location instead of regional abandonment and replacement by new populations (Banning et al., 1994; Gopher and Gophna 1993:304–307).

This argument has been strongly supported with the documentation of a transitional phase [the Pre-Pottery Neolithic C (PPNC)] at large, near "urban" Jordanian settlements, such as 'Ain Ghazal, Wadi Shu'eib, and possibly Basta (Muheisen 1995; Rollefson 1993; Rollefson and Köhler-Rollefson 1993; Simmons et al. 1988). At these sites an unbroken sequence from mid-PPNB (i.e., MPPNB) through PN is documented, with subsequent post-Neolithic abandonment. While the PPNC appears largely restricted to major Jordanian settlements, it may also occur at smaller sites further to the west in Israel (e.g., Galili et al. 1993; Garfinkel 1994). The documentation of the PPNC is important because it supports arguments for local in-situ development rather than abandonments and resettlement by "new" peoples whose material inventory now included pottery. It seems clear that cultural change during the end of the PPN and extending into the PN is much more complex than previously believed.

Bar-Yosef and Meadows have recently questioned the utility of the PPNC, noting that " . . . this phase is chronologically the same as the final PPNB in the northern Levant and the relevant entities in Anatolia, and using the label PPNC suggests a major cultural change where only minor shifts exist" (1995:73). This criticism misses the point in that the PPNC is, indeed, quite distinct. While the PPNC may chronologically overlap with terminal PPNB further to the west and north, it is a strikingly different material phenomenon from both the PPNB and subsequent PN. This distinct character is manifested in artifacts, economy, and burials. Furthermore, at the settlements of 'Ain Ghazal and Wadi Shu'eib, it is part of an unbroken occupational sequence, something rarely documented at other large sites further west or north (Rollefson 1989, 1993; Rollefson et al. 1992; Simmons et al. 1988, 1989).

The Levantine Pre-Pottery Neolithic

Whatever the specifics of individual phases and chronology, the PPN is a relatively well known archaeological entity, with well over 200 sites documented throughout the Near East (Bar-Yosef 1981a; Moore 1973, 1978). In the Levant during the PPNB, Jericho was clearly an important center, covering an estimated 10 acres (*c.* 4 ha) (Kenyon 1957). Although Jericho remains to many the "typical" PPNB site, newer studies have documented a large range of site types. These include small settlements such as Beidha (Kirkbride 1966, 1968), Ghwair I (Najjar 1994; Simmons and Najjar 1996, 1998), Nahal Oren (Noy et al. 1973), and Tel Ramad (de Contenson 1971; de Contenson and Van Liere 1966), to larger villages such as Tel Abu Hureyra (Moore, 1979; Moore et al., 1975) and Beisamoun (le Brun 1969). Recent excavations east of the Jordan River, in fact, have documented huge settlements greatly exceeding Jericho's horizontal extent. These include 'Ain Ghazal and Wadi Shu'eib (Figure 12.1). In fact, Bar-Yosef and Meadows (1995:76, Figure 3.7) and Kuijt (1995:Table 6.2) indicate that sizewise, Jericho tends toward the smaller end of the scale for Neolithic settlements.

In addition to villages, numerous specialized small sites are known throughout the Levant, with many of these located in marginal environmental areas (e.g., Bar-Yosef 1981a; Bar-Yosef and Phillips 1977; Betts 1990; Burian and Friedman 1973; Garrard et al. 1994; Noy 1970, 1971; Rollefson and Frohlich 1982; Simmons,1980, 1981). Most of these are open-air sites, although a cave (Nahal Hemar) containing well-preserved ceremonial PPNB materials has recently been investigated in Israel (Bar-Yosef and Alon 1988). Another cave, El-Khiam (Echegaray 1964, 1966), presumably contains PPNB materials as well. Not all of these sites may have been inhabited by agriculturists, and this variety of site types suggests that a wide array of economic pursuits was followed during the PPNB.

The larger PPNB villages contain all the "classic" elements that have interested researchers working with the Neolithic. These include architecture, abundant lithic materials, burials, ceremonial objects, such as the spectacular 'Ain Ghazal human statues, and economic data in the form of plant and animal remains. There seems little doubt that while PPNB society may have been essentially egalitarian, regional hierarchies between communities existed, as reflected in public structures, artifacts, burials, and site size (see Chapters 5–7 in this volume). Most researchers believe that these material remains reflect changes in social organization toward more complex communities where many decisions were made by extrafamily entities (Bar-Yosef and Meadows 1995:74–82). The operation of a PPNB "interaction sphere" (cf. Bar-Yosef and Belfer-Cohen 1989) indicates a social structure far more complex than what had previously existed.

THE LATE NEOLITHIC IN EASTERN JORDAN: A TENTATIVE SCENARIO

Collectively, archaeological research in the southern Levant has demonstrated a Late PPNB/Early PN cultural record east of the Jordan River that differs significantly from other regions (Rollefson 1989, 1994; Simmons 1995a, 1995b). This has some intriguing implications for the trajectory of social relationships, which we feel were tied to the dramatic economic transformations that highlighted the latter phases of the Neolithic, at least in the eastern Levant. At 'Ain Ghazal, for example, where interdisciplinary investigations concentrated on obtaining paleoeconomic data, a wide range of domesticated and nondomesticated resources characterized the PPN phases—over fifty species were exploited. By the subsequent PN phases, however, diversity was greatly reduced, down to only three major species (Köhler-Rollefson and Rollefson 1990; Rollefson and Köhler-Rollefson 1989).

In fact, data from many PN settlements illustrate a general deterioration in several aspects of material culture. Architecture becomes less substantial, chipped stone technology less sophisticated, and the artistic achievements of the PPNB are not matched (although see Garfinkel 1993:123–126 regarding portable art). Thus, one has an overall impression of PN life reflecting a generalized decline in the standard of living compared to the PPN (Kenyon 1960:60; Mellaart 1975:237–238; Perrot 1968:404–416). Even if Banning, Rahimi, and Siggers (1994:152–154) are correct in their assessment that the PN has been "devalued" by inadequate research and the low archaeological visibility of sites due to geomorphic processes, it still seems apparent that the considerable achievements of the PPNB were not carried over into this last Neolithic phase. Given these turbulent shifts, it is unlikely that extant social organizational systems escaped unscathed, although there has been relatively little direct research as to how it changed.

Recent investigations at the large settlements east of the Jordan River have modified our perception of Neolithic developments in the Levant. We now know that these large Neolithic centers were substantially different from smaller Neolithic settlements situated closer to the Mediterranean coast and that huge settlements such as 'Ain Ghazal and Wadi Shu'eib represented unique desert-edge adaptations (Simmons 1995b). These settlements were initially founded during the MPPNB, with earlier developments (such as PPNA) apparently occurring primarily to the west (although the PPNA sites of Iraq ed-Dubb and Dhra', and a PPNA site near PPNB Ghwair I, are exceptions (Bennett 1980; Kuijt 1995; Kuijt et al. 1991; Kuijt and Mahasneh 1998; Finlayson and Mithen 1999). This population and settlement shift may have been linked to ecological changes and the depletion of local environments, with eastward relocation of communities facilitating the partial exploitation of new areas adjacent to the eastern deserts. Large sites such as 'Ain Ghazal may have attracted populations dealing with these environmental crises, resulting in a dramatic population increase centered within a few nearly urban centers (Simmons 1995b).

This pattern is quite unlike that seen at Neolithic settlements further west, north, or south (see Bar-Yosef and Meadows 1995; Gopher 1994 for summaries). For example, many of the large settlements along the desert ecotone were occupied continuously from PPNB through PN phases and lack the presumed hiatus between aceramic and ceramic phases. Rather, an unbroken sequence exists, as reflected by the previously mentioned transitional PPNC phase. What is equally intriguing about these sites is that after the Neolithic they were abandoned, in stark contrast to Neolithic sites closer to the Mediterranean, where favored locales frequently were reoccupied after a hiatus from terminal Neolithic occupations (Simmons 1995b).

It is useful to characterize these large Jordanian sites. They share several general attributes. First, they all are large. 'Ain Ghazal, for example, covers an estimated 25–35 acres (*c.* 10–14 ha) during the PPNB/PPNC

(Rollefson et al. 1992:444–446), making it roughly three times the size of Neolithic Jericho. Wadi Shu'eib and Basta also exceed Jericho's horizontal dimensions. Although these sites are huge, they are not tells in the proper sense. They do not appear to contain deposits as thick as Jericho's, although test excavations at Wadi Shu'eib have demonstrated a depth exceeding 8 m. These settlements also do not contain early phases of the aceramic Neolithic. That is, there is no PPNA present, and even the PPNB deposits date from the middle to late range of that phase. Many, however, contain the transitional PPNC phase. These sites all appear to be abandoned after the Aceramic Neolithic. This differs from some of the western sites, including Jericho, which was successively reoccupied, even after periods of abandonment. Unlike Jericho, massive architectural features (e.g., large towers or "city walls") are not found at settlements east of the Jordan River. This does not mean that these were "substandard" or impoverished towns, as the material culture at the sites is extraordinarily rich. For example, nothing in the Near East has paralleled the exquisite 'Ain Ghazal statues (Tubb and Grissom 1995). Many of these sites are located in relatively fragile ecological zones, or at least adjacent to what might be termed marginal environments. As such, they just barely occur in what Bar-Yosef and Meadows (1995:73–74) refer to as the "Levantine Corridor," a relatively favorable environment. 'Ain Ghazal, for example, is in a somewhat precarious ecological setting. Despite its proximity to a major drainage (the Wadi Zarqa), it is situated along the minimum (250 mm) rainfall isoheyt for nonirrigation agriculture, and desert environment is rapidly encountered just to the east. This may have been a crucial factor for peoples who subsisted to a large degree on domesticated plants. We can observe generally similar patterns at the other large Neolithic sites in Jordan, illustrating the importance of selected microhabitats along desert ecotones.

This patterning stands in stark contrast to what occurred elsewhere during the PPNB and challenges us to explain this important regional shift. As an exploratory model, the following interpretative scenario may be proposed. If initial south Levantine PPNA development primarily occurred west of the desert ecotone, the natural population increase associated with Neolithic economies could ultimately have depleted local microenvironments and stretched carrying capacities to their maximum. A subsequent adaptive response may have been to expand further east, to the edges of where a "traditional" Neolithic economy could still be practiced. This does not mean the western Levant was abandoned; clearly it was not. I do not believe that all MPPNB and LPPNB populations moved to the desert fringes because the existence of large sites west of the Jordan River, such as Beisamoun, belie such a simplistic explanation. Furthermore, I do not believe that "abandonment" at sites such as Jericho was necessarily immediate and complete. Certainly the PPNB component at Jericho partially overlaps with that of 'Ain Ghazal, for example. Given that a partial population reshifting could have eased pressure in the western region, the "gap" between Aceramic and Ceramic Neolithic phases may be linked to this event and represents a readjustment to new economic parameters. Large sites such as 'Ain Ghazal, for example, may have served as population "magnets" from places like Jericho (see also Kuijt 1995). The consequence of this was that the former expanded at the expense of the latter, perhaps resulting in temporary abandonments at some of the western Levantine sites.

A move to the east was not without difficulties, however. Such regions were more environmentally precarious. One adaptation to this ecological constraint may have been the forced consolidation of populations into larger settlements, such as 'Ain Ghazal or Wadi Shu'eib. This could have allowed for scarce resources to be pooled. Such a situation would have created social organization pressures previously unfelt but may have had the advantage of requiring more cooperation and the development of more efficient exploitation technologies and land use patterns focused on regional population centers. Certainly around sites such as 'Ain Ghazal, there is little evidence for the existence of smaller "hamlets" or support sites (Simmons and Kafafi 1988).

This does not mean, however, that a similar situation existed in all areas of the southern Levant. In the southern portion of Jordan, where one might argue for an even more marginal environment, it seems that a wider range of site types existed. These included huge settlements such as Basta (Muheisen 1995; Nissen 1990) and Es-Sifyia in Wadi Mujib (Mahasneh 1995), to more modest villages such as Beidha (Kirkbride 1966, 1968), Khirbet Hamman (Rollefson and Kafafi 1985), and several in the Wadi Feinan (Adams 1991; Raikes 1980), including the superbly preserved Ghwair I (Najjar 1994; Simmons and Najjar 1996, 1998). In addition, several nonvillage Neolithic sites are known, such as those in the Hisma Basin (Jobling and Tangri 1991; Vianello 1985), the adjacent Wadi Hafir (Borzatti von Löwenstern 1984), the Wadi Rumm (Kirkbride 1978), and between Aqaba and Ma'an, where nearly twenty stone circle sites datable to the PPN have been recorded (Jobling and Tangri 1991:147). Clearly, very different events were occurring in the south.

Thus, the model I have developed for population aggregation is more applicable to west-central Jordan, where the diversity in Neolithic site types does not appear to be as great as it is further to the south. Confirmation of this, however, will have to await further research, since most of the southern sites (excepting Basta and Beidha) have not been thoroughly investigated (cf. Rollefson and Kafafi 1985:69). In particular, refinement of the chronological sequence is critical in order to evaluate settlement shifts thorough time during the Neolithic in the south. It is interesting to note, however, that at least some of the large southern sites appear to contain PPNC components. PN sites, however, are rare in southern Jordan, and most of those known appear to be related to either the Jericho IX PN or the Qatifian variant of the Wadi Raba PN (cf. Gilead 1990; Gopher and Gophna 1993:318, 337; Goren 1990). It is apparent that the entire range of Neolithic occupation, from PPNA through PN, occurs in southern Jordan. As with central Jordan, however, only limited evidence exists for the earliest (PPNA) occupation. What is particularly intriguing about the south is that both villages and more limited activity sites occur during the PPNB. This has prompted some authors (e.g., Jobling and Tangri 1991:147) to suggest that the PPNB adaptation here was one characterized by seasonally mobile hunters and gatherers, as suggested for similar sites in the Sinai (e.g., Bar-Yosef 1981b; Tchernov and Bar-Yosef 1982). Many such sites, however, may also represent logistical camps away from villages or pastoral activity. Garrard et al. (1996:221) note that may of the Late Neolithic inhabitants in the more arid portions of the southern Levant made use of a "pastoral package" that " . . . would have provided a useful risk-buffer for those engaged in marginal farming and hunting." Certainly the identification of large PPNB (e.g., Basta, Es-Sifyia) and small (e.g., Ghwair I) villages clearly does not support such a model. In any event, it is clear in the PPN/PN of the southern Levant that population movement was dynamic, complex, and must be interpreted in a regional context. It is equally clear that the huge settlements, such as 'Ain Ghazal and Wadi Shu'eib, changed dramatically at the start of the PPNC and PN and were, ultimately, abandoned. It is useful to examine some possibilities as to why this may have occurred.

PALEOECOLOGICAL PERSPECTIVE

Although there is little direct evidence that preagricultural populations in the Levant had a serious effect on the landscape, by the PPN there are considerable data to suggest that human actions drastically, if not irreversibly, impacted the local/regional environment (cf. Hershkovitz 1989; Köhler-Rollefson 1988; Köhler-Rollefson and Rollefson 1990; Simmons et al. 1988). It was during this time that we see the first settled villages and exploitation strategies based on agriculture, herding, and wild resources. With the domestication of plants and animals, human settlement on the landscape changed dramatically. Debate continues to exist, however, as to whether these activities were the principal drivers behind environmental decay or whether this deterioration was caused by climatic variables in the form of increasing aridity. Köhler-Rollefson and Rollefson (Köhler-Rollefson 1988;

Köhler-Rollefson and Rollefson 1990; Rollefson and Köhler-Rollefson 1989, 1992) clearly prefer a model where local environmental degradation and settlement abandonment were culturally induced. However, the role of climatic change must be seriously considered. Certainly, climatic variations have long been invoked by archaeologists as being linked to cultural change in the Near Eastern Neolithic. Most models, however, focused on climatic variation as a device for stimulating agricultural origins (e.g., McCorriston and Hole 1991; Moore and Hillman 1992) and have not explored how this might have impacted later Neolithic developments.

We have examined this issue elsewhere, suggesting that these cultural developments occurred against a backdrop of steadily deteriorating climatic conditions (cf. Davis et al. 1990; Simmons 1995a). Based on regional archaeological, climatic, and geological data, we believe that the combination of drought in the Levant, ironically coupled with increased precipitation brought on by summer monsoons (cf. Kutzbach and Guetter 1986; Kutzbach et al. 1993; Street and Grove 1979), the impacts of expanded human population, intensive agriculture and herding, and deforestation for fuel caused an environmental crisis that has dominated human adaptation in the region ever since (Davis et al. 1990; Simmons 1995a). Specifically, by *ca.* 9,000 bp, a Mediterranean climatic regime apparently was well established in the Levant (Roberts and Wright 1993:201). At roughly the same time, however, summer precipitation may have declined and ceased altogether by about 7,000 years ago. Importantly, this occurs simultaneously with increased human-induced environmental degradation brought about by intense farming and herding and the eventual abandonment of major PPN/PN centers. It is important to realize, however, that increased precipitation, as proposed here, does not necessary translate into increased effective moisture. It is thus possible to have short-term and intense seasonal precipitation in conjunction with overall increasing aridity.

We have suggested that loss of agricultural productivity may have begun with erosion accelerated by torrential summer rains (Davis et al. 1990). Previously forested areas cleared for fields, pastures, fuel, and settlements during the PPNB would have been especially prone to erosion by the PPNC/PN. Possible evidence for the erosion occurs at many of the Jordanian sites as layers of comparatively well sorted cobbles. These may be attributed to debris flow caused by torrential summer thunderstorms, the physical manifestation of the increased monsoonal pattern. Erosion could have been increased if surrounding vegetation had been reduced by human activity and increasing aridity.

This constellation of cultural and climatic activities had substantial ecological consequences, ultimately straining the landscape. This was particularly acute at sites such as PPN/PN settlements situated in ecologically fragile zones. While the pattern of population aggregation and pooling of scarce resources were initially adaptive, they ultimately began to deplete the environment. This resulted in critical resource shortages that may have required an economic split between farmers and pastoralists, and resulted in the abandonment of the large PPN/PN communities and led to the establishment of the famed Near Eastern dichotomy between the "desert" and the "sown," or between village farmers and pastoral nomads (Köhler-Rollefson 1988, 1992; Köhler-Rollefson and Rollefson 1990; Rollefson and Köhler-Rollefson 1989; Simmons et al.1988).

A SPECULATIVE MODEL OF SOCIAL IMPLICATIONS

Although speculative, consideration of the interplay between population oscillations, environmental change, and regional settlement shifts aids us in understanding social changes that may have occurred between the PPN and PN. These can be viewed as having occurred in at least three stages, each of which will be discussed.

Stage 1: Aggregation into Large Regional Centers

The florescence of the Neolithic was during the PPNB. The experiments with village, and presumably sedentary, life that commenced during the earliest Neolithic, the PPNA, and perhaps even the Natufian, culminated during the PPNB. Not only were large settlements occupied, but numerous smaller sites also constituted the PPNB settlement system. All of these events, from the Late Natufian through the LPPNB, occurred over but a few thousand years and brought about monumental and irreversible changes from the relatively simple social structures of hunters and gatherers to more stratified organizations required for the day-to-day existence of village life. It is likely that simple tribal societies were no longer viable social structures within these communities, as the PPN required more complex organizational structures. This is not to say that the emergence of extremely complex social hierarchies developed during the Neolithic. Indeed, many researchers believe that these societies still maintained a degree of egalitarian organization (cf. Kuijt 1995, 1996; Moore 1985:61). The point, however, is that previous methods of maintaining social control were no longer adequate. New structures were required to cope with the increasingly complex life imposed by Neolithic society. In particular, aggregation of large populations would have required the emergence of more elaborate social cohesion and control.

As successful as the PPNB was, turbulent changes were looming on the horizon. By the end of the PPNB, severe population resettlements occurred. While the old explanation of abandonment, hiatus, and replacement by "new" peoples bearing ceramics has been largely disregarded, there is no question that major changes were under way. For whatever reasons, be they climatic, ecological overexploitation, or social, the MPPNB and LPPNB witnessed the establishment of major centers in the eastern Levant, such as 'Ain Ghazal, Basta, and Wadi Shu'eib. Many of these were occupied at the time that major western centers were being depopulated.

The generally marginal Jordanian Plateau and adjacent areas may not have allowed for the luxury of smaller villages and larger towns spread throughout the region (although a different pattern seems to have existed in southern Jordan). Neolithic communities were faced with a decision: either live in smaller settlements or congregate into larger towns such as 'Ain Ghazal. Either response could have been taken, with perhaps different outcomes. I suggest that the latter occurred, resulting in the development of large, spread out settlements like 'Ain Ghazal. Massive defensive structures would not have been necessary, since previous smaller social groups had by now aggregated into a large unit, thereby lessening the change of intercommunity conflict.

Initially, this pattern was adaptive, although it required a modification of existing subsistence and social strategies. For example, throughout 'Ain Ghazal's aceramic development, we see a remarkably wide range of resources being exploited, both domestic and wild; this pattern opposes many of our assumptions of the Neolithic. In point of fact, this strategy was a reflection of a more efficient economic organization in which it was necessary to exploit as wide an array of resources as possible in order to feed a large population located in or near a marginal environment.

With these dramatic population and subsistence fluctuations, it is certain that social organization also had to be reoriented. How were these large villages managed? Was there an overall central authority, or were portions of each settlement run by family or clan units, with central authority minimized or only necessary for certain activities? These issues, of course, are difficult to document archaeologically, but there seems little question that a relatively efficient organizational structure with some degree of centralized authority must have been in place to have allowed the prosperity and growth of settlements that probably contained several thousands residents.

While difficult to directly assess, there is an array of archaeological evidence that indicates the existence of a more complex social interactive system. Architecture has long been used as a marker of social hierarchy and complexity in the Neolithic (cf. Banning and Byrd 1989; Byrd 1994). Large "public rooms" occur at many PPNB settlements, even if the elaborate ceremonial structures seen in Anatolia seem absent in the southern Levant. Of interest, however, are a recently reported LPPNB "temple" and other possible cult structures at 'Ain Ghazal (Rollefson et al. 1994:23-24; Rollefson and Kafafi 1996:20–22). Ritual objects are another social indicator. By the PPNB there is ample evidence for cult objects. Female figurines, for example, are often cited as "mother goddesses" (cf. Cauvin 1985) and may reflect changes in traditional gender roles. Another group of cultic objects contains the elaborate statutes of 'Ain Ghazal (Tubb and Grissom 1995); these undoubtedly had major social significance, possibly related to clan structures within the larger settlement. While most of these remarkable statues are asexual, a few are clearly female (Figure 12.1) and undoubtedly relate to the changing roles of women in Neolithic society (see Chapters 11 and 12, this volume). Certainly the establishment of permanent villages relying heavily on farming required a shifting of traditional gender roles (cf. Bar-Yosef and Meadows 1995:93). Regardless of specific interpretations, there seems little doubt that ritual objects functioned as a form of information exchange, and their abundance in the PPNB suggests a degree of social complexity previously unseen.

Another method of examining social structure is through burials. Researchers have long held that " . . . the deceased are given a set of representations of his or her various social identities or roles when alive so that their status or social position may be given material form after death (e.g., grave-goods, monuments, place of burial, etc.) . . . Consequently the social organization of any society may be reconstructed . . . " (Pearson 1982:99). At PPNB 'Ain Ghazal, as at other contemporary sites, there are burial data supporting status distinctions, although there is little evidence for individual high-status burials. Essentially it appears that "normal" inhabitants were buried in a highly patterned manner, in single interments and in flexed positions with their skulls removed and with few burial goods. At least some others were discarded much more casually; these individuals are found in trash deposits in extended positions with their skulls intact. Children appear to have had little status, as witnessed by scattered remains of adolescents and subadolescents.

At sites such as Wadi Shu'ieb, however, there is tantalizing evidence for different mortuary patterns. Although only a tiny portion of the settlement was sampled and the burial population is small, interments consisting of up to four individuals were present in addition to single interments. Both types were subfloor and flexed, with skulls removed. Interestingly, burial goods, in the form of plaster beads and in one case a small plaster statuette, were recovered with some of these interments (Roler 1991). This pattern possibly reflects a different social structure than that observed at 'Ain Ghazal. Overall, though, the patterns at 'Ain Ghazal and Wadi Shu'eib are very similar to other PPNB sites, including Jericho. There appears to have been a widespread mortuary unity among many PPNB populations, and secondary mortuary practices may have been an important means of increasing community cohesion and limiting social differentiation (Kuijt 1995, 1996, Chapter 6 this volume).

Stage 2: Disharmony During the PPNC

While the PPNB may reflect the florescence of Neolithic life, this pattern of prosperity was not to continue. At around 8,000 bp, as exemplified by the PPNC, dramatic economic and, presumably, social, shifts occurred (cf. Rollefson 1993). The advantages initially offered by population aggregation during the Late PPNB were becoming increasingly fragile by the PPNC, probably due to ecological degradation. Whether it was solely culturally induced or a combination of human mismanagement coupled with deteriorating climatic conditions is, in a sense, a moot question. The point is that by *ca.* 8,000 bp, the largess of the environment was diminishing. The huge populations of centers such as 'Ain Ghazal and Wadi Shu'eib had diminished the local environment, and the occupants of these sites were finding it increasingly difficult to maintain the viable and diverse economy they had enjoyed during the PPNB.

This was occurring at a time when social organization was sophisticated enough to manage the daily operation of the large settlements. The intergroup stress brought about by the increasingly difficult farming and herding conditions would have required careful resolution. Several outcomes were possible, not the least of which may have been interpersonal violence. There is, however, little data to suggest conflict within or between Neolithic settlements. Some evidence exists at 'Ain Ghazal (see Rollefson Chapter 7 this volume), but the overall lack of violence in Late Neolithic society may be a testament to the efficiency of the social organization that had been forged during the PPNB. It apparently was robust enough to deal with deteriorating economic conditions in a nonconfrontational manner. Nonetheless, there would have been consequences of a weaker economy, and these would have had social manifestations. Unfortunately, our database from the PPNC is relatively scant.

While the PPNC is still poorly understood in the southern Levant, research at 'Ain Ghazal, where considerable exposures relating to that period have been excavated, outline profound changes in architecture, artifacts, and the reduced size of the settlement compared to the LPPNB. For example, two different building styles suggest that seasonal nomadism may have been practiced by some of the population (Rollefson and Köhler-Rollefson 1993), lime plaster on floors was no longer used, and houses were smaller in the PPNC. Changes in burial practices also provide evidence for a major change of social behavior during the PPNC. While still interred in flexed positions, none of the thirty-four PPNC burials recovered from 'Ain Ghazal had their skulls removed (Rollefson, personal communication). What the social implications of this are is unclear, but obviously a change in treatment of the dead reflects changing social mores. Based on present evidence, during the PPNC sites such as 'Ain Ghazal and Wadi Shu'eib continued to function as less prosperous versions of the LPPNB. Overall, the PPNC is perhaps best viewed as a true transitional period leading into the PN.

Stage 3: The Solution: The Return to Tribal Society and Consequent Social Fragmentation

During the PN the deteriorations that begun during the PPNC continued. PN sites frequently contain less substantial structures than during the PPNB. Often these are no more than circular huts, and while more complex architecture does exist at 'Ain Ghazal (Rollefson et al. 1992:450–452), Jebel Abu Thawwab (Kafafi 1988), Wadi Ziqlab (Banning 1995; Banning et al. 1989, 1994), Shar'ar Hagolan (and possibly other sites) (Garfinkel 1993:127–129), it is not as elaborate or well constructed as previously. Regardless of variations in architecture, however, major sites such as 'Ain Ghazal and Wadi Shu'eib shrank in size during the PN (cf. Rollefson 1996; Rollefson et al. 1992:466–468). This clearly has social implications relating to family and clan arrangements. The early change (PPNA to PPNB) from circular to rectangular structures and possible social implications has been discussed by Flannery (1972, 1993), but this issue has not been addressed for the PN despite the widespread "return" to circular structures.

Some of the major shifts that occurred during the PN are best exemplified by the dramatic economic transformations occurring at 'Ain Ghazal and Wadi Shu'eib. At these sites (particularly well represented at 'Ain Ghazal), principal subsistence strategies were now focused on a few select species, specifically sheep/goat. Although farming still occurred, a new emphasis on pastoralism developed. As detailed elsewhere (Köhler-Rollefson 1988; Rollefson and Köhler-Rollefson 1989), what had started out as mutually compatible economic systems of agriculture and animal husbandry turned into a mutually exclusive system. No longer could the already marginal environment, stretched to its limits by large aggregated groups of humans, allow for both strategies to be viable in the same locality. This situation was exacerbated by climatic conditions leading to a steady increase of arid conditions (cf. Davis et al. 1990). Thus the previously underutilized desert areas were now more intensively occupied by Neolithic pastoralists, while the better watered, but still marginal, core area maintained an agricultural focus.

The cultural response to this was the evolution of two economic strategies during the PN, one based on farming and the other on pastoralism. Sites such as 'Ain Ghazal continued to function, but we can assume a deterioration in the organizational structure previously extant. If a substantial proportion of the population now practiced pastoralism, then it seems logical that social structure must have been reoriented toward smaller group sizes, at least for those engaged in this economic pursuit, and no longer required the elaborate social controls of the PPNB.

Unfortunately, no burials have been uncovered from PN 'Ain Ghazal. Indeed, only three PN burials are documented for the entire Levant (Garfinkel 1993:127). Additionally, information on gender roles, so tantalizingly hinted at by PPNB ritual objects, is rare in the PN. Portable art, however, continues in the PN, and many figurines, including the intriguing "coffee-bean" eye forms, appear to be female. Others, however, are clearly male; even though no complete examples have been found, there are headless examples and torsos containing male genitalia (Garfinkel 1993:124). The presence of clear male figurines, generally absent in the PPN, may suggest some rather dramatic social changes that could be linked to traditionally male activities such as animal husbandry. One could, albeit tentatively, posit a reduction in the role of females during the PN, when pastoralism assumed more importance (see Chapters 11 and 12, this volume, for further discussion).

Ultimately, as aridity increased and agriculture became even more difficult, the pastoralists may have had the adaptive edge. This resulted in the eventual abandonment of the large Neolithic towns. Subsequent cultural evolution during the Chalcolithic and later periods witnessed a new type of adaptation and settlement type. Social organization must have particularly suffered with such a radical change. From a period of but several hundred years, the fabric of life had gone from elaborate and large population aggregations focused on villages to one consisting of a splintering of populations, many of whom were nomadic, at least on an annual basis. Dramatic social reorientations would have been necessary to accommodate these changes in settlement and subsistence. The degree of centralized authority necessary for controlling a large settlement such as LPPNB 'Ain Ghazal would no longer have been necessary. A return to more clan- or tribal-oriented authority better adapted to pastoral life would now have been a more efficient control mechanism.

CONCLUSIONS

Much research on the Neolithic of the Levant has justifiably concentrated on the earlier, PPN, stages, during which settled village life became a standard practice. It is only in recent years that a concentrated and systematic attempt has been made to thoroughly examine the later Neolithic, as exemplified by the PN. While we now know that older scenarios of a deterioration of culture during the PN are undoubtedly overly simple, there is no denying that major changes occurred by the end of the Neolithic. The expiry of the "PPNB interaction sphere" and subsequent development of ceramic-producing cultures was a major readjustment that contrasts sharply to the prosperity of the Late PPN. This clearly was a time of significant economic and settlement changes, and social systems also must have undergone radical modifications.

REFERENCES

Adams, R., 1991, The Wadi Fidan Project, Jordan. *Levant* 24:181–183.

Banning, E., 1995, Herders . . . or Homesteaders? *Biblical Archaeologist* 58:2–13.

Banning, E., and Byrd B., 1989, Renovations and the Changing Residential Unit at 'Ain Ghazal, Jordan. In *Households and Communities,* edited by S. MacEachern, D. Archers, and R. Garvin, pp. 525–533. Chacmool, the Archaeological Association of the University of Calgary, Calgary.

Banning, E., Rahimi, D., and Siggers, S., 1994, The Late Neolithic of the Southern Levant: Hiatus, Settlement Shift or Observer Bias? The Perspective from Wadi Ziqlab. *Paléorient* 20:151–164.

Banning, E., Dobs, R., Field, J., Kuijt, I., McCorriston, J., Monckton, S., Rubenstein, R., and Sheppard, P., 1989, Wadi Ziqlab Project 1987: A Preliminary Report. *Annual of the Department of Antiquities of Jordan* 33:43–58.

Bar-Yosef, O., 1981a, The "Pre-Pottery Neolithic" Period in the Southern Levant. In *Préhistoire du Levant,* edited by J. Cauvin and P. Sanlaville, pp. 551-569. Colloques Internationaux du Centre National de la Recherche Scientifique, No. 598. Editions du Centre National de la Recherche Scientifique, Paris.

Bar-Yosef, O., 1981b, Neolithic Sites in Sinai. In *Beitrage zur Umweltgeschichte des Vorderen Orients,* edited by W. Frey and H. Uerpmann, pp. 217–233. Dr. Ludwig Reichert Verlag, Wiesbaden.

Bar-Yosef, O., and Alon, D., 1988, *Nahal Hemar Cave.* 'Atiqot 18. Graphpress, Jerusalem.

Bar-Yosef, O., and Belfer-Cohen, A., 1989, The Levantine "PPNB" Interaction Sphere. In *People and Culture in Change,* edited by I. Hershkovitz, pp. 59–72. British Archaeological Reports, International Series 508, Oxford.

Bar-Yosef, O., and Meadows, R., 1995, The Origins of Agriculture in the Near East. In *Last Hunters, First Farmers,* edited by T. D. Price and A. Gebauer, pp. 39–94. School of American Research Press, Santa Fe, New Mexico.

Bar-Yosef, O., and Phillips, J. (eds.), 1977, *Prehistoric Investigations in Gebel Maghara, Northern Sinai.* Qedem 7, Monographs of the Institute of Archaeology, Hebrew University, Jerusalem.

Bennett, C., 1980, Soundings at Dhra', Jordan. *Levant* 12:30-39.

Betts, A., 1990, The Pre-Pottery Neolithic in Eastern Jordan. *Paléorient* 15: 147–153.

Borzatti von Löwenstern, F., 1984, Pitture Schematiche in Ripari Sotto Roccia nel Deserto della Giordania. *Studi per l'Ecologia del Quaternario* 6:71–80.

Braidwood, L., Braidwood, R., Howe, B., Reed, C., and Watson, P., 1983, *Prehistoric Archaeology along the Zagros Flanks.* Oriental Institute Publications, Volume 105. University of Chicago Press, Chicago, Illinois.

Braidwood, R. and Howe, B., 1960, *Prehistoric Investigations in Iraqi Kurdistan.* Studies in Ancient Oriental Civilization, No. 31. University of Chicago Press, Chicago, Illinois.

le Brun, A., 1969, *Beisamoun.* Israel Exploration Journal 19.
Burian, F., and Friedman, E., 1973, Prehistoric Hunters in the Halutza Dunes. *Mitekufat Haeven* 13:58–69.
Byrd, B., 1994, Public and Private, Domestic and Corporate: The Emergence of the Southwest Asian Village. *American Antiquity* 59:639–666.
Cauvin, J., 1985, La Question du "Matriarcat Préhistorique" et le Rìle de la Femme dans la PrÇhistoire. In *La Femme dans le Monde Mediterranéen,* edited by A. Vérilhac, pp. 7–18. Travaux 10, Maison de l'Orient, Lyon.
de Contenson, H., 1971, Tel Ramad, a Village of Syria of the Seventh and Sixth Millennia B.C. *Archaeology,* 232:278–285.
de Contenson, H., and Van Liere, W., 1966, Seconde Campagne Ö Tell Ramad 1965: Rapport Préliminaire. *Annales Archéologiques Arabes Syriennes* 16: 167–176.
Davis, J., Simmons, A., Mandel, R., Rollefson, G., and Kafafi, Z., 1990, A Postulated Early Holocene Summer Precipitation Episode in the Levant: Effects on Neolithic Adaptations. Presented at the 55th annual meeting, Society for American Archaeology, Las Vegas.
Echegaray, J., 1964, *Excavaciones en la Terraza de 'El Khiam* (Jordania), Volume 1. Biblioteca Prehistorica Hispana, Volume IV, Madrid.
Echegaray, J., 1966, *Excavaciones en la Terraza de 'El Khiam (Jordania),* Volume II. Biblioteca Prehistorica Hispana, Volume V, Madrid.
Finlayson, B., and Mithen, S., 1999, The Dana-Faynan-Ghuwayr Early Prehistory Project: Second Interim Report—1998 Field Season. Unpublished manuscript, on file at Centré for Field Archaeology, University of Edinburgh, Scotland.
Flannery, K., 1972, The Origins of the Village as a Settlement Type in Mesoamerica and the Near East: A Comparative Study. In *Man, Settlement, and Urbanism,* edited by P. Ucko, R. Tringham, and G. Dimbleby, pp. 23–53. Duckworth, London.
Flannery, K., 1993, Will the Real Model Please Stand Up: Comments on Saidel's "Round House or Square?" Journal of Mediterranean Archaeology 6:109–117.
Galili, E., Weinstein-Evron, M., Hershkovitz, I., Gopher, A., Kislev, M., Lernau, O., Kolska-Horwitz, L., and Lernau, L., 1993, Atlit-Yam: A Prehistoric Site on the Sea Floor off the Israeli Coast. *Journal of Field Archaeology* 20: 133–157.
Garfinkel, Y., 1993, The Yarmukian Culture in Israel. *Paléorient* 19:115–134.
Garfinkel, Y., 1994, The PPNC Flint Assemblage from Tell 'Ali. In *Neolithic Chipped Stone Industries of the Fertile Crescent,* edited by H. Gebel and S. Kozlowski, pp. 543-562. Studies in Early Near Eastern Production, Subsistence, and Environment I. Ex Orient, Berlin.
Garrard, A., Baird, B., Colledge, S., Martin, L., and Wright, K., 1994, Prehistoric Environment and Settlement in the Azraq Basin: An Interim Report on the 1987 and 1988 Excavation Seasons. *Levant* 26:73–109.
Garrard, A., Colledge, S., and Martin, L., 1996, The Emergence of Crop Cultivation and Caprine Herding in the "Marginal Zone" of the Southern Levant. In *The Origins and Spread of Agriculture and Pastoralism in Eurasia,* edited by D. Harris, pp. 204–226. University College London Press, London.
Gilead, I., 1990, The Neolithic-Chalcolithic Transition and the Qatifian of the Northern Negev and Sinai. *Levant* 22:47–63.
Gopher, A., 1994, *Arrowheads of the Neolithic Levant.* Dissertation Series 10, American Schools of Oriental Research. Eisenbrauns, Winona Lake, Indiana.
Gopher, A., and Gophna, R., 1993, Cultures of the Eighth and Seventh Millennia BP in the Southern Levant: A Review for the 1990s. *Journal of World Prehistory* 7:297–353.
Goren, Y., 1990, The "Qatifian Culture" in Southern Israel and Transjordan: Additional Aspects for its Definition. *Mitekufat Haeven* 23:100–112.
Hershkovitz, I. (ed.), 1989, People and Culture Change: Proceedings of the Second Sympo-

sium on Upper Paleolithic, Mesolithic and Neolithic Populations of Europe and the Mediterranean Basin. British Archaeological Reports, International Series 508, Oxford.

Jobling, W., and Tangri, D., 1991, A Pre-Pottery Neolithic Site in the Hisma Basin, Southern Jordan. *Paléorient* 17: 141–148.

Kafafi, Z., 1982, The Neolithic of Jordan (East Bank). Ph.D. dissertation, Freien Universitat, Berlin.

Kafafi, Z., 1988, Jebel Abu Thawwab: A Pottery Neolithic Village in North Jordan. In *The Prehistory of Jordan,* edited by A. Garrard and H. Gebel, pp. 451–471. British Archaeological Reports, International Series 396, Oxford.

Kenyon, K., 1957, *Digging up Jericho.* Ernest Benn, London.

Kenyon, K., 1960, *Archaeology in the Holy Land.* Praeger, New York.

Kirkbride, D., 1966, Five Seasons at the Pre-Pottery Neolithic Village of Beidha in Jordan: A Summary. *Palestine Exploration Quarterly* 98:8–72.

Kirkbride, D., 1968, Beidha: Early Neolithic Village Life South of the Dead Sea. *Antiquity* 42:263–274.

Kirkbride, D., 1978, The Neolithic in Wadi Rumm: 'Ain Abu Nekheileh. In *Archaeology in the Levant,* edited by R. Moorey and P. Parr, pp. 1–10. Aris and Phillips, Warminster.

Köhler-Rollefson, I., 1988, The Aftermath of the Levantine Neolithic Revolution in Light of Ecologic and Ethnographic Evidence. *Paléorient* 14:87–93.

Köhler-Rollefson, I., 1992, A Model for the Development of Nomadic Pastoralism on the Transjordanian Plateau. In *Pastoralism in the Levant,* edited by O. Bar-Yosef and A. Khazanov, pp. 11–18. Monographs in World Archaeology No. 10, Prehistory Press, Madison, Wisconsin.

Köhler-Rollefson, I., and Rollefson, G., 1990, The Impact of Neolithic Subsistence Strategies on the Environment: The Case of 'Ain Ghazal, Jordan. In *Man's Role in the Shaping of the Eastern Mediterranean Landscape,* edited by S. Bottema, G. Entjes-Nieborg, and W. Van Zeist, pp. 3–14. Balkema, Rotterdam.

Kuijt, I., 1995, *New Perspectives on Old Territories: Ritual Practices and the Emergence of Social Complexity in the Levantine Neolithic.* Ph.D dissertation, Harvard University. University Microfilms, Ann Arbor, Michigan.

Kuijt, I., 1996, Negotiating Equality through Ritual: A Consideration of Late Natufian and Pre-Pottery Neolithic A Period Mortuary Practices. *Journal of Anthropological Archaeology* 15(4):313–336.

Kuijt, I., and Bar-Yosef O., 1994, Radiocarbon Chronology for the Levantine Neolithic: Observations and Data. In *Late Quaternary Chronology and Paleoclimates of the Eastern Mediterranean,* edited by O. Bar-Yosef and R. S. Kra, pp. 227–245. Radiocarbon, Tucson, Arizona.

Kuijt, I., Mabry, J., and Palumbo, G., 1991, Early Neolithic Use of Upland Areas of Wadi El-Yabis: Preliminary Evidence from the Excavations at 'Iraq ed-Dubb, Jordan. *Paléorient* 17:99–108.

Kuijt, I., and Mahasneh H., 1998, Dhra': An Early Neolithic Village in the Southern Jordan Valley. *Journal of Field Archaeology* 25:153–161.

Kutzbach, J., and Guetter P., 1986, The Influence of Changing Orbital Parameters and Surface Boundary Conditions on Climate Simulations for the Past 18,000 years. *Journal of Atmospheric Science* 43:1726–1759.

Kutzbach, J., Guetter, P., Behling, P., and Selin, R., 1993, Simulated Climatic Changes: Results of the COHMAP Climate-Model Experiments. In *Global Climates Since the Last Glacial Maximum,* edited by H. Wright, J. Kutzback, T. Webb, W. Ruddiman, F. Street-Perrott and P. Bartlein, pp. 24–93. University of Minnesota Press, Minneapolis, Minnesota.

Mahasneh, H., 1995, Es-Sifyia, a Pre-Pottery Neolithic Site in Wadi Mujib, Jordan. Paper presented at the Sixth International Conference on the History and the Archaeology of Jordan, June 5–10, Turin, Italy.

McCorriston, J., and Hole, F., 1991, The Ecology of Seasonal Stress and the Origins of Agriculture in the Near East. *American Anthropologist* 93:46–69.
Mellaart, J., 1975, *The Neolithic of the Near East*. Thames and Hudson, London.
Moore, A., 1973, The Late Neolithic in Palestine. *Levant* 5:38–68.
Moore, A., 1978, *The Neolithic of the Levant*. Ph.D. dissertation, University of Oxford, Oxford.
Moore, A., 1979, A Pre-Pottery Neolithic Farmers' Village on the Euphrates. *Scientific American* 241:62–70.
Moore, A., 1985, The Development of Neolithic Societies in the Ear East. In *Advances in World Archaeology,* edited by F. Wendorf and A. Close, pp. 1–69. Academic Press, New York.
Moore, A., and Hillman, G., 1992, The Pleistocene to Holocene Transition and Human Economy in Southwest Asia: The Impact of the Younger Dryas. *American Antiquity* 57:482–494.
Moore, A., Hillman, G., and Legge, A., 1975, The Excavation of Tell Abu Hureyra in Syria: A Preliminary Report. *Proceedings of the Prehistoric Society* 41:50–72.
Muheisen, M., 1995, Basta Excavations (1986–92). Paper presented at the Sixth International Conference on the History and the Archaeology of Jordan, June 5–10, Turin, Italy.
Najjar, M., 1992, Tell Wadi Feinan/Wadi Araba: A New Pottery Neolithic Site from Jordan. *The Near East in Antiquity* 3: 9–28.
Najjar, M., 1994, Ghwair I, a Neolithic Site in Wadi Feinan. *The Near East in Antiquity* 4:75–85.
Nissen, H., 1990, Basta: Excavations of 1986–89. *The Near East in Antiquity* 1:87–94.
Noy, T., 1970, Prehistoric Sites in the Halutza Dunes. *Mitekufat Haeven* 10:1–10.
Noy, T., 1971, *Halutza Dunes: A Prehistoric Survey*. Israel Museum, Jerusalem.
Noy, T., Legge, A., and Higgs, E., 1973, Recent Excavations at Nahal Oren, Israel. *Proceedings of the Prehistoric Society* 39: 75–99.
Pearson, J., 1982, Mortuary Practices, Society and Ideology: An Ethnoarchaeological Study. In *Symbolic and Structural Archaeology,* edited by I. Hodder, pp. 99–113. Cambridge University Press, Cambridge.
Perrot, J., 1968, La Préhistorire Palestinienne. *Dictionnaire de la Bible—Supplement* 8:286–446.
Raikes, T., 1980, Notes on Some Neolithic and Later Sites in Wadi Araba and the Dead Sea Valley. *Levant* 12: 40–60.
Roberts, N., and Wright, H., 1993, Vegetational, Lake-Level, and Climatic History of the Near East and Southwest Asia. In *Global Climates Since the Last Glacial Maximum,* edited by H. Wright, J. Kutzback, T. Webb, W. Ruddiman, F. Street-Perrott, and P. Bartlein, pp. 194–220. University of Minnesota Press, Minneapolis, Minnesota.
Roler, K., 1991, Human Remains from Wadi Shu'eib, Jordan. Ms. on file, Department of Anthropology, University of Nevada, Las Vegas.
Rollefson, G., 1989, The Late Aceramic Neolithic of the Levant: A Synthesis. *Paléorient* 15:168–173.
Rollefson, G., 1993, The Origins of the Yarmoukian at 'Ain Ghazal. *Paléorient* 19:91–99.
Rollefson, G., 1994, La Néolithique de la Vallée du Jourdain. *La Recherche* 271:1254–1259.
Rollefson, G., 1996, The Neolithic Devolution: Ecological Impacts and Cultural Compensation at 'Ain Ghazal, Jordan. In *Retrieving the Past: Essays on Archaeological Research and Methodology in Honor of Gus W. Van Beek,* edited by J. Seger, pp. 219–229. Eisenbrauns, Winona Lake, Indiana.
Rollefson, G., and Frohlich, B., 1982, A PPNB Burin Site on Jabal Uweinid, Eastern Jordan. *Annual of the Department of Antiquities of Jordan* 26:189–198.
Rollefson, G., and Kafafi, Z., 1985, Khirbet Hammam a PPNB Village in the Wadi el-Hasa, Southern Jordan. *Bulletin of the American Schools of Oriental Research* 258:63–69.
Rollefson, G., and Kafafi, Z., 1996, The 1995 Season at 'Ayn Ghazal: Preliminary Report. *Annual of the Department of Antiquities of Jordan* 40:28.
Rollefson, G., Kafafi, Z., and Wada, H., 1994, The 1993 Season at 'Ain Ghazal: Preliminary Report. *Annual of the Department of Antiquities of Jordan* 38:11–32.

Rollefson, G., and Köhler-Rollefson, I., 1989, The Collapse of Early Neolithic Settlements in the Southern Levant. In *People and Culture Change: Proceedings of the Second Symposium on Upper Paleolithic, Mesolithic and Neolithic Populations of Europe and the Mediterranean Basin,* edited by I. Hershkovitz, pp. 59–72. British Archaeological Reports, International Series 508, Oxford.

Rollefson, G., and Köhler-Rollefson, I., 1992, Early Neolithic Exploitation Patterns in the Levant: Cultural Impact on the Environment. *Population and Environment* 13:243–254.

Rollefson, G., and Köhler-Rollefson, I., 1993, PPNC Adaptations in the First Half of the 6th Millennium B.C. *Paléorient* 19:33–42.

Rollefson, G., Simmons, A., and Kafafi, Z., 1992, Neolithic Cultures at 'Ain Ghazal, Jordan. *Journal of Field Archaeology* 19:443–470.

Simmons, A., 1980, Early Neolithic Settlement and Economic Behavior in the Western Negev Desert of the Southern Levant. Ph.D. dissertation, Southern Methodist University, Dallas, Texas.

Simmons, A., 1981, A Paleosubsistence Model for Early Neolithic Occupation of the Western Negev Desert. *Bulletin of the American Schools of Oriental Research* 242:31–49.

Simmons, A., 1995a, Town Planning in the Neolithic—Is 'Ain Ghazal "Normal?" In *Studies in the History and Archaeology of Jordan, V: Art and Technology throughout the Ages,* edited by K. 'Amr, F. Zayadine, and M. Zaghloul, pp. 119–122. Jordan Press Foundation, Amman.

Simmons, A., 1995b, Landscape Archaeology and the Exploitation of Natural Resources in the Eastern Levant. Paper presented at the Sixth Conference on the History and Archaeology of Jordan, June 5–10, Turin, Italy.

Simmons, A., and Kafafi, Z., 1988, Preliminary Report on the 'Ain Ghazal Archaeological Survey. *Annual of the Department of Antiquities of Jordan* 32: 27–39.

Simmons, A., and Najjar, M., 1996, Current Investigations at Ghwair I, A Neolithic Settlement in Southern Jordan. *Neolithics* 2:6–7.

Simmons, A., and M. Najjar, M., 1998, Al-Ghuwayr I. A Pre-Pottery Neolithic Village in Wadi Faynan, Southern Jordan: A Preliminary Report on the 1996 and 1997/98 Seasons. *Annual of the Department of Antiquities of Jordan,* 42:91–101.

Simmons, A., Rollefson, G., Kafafi, Z., and Moyer, K., 1989, Test Excavations at Wadi Shu'eib, a Major Neolithic Settlement in Central Jordan. *Annual of the Department of Antiquities of Jordan* 33:27–42.

Simmons, A., Rollefson, G., Köhler-Rollefson, I., Mandel, R., and Kafafi Z., 1988, 'Ain Ghazal: A Major Neolithic Settlement in Central Jordan. *Science* 240(4848):35–39.

Stekelis, M., 1973, *The Yarmukian Culture of the Neolithic Period.* The Magnes Press, Hebrew University, Jerusalem.

Street, F., and Grove, A., 1979, Global Maps of Lake-level Fluctuations Since 30,000 Years BP. *Quaternary Research* 12:83–118.

Tchernov, E., and Bar-Yosef, O., 1982, Animal Exploitation in the Pre-Pottery Neolithic B Period at Wadi Tbeik, Southern Sinai. *Paléorient* 8:17–37.

Tubb, K., and Grissom, C., 1995, 'Ain Ghazal: A Comparative Study of the 1983 and 1985 Statuary Caches. In *Studies in the History and Archaeology of Jordan, V,* edited by K. 'Amr, F. Zayadine, and M. Zaghloul, pp. 437–447. Jordan Press Foundation, Amman.

de Vaux, R., 1966, Palestine during the Neolithic and Chalcolithic Periods. In *Cambridge Ancient History,* Volume I, pp. 498–520. Cambridge University Press, Cambridge.

Vianello, F., 1985, La Preistoria nel Deserto. *L'Universo* 65:72–89.

Part IV

Social Relations and Material Culture

Symbolism and Meaning

As with other recent considerations of the nature of social relations in middle-range societies, these chapters focus on exploring some of the relationships between Neolithic social relations, material culture, and symbolism and meaning. One of the relatively recent developments in the study of Neolithic social systems is that researchers have begun to employ a wider range of material data sets to explore the nature of social relationships within and between different social units. In this section, three researchers explore the function, role, and possible social messages developed by the construction, use, and destruction of anthropomorphic and zoomorphic clay figurines, as well as ceramics, of the Pottery Neolithic. Moving between consideration of data from individual settlements to discussion of the broader existence of shared material systems in Near Eastern Neolithic communities, these authors explore the social context of a wide range of material objects, their possible meanings, and how archaeological data help us to understand how groups and individuals may have employed figurines and ceramics to negotiate relationships, differentiate select members of the community, and/or link individuals across the household, community, and intercommunity level.

Based on his long-term interest in ideology, ritual, and social change, Jacques Cauvin attempts to intellectually bridge between broader regional changes in figurine use in the Neolithic and past belief systems. Cauvin provides a detailed consideration of the cognitive and symbolic nature of Neolithic cultures in the Near East, arguing that the Neolithic revolution represents, above all else, a profound mental and social transformation.

Adopting a comparative perspective for the entire Near East emphasizing the symbolic uses of material culture as a means of constructing social meaning, he argues that the emergence of food production is not a reaction to a necessity for food but is a by-product of a new way of looking at the environment by humans, and that this is expressed through the symbolic and physical use of figurines and skulls. In brief, he argues that the transition from the Natufian to Neolithic was one in which the nature of symbolic systems was transformed from the Natufian emphasis on animals to that of the Neolithic emphasis upon women and bulls. Based on a consideration of the environmental context within which the revolution of symbols occurred, Cauvin rejects the argument that changes in social organization were linked to overexploitation of the environment; he suggests that ideological changes were indigenous and reflect new ways of conceptualizing the relationship between the environment and human culture. Noting that "religion," as is the case in traditional societies, is both cognitive and dynamic and explains the cosmos as well as simultaneously managing social tensions, he argues that the widespread appearance of clay and stone figurines in the PPNA reflects a ritual and religious focus on collective ancestors and divinities as part of organized ritual life. Cauvin places the Neolithic transition of symbols within the scope of the emergence of institutionalized inequality and changes in social relations within communities and in how humans conceptualized their relationship with the physical environment.

Mary Voigt synthesizes published archaeological data on figurines from the Neolithic settlements of Hajji Firuz Tepe, Gritille Hüyük, Nevali Çori, and Çatal Höyük and explores how these data inform us as to how and why ritual practices were employed during the Neolithic in the northern areas of the Near East. Expanding Ucko's (1968) functional framework for figurines, the author illustrates how figurines from Hajji Firuz Tepe and Gritille and the analysis of breakage and wear patterns on the figurines demonstrate patterning consistent with ethnographic uses of figurines as "vehicles of magic" and symbols of worship in which figurines were used in rituals intended to convey social and spiritual meanings. She argues that Neolithic clay figurines were often intentionally incorporated into roasting or ash pits as part of household ritual practices. Based on contextual information, she notes that animal figurines (specifically cattle) were often recovered from domestic areas and were, therefore, most likely symbolically associated with herd fertility and production rather than hunting. In contrast, the frequent occurrence of human figurines with protruding stomachs, navels, and breasts (presumably female) appear to be associated with the well-being of females and children. Moving to a detailed discussion of changing use of figurines during the Neolithic occupation of Çatal Höyük, the author argues that stone and large clay figurines were employed as cult statues. She interprets their changing use from the position of form, archaeological context, and the condition of the artifacts at the time of deposition, and develops generalizations based on ethnographic parallels. Voigt argues that rituals were employed as a means of alleviating social stresses, ensuring the well-being of the community. In her conclusion, the author briefly addresses the implications for such a strong regional patterning, how these collective data inform us of social relationships within and between settlements, and how specific ritual practices were employed to craft social relationships.

Finally, Estelle Orrelle and Avi Gopher address several aspects of the meaning, symbolism, and decoration of pottery from several periods of the Pottery Neolithic. Placing this discussion in the context of the entrenchment and expansion of food production, the authors explore how the social manipulation of symbolic material culture may inform researchers as to the nature and mechanisms of social and economic change in Pottery Neolithic communities as materially expressed through figurines and decorated and undecorated pottery vessels. Moving beyond functional considerations, they regard the first pottery assemblages as fulfilling a social role, focus on describing the main elements of these symbolic assemblages, and highlight some of the possible implications of these artifacts. In the context of the earliest stages of the Pottery Neolithic period, they argue that specific motifs stood as symbols for a social system governing rules of access to women. Dealing with material systems in the later Pottery Neolithic, the authors argue that vessels of clay traditionally identified with the female body and social body acted as a kind of blueprint onto which the norms and beliefs of society were displayed and which were expressions of restrictions and guidelines for social behavior in the Pottery Neolithic. This chapter, as with the previous two, provides a provocative and thought-provoking exploration of the possible interconnections between some aspects of Neolithic material culture, symbolism, meaning, and the way social relations were structured at the household, community, and regional scale.

Chapter 10

The Symbolic Foundations of the Neolithic Revolution in the Near East

JACQUES CAUVIN

INTRODUCTION

One of the main contributions of "processual" archaeology in the 1960s was to emphasize demography and the functioning of societies, beyond a consideration of artifact typology, subsistence practices, and ecology. Although Binford anticipated, in the general program of the "New Archaeology", the study of "ideotechnical" systems, the question of prehistoric ideologies remains relatively unexplored. From the functionalist perspective of Binford, culture functions to adapt human societies in extrasomatic ways to their natural environment. This is the materialism common to both Marxist and capitalist thought, which is itself derived from positivist scientific positivism. It is in fact none other than a "paradigm," in the Kuhnian sense of the term (Kuhn 1977)—that is, a basic assertion considered to be so obvious that it is not even necessary to justify it. This is in reality much less rational than one might believe: Nietszche had already denounced the "Occidental myth of

JACQUES CAUVIN • Centre National de la Recherche Scientifique, Institute de Prehistoire Orientale, Jalès, France.

Life in Neolithic Farming Communities: Social Organization, Identity, and Differentiation, edited by Ian Kuijt. Kluwer Academic/Plenum Publishers, New York, 2000.

Science," and an epistemologist such as Feyerabend (1979) does not hesitate to suggest that today's science may be tomorrow's fairy tale. This is why, in the 1980s and 1990s, through the impetus of Ian Hodder and a few others, the school of "symbolic" or "contextual" archaeology reacted against this systematic reductionism by advocating a more interpretive and hermeneutic approach (Bintcliff 1991), to the point of considering not only the ritual documents (funeral practices and art) but the artifacts themselves, their decoration, and the architectural structures as "symbols" of an underlying way of thinking to be reconstructed. While a fruitful approach, in practice prehistorians have focused on that nature of social structures, in particular the first hierarchies of the end of the Neolithic and of the Chalcolithic foreshadowing the urban revolution (Hodder 1989).

What about the earlier Neolithic, before societies with elites? Hodder's interesting ideas (1990) on the European Neolithic can sometimes give an impression of arbitrariness and subjectivity. Arguing that modern science is focused too much on the positivist and behaviorist ideologies that influenced processual archaeology, this does not get in the way of the "new school," which seeks to be resolutely relativist, antiscientific, and postmodern. The recent brainstorming concerning scientific epistemology itself (e.g., Feyerabend 1979; Kuhn 1977; Morin 1991; Popper 1959) allows that human sciences may be practiced without being necessarily positivist or behaviorist (Peebles 1992). What is important is to widen the field of the science itself to include areas so far neglected. The advent of "cognitive archaeology" is, for example, one of the principal avenues toward which this broadening can occur (Renfrew and Bahn 1991; Renfrew 1994a). Here again, however, ambiguity exists. In principle it is a matter of reconstructing prehistoric "cognition," but in reading the publications related to this trend one realizes that they are more often concerned with analyzing the cognitive methods and practice of the archaeologist in crafting prehistory rather than penetrating the actual functioning of prehistoric thinking. As Gardin has said, "the two programs are not mutually exclusive, but it is not a reason to confound them" (Gardin 1992:99).

The problem is, then, to know whether, independently of any reducing positivism, it is possible to have discussion on the prehistoric mentality that is sufficiently rigorous and coherent for a minimal consensus among the scholars of our discipline, an element that Popper holds to be the essential virtue of a scientific theory. The undertaking is a difficult one. There is obviously a danger of projecting on the past our cultural conceptions of the present (Renfrew 1994b), not only in the matter of religion but even for the most trivial activities. One will note, for example, the modern tendency to treat the new food strategies of the Neolithic chiefly in terms of production–consumption, as would a present-day economist. Perhaps it is the same in technology for the French method of "chaînes opératoires," given as an eminent example of cognitive archaeology: just because we observe in a prehistoric technique a rational linkage of gestures and inventions does not mean these were inevitably conceptualized in the same way by prehistoric craftsmen. Technological analysis today is related to that which Popper has called "the world of linguistically formulated human knowledge"; it is a relatively late stage in the evolution of our species. There could, however, have been an unformulated practical rationality that could have been accompanied by "thinking" of a very different nature.

Beyond the overly theoretical controversies between "processual archaeologists" and "contextual archaeologists," we are striving (Cauvin 1994) to combine clearly the various methods with regard to the Neolithic in the Near East. The first step in this process is to construct a classic synthesis on the basis of our own excavations and those of our colleagues, using available architectural, technological, environmental, and subsistence data, then situating these chronologically in relation to other more specifically cultural information, such as art and religious and funerary practices. This multidimensional analysis reveals anomalies that are counter to current interpretive models and allows us to enter into a "palaeopsychological" interpretation of the past, in spite of the somewhat pejorative meaning of the word in the eyes of the new archaeology.

Contrary to other regions of the world where the Neolithic arrived abruptly, such as Europe, in the Near East it was an indigenous process, spread over three millennia. Its different components (economic, technological, ideological) occurred progressively in a determined order. This order, which is an irrefutable stratigraphic fact, is in itself significant for it automatically excludes certain explanations since certain phenomena occurred chronologically *later* than those that they were supposed to have caused. For example, Gordon Childe's argument that the move toward sedentarism was the consequence of food production was undermined with the discovery of preagricultural Natufian villages such as 'Ain Mallaha (Perrot 1966a).

One of the components of the Neolithic revolution is that which we call the "revolution of symbols," manifested preeminently in Neolithic art. Although there are rare anthropomorphic silhouettes, still schematic and asexual, the great majority of Natufian figurines depict small ruminants (gazelles, perhaps deer) carved in bone or stone. In contrast, two new figures accompanied, and continue after, the Neolithic revolution: the representation of women and of bulls, which henceforth dominated all other representations. These representations first appeared between 10,000 and 9,500 BP in the Khiamian culture, intermediate between the Natufian and the PPNA. Female statuettes are found almost exclusively from settlements in the Jordan valley (Bar-Yosef 1980) as well as on the Euphrates (at Mureybet II). The theme of the bull, however, was limited to the Euphrates valley and was expressed through the burial of bulls' skulls in clay benches. These devices are obviously symbolic, for very little hunting of the wild bull itself took place. Between 10,500 and 10,000 BP the early Mureybetian of the Euphrates (phase IIIA of Mureybet) was to see a proliferation of female figurines in stone and in fired clay (Cauvin 1977, 1994), along with the horns of aurochs imbedded within walls. Paradoxically, there are no perceptible traces of agriculture or of animal domestication. Representation of bulls do not reach the southern Levant until the middle PPNB (about 9,500 BP).

These two figures, the woman and the bull, were destined to represent the divine couple, mother-goddess and bull-god, which were to persist in the Near East and the eastern Mediterranean from the Neolithic until the classical period. In the Khiamian and Mureybetian, where we see them emerging as "dominant symbols," they may have already been viewed as divinities. Although we have no formal proof for this early period, 2,000 years later in the Ceramic Neolithic of Çatal Höyük (Mellaart 1967) this is the case where the monumental representations of the goddess are found on relief on the walls and the well-known statuette depicting the mother-goddess giving birth on a throne of panthers—attributes unequivocally royal and maternal. The obsessive and oversized representations of the bull (frescos and clay bucrania), and particularly a statuette of a bearded man riding a bull, foreshadow the Phoenician and Hittite god Hadad, also represented as riding a bull. But this certainty is based on the abundance, the variety, and, especially the exceptional preservation of the material at Çatal Höyük, particularly its architecture. Although there is limited information on the subject, the interpretation that the representations of women and bulls in the Near East by 9,000 BP were linked to "divinities" remains a theoretical probability, one that is worthy of further consideration (see also Voigt Chapter 11, this volume).

What was the environmental context of this revolution of symbols? The Khiamian period, like the end of the Natufian that preceded it, occurred during the dry climatic episode of the Younger Dryas, which lasted from 12,000–11,500 BP. At Mureybet, phases I and II (Final Natufian and Khiamian) are both characterized palynologically by the abundance of chenopodiaceae and the relative rareness of tree species and grasses (Leroi-Gourhan 1974). Although the utility of palynological analysis is limited when applied to ells in the Near East, these results are unquestionably confirmed by the dating of the Younger Dryas in the eastern Mediterranean (Rossignol-Strick 1993), as well as the archaeobotanical results (macroremains) from Abu Hureyra situated on the Euphrates, where the climatic dryness began at the end of the Late Natufian ("Mesolithic"), about 12,000 BP (Moore and Hillman 1992).

Thus, the change in the symbolic material occurred in the middle of an arid phase. Although it is difficult to say whether there is a connection between the two phenomena, what is certain is that the arid phase and period of food stress occurred much earlier than the appearance of an agricultural economy which was not linked to climatic stress. In fact, the Khiamian diet was no different from that of the Final Natufian: at Mureybet it was a broad spectrum economy where the gathering of *Polygonum* (with some wild cereals), the hunting of gazelle and small game, and fishing were predominant.

The first traces of agriculture, contemporary with an increasingly moist climate throughout the Levant, did not occur before 10,000 BP, as seen in the PPNA (Sultanian) of Jericho and especially Tell Aswad (Aswadian) in the oasis of Damascus. For these two villages, the morphological "domestication" of cereals has been proposed, but continues to be debated. At Netiv Hagdud (Bar-Yosef et al. 1980) in the Jordan Valley, the "domestication" of barley is only partial, which has led to doubt on the part of some archaeobotanists (Kislev 1992), though not of others (Zohary 1992), as to the presence of true agriculture. On the Middle Euphrates, the agricultural economy appears to be established at the same time as phase IIIB at Mureybet (Late Mureybetian), but in an entirely predomestic form, indicated by a strong and sudden rise in the quantity of cereals probably due to artificial human intervention. Here, again, debate exists (Van Zeist and Bakker-Heeres 1984; Willcox 1995). It should be noted, however, that G. Hillman's proposed "pre-domestic phase" has been demonstrated through experimental prehistoric cultivation at Jalès (Anderson-Gerfaud et al. 1991). Moreover, the selection for domesticated cereals could have been delayed if spontaneously sown seed was not kept apart from cultivated seed. The exact content of this first agriculture still remains to be defined from an archaeobotanical point of view, but a first phase of a more active management of the environment is certainly supported by the deliberate selection of certain species (cereals among them) requiring increase in technical investment, whereas other resources were abandoned (Cauvin 1977, 1978). At Mureybet III, we see the end of fishing and a progressive decrease in the hunting of gazelle and small game, replaced by the hunting of large herbivores (cattle, equines), which becomes dominant (Ducos 1972). Interestingly, this pattern indicates that this new economy was not the reaction to a lack of food resources, such as proposed by Flannery and Binford, but a deliberate cultural choice previously suggested by Braidwood.

This choice caused the human settlements to enlarge quickly; Mureybet and Jericho grew to larger than 2 ha. There were then *local* demographic increases, in contrast to the small settlements of the Natufian and the Khiamian. As Flannery suggested (1972), such an enlargement necessarily implies a new type of social organization and probably led to the collective construction of monumental buildings such as the PPNA tower of Jericho and a general reorganization of food strategies. But this new organization, whose exact nature is still quite difficult to define, was not a reaction to an environmental stimulus, for as we have seen the environment was not overexploited but, on the contrary, abounded in unexploited resources. This is definitely not adaptive behavior in Binford's sense but an adaptation of human society to itself and to its own maturing process.

Since it was not an environmental stimulus, it must have been cultural. Thus, it is necessary to look for the reason for these changes in the "interior environment" (Leroi-Gourhan 1965). We have seen that a spectacular change has already occurred that foreshadows the economic upheaval to come: the Khiamian revolution of symbols. Moreover, given that this chronological sequence is now indisputable, it is necessary to challenge and dismiss former materialist theories in which symbolic constructions were only derived "superstructures." We must note, however, that the true logical links between the two events, that of ideology and economy, are not *a priori* obvious. We have thus presented a theoretical explanation for this relation between religious beliefs and economic and sociological changes (Cauvin 1987, 1994).

THE ADVENT OF DIVINITIES

The pre-Natufian food gatherers of the Near East lived, as do all animal species, in an environment of plants and animals that they exploited but did not think to modify in any economically significant way. At the very most they varied their acquisition strategies according to the seasons, often moving from place to place to follow and exploit this or that edible species. This is why the first sedentary Natufian villages were established in rich environmental zones permitting a diversified (broad spectrum) economy where food resources were seasonably available in the same area.

Thanks to the painted caves of the "Franco-Cantabrian" region we know that during the Upper Paleolithic animals were not only prey for the hunters but already a "spectacle" subject to being reproduced in works of art (Leroi-Gourhan 1965). It is generally agreed that these caves were sanctuaries and that the profusion of animals represented on their walls were not drawn by accident, but form structured symbolic wholes with cognitive and religious meaning and value. According to the region, certain species are represented more frequently than others: for example, horses and bison in southwest France and lions and rhinoceros in southeast France (Chauvet et al. 1995). Importantly, these are almost always arranged in a way that is collective and "horizontal"—that is, to say no individual animal that could represent a divinity ever stands out in these groups. This is clearly deliberate, since it would have been very easy in these spatially composed cave paintings to have emphasized one or another individual animal over others if it had a truly dominant religious value.

One wonders (Delporte 1979) whether the female figurines of the Upper Paleolithic, sometimes called "Aurignacian Venuses," could have foreshadowed the future Neolithic goddess. Leroi-Gourhan and Delporte have pointed out how much the deformation of these figures (the distention of the hips and breasts) emphasizes that which is least specifically human in the idea of fecundity, whereas the extremities and, particularly, the head, are usually atrophied. One thinks more of a simple symbol of fecundity than a truly divine personage. The only exception is perhaps the clay relief in the cave of Laussel that represents a woman holding a horn in her hands. This figure of a woman with an aurochs horn suggests some mythical figure, perhaps the germ of a female divinity, already associated with the evocation of an ox. One may consider in any case that in the whole of the Upper Paleolithic a symbolic classification of the natural world was well established but that it was very probably a world without gods.

In contrast, excavations at Çatal Höyük have unquestionably documented the existence by 9,000 BP of a sort of divine couple with a female dominance. Although the information available is less forthcoming, it is tempting to push back its emergence to the Khiamian culture at about 10,000 BP when the first representations of women and (on the Euphrates) of bulls appeared and began to dominate assemblages. In fact this patterning may outline the actual mental transformation that provided the initial impetus to the Neolithic revolution. The appearance of divinities indicates a new distinction in the religious imagination between a High and a Low, between a perfect divine power, experienced henceforth as removed from ordinary man, and this man himself, who consciously feels himself to be a finite and imperfect creature. This new tension puts an end to an equilibrium, which is not at all an ecological balance between population and resources as seen from processual archaeology, but an entirely psychic one: a harmonious insertion of humanity in a natural world that had yet to be transformed or manipulated.

On the cognitive level the perception of the sacred would have taken on a hierarchical form, with the goddess as the keystone of the system. This is, however, in the realm of the religious, where theoretical representations of the world are inseparable from specific emotional attitudes and from a specific active dynamic. Belief in divinities is often associated on the iconographic level with the theme of the "orant," that is, a human figure in a state of prayer, arms raised. In the Near East this iconographic theme was not to be formally depicted until the Bronze Age on Mesopotamian seals. Elsewhere, for example on the decorated shelters of the Saharan region (so-called phase of the "Round heads"), it is explicit from the beginning of the Neolithic. The intimate tension represented by prayer is obviously evidence of a certain existential dissatisfaction that could have emerged at the same time as the Neolithic. Moreover, I argue that there is a probable correlation between this uneasiness and a compensating new dynamism in the face of the material world, attested by the beginning of food production. In this intellectual context the original invention of agriculture and herding is not a reaction to a necessity for food, but resulted from a new perception of humans toward nature, toward themselves, and to the role they played in this relationship. It must not be forgotten that the Neolithic revolution brought not only a new conception of the world but also the beginning of its transformation by humans with all the novel behavior and activities that resulted from this conception. "Religion," as is the case generally in traditional societies, is cognitive and dynamic, in that it explains the cosmos but at the same time manages psychic tensions and releases energies. This profound mental unity between symbolic construction, emotion, and action is a familiar notion to the psychoanalyst today when dealing with the individual. Perhaps it would be appropriate not to be stopped here by Popper's views, for whom psychoanalysis is not truly scientific, and to extend this approach as a sort of collective paleopsychoanalysis. Given that the goal of archaeology is to reconstruct human history as a whole and not only its economic and social practices, it is necessary to use all methods available to us, whether or not they are "scientific" according to one or another school of thought. In this respect, then, there is an indisputable contribution from contextual archaeology to which we will return later.

We should not forget, however, that the "mentalist" model presented here in relation to the Khiamian and the PPNA still rests upon limited data, especially given the relative scarceness of PPNA villages excavated compared to those that preceded and followed them. In particular the actual "divine" nature of female and bull representations remains problematic, as we have seen, until the seventh millennium, even though their evolution was certainly moving in this direction. It remains possible that their transformation from "dominant symbols" into real representations of divinities was progressive, as the simultaneous emergence of "agriculture" itself occurred in successive stages, the details of which are far from clear. What is important is that the two transformations occurred side by side, at least according to the information we have, while the transformation of symbolic data appears to have begun earlier than the new economy.

It is also possible, given the demonstrated delay of the morphological proof of "domestication," that incipient cultivation occurred during the Khiamian in a form still imperceptible to us. As it was not yet an "agricultural economy" in which cultivated plants were treated as an essential food source, there may not have been the quantity of materials found at settlements such as Mureybet IIIB. Thus, the first experiments could have escaped us completely. Similar to Isaac's ideas (1962) of a "religious" origin of animal domestication, we can model that the first agricultural experiments were contemporary to the revolution of symbols, expressing simultaneously in the *praxis*, as a sort of "ritual," the mental transformations of human communities in the Levant.

THE ASCENDANCY OF MASCULINE SYMBOLS AND THE NEOLITHIC TRIUMPHANT

In comparison to the PPNA, archaeological excavation and survey of PPNB settlements of the Near East have furnished a much more detailed understanding of a clear overall demographic increase, the appearance of more numerous villages, and the expansion of communities into new regions such as Anatolia. On the symbolic level, female representations continued to dominate, but a few explicitly masculine anthropomorphic figurines also appeared, and the theme of the bull spread throughout the Near East (Levant and eastern Anatolia). Other artistic and funerary phenomena, discussed later, enrich our understanding of the religious data. In addition to architectural, technological, and economic data, consideration of the PPNB illustrates the utility and importance of symbolic data in addressing the mental structures of a period and outlines that a profound unity of thought underlies in reality all the manifestations of PPNB culture from the most spontaneous to the most utilitarian.

Despite local variations the cultural tradition of the PPNB comprises from beginning to end a definite number of stable features, some of which had their origin in the end of the PPNA of the Euphrates. Besides the symbols of the woman and the bull, there is rectangular architecture and flint knapping on bipolar nuclei (naviform) to produce fine regular blades; these blades were intended particularly for arms (projectile points and daggers) of which the quantitative importance and the aesthetic are already perceptible in the earlier Mureybetian. Other features appeared only in the PPNB: the use of a flat lamellar retouch on flint or obsidian blades gives rise to a specific weapons typology (Byblos points and Amuq points). These weapons rise in number and quality, suggesting that they were prestigious items in these communities.

While the use of domesticated plants and agriculture spread throughout the Near East by the start of the PPNB, the beginnings of animal domestication occurred only in the middle PPNB, about 10,000 BP. The first domesticated animal was the goat, simultaneously in the northern and in the southern Levant. It was followed in the late PPNB (about 9,500 BP) by sheep, then at the end of the eighth millennium by cattle and pigs (Helmer 1992). About 9,000 BP, humans had completed their mastery over the main food animals; among these the domestication of the sheep in the late PPNB led to pastoral nomadism, which gave to certain human groups a new freedom of movement (Cauvin 1980; Stordeur 1993).

Another important feature to keep in mind is the speed and encompassing nature by which the PPNB expanded into other regions of the Near East. From the late PPNB, perhaps even from the end of the Late Mureybetian that gave rise to it, southeast Anatolia (Çayönü, Cafer Höyük, Nevali Çori) became Neolithic, with the appearance of the "PPNB of the Taurus" (Cauvin 1988), which combined local elements to the main features of the PPNB as a whole, including agriculture, rectangular houses, importance of arms, cult of the goddess and of the bull, and, later, animal domestication. As discovered by Kenyon (1957), it is in the Middle PPNB that Palestine was quickly taken over by PPNB population from the north. It is, moreover, during the late PPNB that there occurs the "great exodus" when Neolithic occupation with pastoral nomadism takes place in zones peripheral to the central area of the Levant, such as western Syria and Lebanon, northern Iraq (Magzalia), and the semidesert zones of the Levant (the Syrian, Jordanian, and Sinai deserts). While these regional cultures are distinct facies of the PPNB, the general traits of this culture are maintained, with the exception of the round houses characteristic of the nomadic PPNB.

I have suggested elsewhere that these diverse cultural or economic domains should not be looked on as so many separate "chapters" without relation to each other, but that they were in fact "homologous" in the sense of Levi-Strauss. That is, they all presented on various planes of reality a single mental structure, directly intelligible at the level of religious symbolism. In this symbolism there is evidence of the beginning of an ascendancy of male artistic representations, such as the masculine figurines at Mureybet IV, at Cafer Höyük (Cauvin 1989:Figure 11), and in Palestine at Munhata, with ithyphallic figures (Perrot 1966b). Bull figurines especially spread everywhere from the Taurus to the Dead Sea. A particularly expressive example is that of 'Ain Ghazal (Rollefson 1983), where zoomorphic figurines are abundant and, at times, found with a bladelet of flint stuck in the head or chest, a motif that could be an early allusion to the bullfighting theme found in the painted frescoes of the seventh millennium at Çatal Höyük and in the eastern Mediterranean of the Bronze Age.

There also appears to be a relationship between this ascendancy of masculine symbols and the broader patterns of PPNB material culture previously discussed. The angularity of PPNB residential architecture, for example, where the straight line dominates, contrasts with the matrix-like roundness of the pithouses of the Natufian–PPNA tradition. The abundance of arms and the aesthetic perfection of certain specimens do not appear to be linked to any new need; on the contrary, at this period hunting gives way to more peaceful herding; this could only indicate a "love of arms" as a social value with virile significance. It is also important to note that the origins of animal domestication to not appear to be linked to securing reliable meat resources. Following Legge, Helmer (1992) has shown that animal domestication at Middle PPNB Abu Hureyra with the first modifications in size of domesticated animals concerns only a small number of animals of limited food importance, and does not become a vital economic factor until the Late PPNB. Digard argues within present-day ethnographic cases that "the stupefying domesticating zeal of man" doubtless had as an end simply "human domination over creatures and things" (Digard 1990:215). Contrary to the traditional conceptions, pastoral economies are the practical and already secularized result of a confrontation of man with the animal kingdom, at first lived in a mythic and ritual form. The bull was felt to be the most expressive symbol of this irrational and sometimes destructive power of wild nature, to master in order to be truly "a man." This primordial confrontation could be represented by the bull–man combat, which goes back to the Neolithic and is maintained in present-day in Spain due to the persistent wildness of uncastrated bovine males. It is present at Çatal Höyük where cattle were, according to the archaeozoologists, in the process of being domesticated, whereas the PPNB of 'Ain Ghazal had mastered only sheep and goats, obviously more easily approachable than wild cattle.

This interpretation certainly requires on our part a certain "empathy" (Colingwood), in accordance with contextual archaeology, but it is not arbitrary. The only license the prehistorian has here, in order to understand Neolithic symbolic system and imagery, is to incorporate models based on later and literary civilizations, such as those furnished by the Bronze Age civilizations occurring in the *same* cultural area in conjunction with the *same* images. We have evoked the formal identity, already noted by Mellaart, between the masculine figure at Çatal Höyük riding a bull and the Phoenician Hadad. Similarly, the "Baal" of Ugarit, described as riding the heavenly bull, is viewed as symbolizing the storm and war (his destructive power), but he is also the master of cultivated fields and the civilizing hero (Caquot et al. 1974). This dialectic relationship, previously illustrated by the bullfighting image, exists between the raging force and controlled virility, between "bestial" violence (outside man or within him) and the hero's mastery. It is also, collectively, the passage from nature to culture. The earliest history of this symbolism teaches us that the great civilizing changes of the Neolithic were at first anticipated and played out within religious and ritual contexts and that only then men themselves achieved that which was first attributed to the gods.

Where are social structures situated in this sequence? Funeral rites provide precious information for this question, but so far most archaeologists have felt at ease only when addressing dimensions of social hierarchy and the emergence of social differentiation. Burials and the richness of their offerings, for example, are a highly visible means by which inequalities are expressed. Inequality occurs among all human groups for natural abilities are never divided evenly. For the most part, however, archaeologists have been more concerned with institutional inequality—something that only emerges at the end of the Near Eastern Neolithic when social differentiation in urban civilizations is foreshadowed. Nevertheless, study of burials, sacred places and mortuary ceremonies in the egalitarian Pre-Pottery Neolithic provides a rich source of information for understanding existing social organization. It is to be recalled that for Gordon Childe the "temple" and the "palace" were the principal indicators of urbanization. Interestingly, recent archaeological research has demonstrated that by the PPNB—that is, long before this urbanization—exceptional nonresidential buildings existed, buildings that probably served as sanctuaries constructed for religious use. These are seen from the early PPNB at Çayönü (Özdogan and Özdogan 1990) and at Nevali Cori (Hauptmann 1988) in Anatolia, to which should probably be added a large Middle PPNB building at Beidha in Jordan (Kirkbride 1966). The sanctuary of Çayönü contains an enormous concentration of burials and human skulls (more than 400 individuals), which make it a veritable "House of the Dead." Although different, one structure at Nevali Cori has in its center a large stone stele, roughly anthropomorphic in design. The three sanctuaries each have a large quadrangular stone slab; that of Çayönü was discovered by microbiologists to have traces of blood on its surface (Ozbek 1988), which would apparently indicate its use as a sacrificial stone.

These sanctuaries were, then, places of assembly and religion. We should compare the many lime plaster statues found in Palestine at Jericho and 'Ain Ghazal and the stone masks found in the south Levant in several places, in particular at Nahal Hemar (Bar-Yosef 1985). The statues are light and flat, obviously intended to be carried or exhibited; like the masks and the sanctuaries themselves, they indicate a socially organized religious life, with ceremonies that were probably recurring and directed by a few charismatic individuals rather than a true clergy. Examination of the architectural context of these villages provides no evidence for "palaces," the more secular version of exceptional buildings. The residential structures remain homogenous, without discontinuities in size suggesting social divisions. The fact that the first exceptional constructions were religious buildings is thus significant. The PPNA tower of Jericho, for example, was already an exceptional construction the role of which was probably more symbolic than defensive, as demonstrated by its last reuse as a collective burial place. But it was not, however, a place of assembly. The great sanctuaries of the PPNB introduce for the first time architectural evidence for the possibility of organized secular and religious meetings.

Consideration of the evolution of Near Eastern Neolithic funerary customs, and specifically the "cult of skulls," illustrates that these played a fundamental role in structuring broader social life itself. It first appears in the PPNA in the custom of burying human skulls separately from the bodies, as for example at Jericho (Kenyon 1957:72) and in more recent discoveries at Cheikh Hassan (Cauvin 1980) and Jerf el Ahmar. This is not only an indicator of secondary burial but also of a marked preference for the head in the evocation of the reburied dead. In the same way, I argue elsewhere (Cauvin 1987:1478) that, contrary to the Paleolithic "Venuses," from the PPNB onward anthropomorphic figures of the Near East more often emphasized the head and the eyes, often painted or represented by clay globules or shells. The eyes and their expression evoke that which is most subjective and specifically psychic in the human being. In the case of the figures of the goddess, it is her power of fascination that is probably indicated by the emphasis on the eyes.

Two facts appear particularly significant when considering the PPNB "cult of skulls" (see Bienert 1991; Kuijt 1995). The first concerns their location: accurate observations of excavations have shown that the skulls were possibly not buried, but stored in cells or fabricated containers (for example at Çayönü or Ramad) or discovered in situ on the very floor of the houses (and not beneath), as for example at Mureybet IV. At Mureybet, they were placed on clay supports, which probably also existed, roughly anthropomorphic, at Ramad (Contenson 1992). In other words, these are a kind of cult object intended to be visible (continuously, or intermittently and ceremoniously if they were stored) to the living. This "object" aspect is confirmed by a second fact: this is the "artistic" application to which the skulls could be subject. For example, traces of paint remain on a skull from Abu Hureyra, and in most cases of lime-plaster modeling of skulls in the central and southern Levant extensive attempts were taken to reconstruct the face. One finds the same emphasis on the facial expression, through the use of painting or shells for the eyes, with the statues of Jericho or 'Ain Ghazal.

The fact that the heads were preserved and kept in condition for viewing by the broader village community was a new phenomenon that contrasts to earlier simple funerary piety, in which the dead were buried in the earth, leaving their memory to become fragile. This indicates, in the sedentary communities of the PPNB, a carefully maintained memory of ancestors through ritual ceremonies. Thus, there was an increased sense of human lineage, a strengthening of social links, and an affirmation of collective property in definite territories. This does not necessarily mean, however, that this ideology was only the outcome of a new organization of society. Although it is difficult to define which came first, it is more probable that we have two simultaneous sides—the one interior, the other exterior—of the same overall transformation of Near Eastern populations. The projection of the human image in religious art as well as in funeral practices unquestionably constitutes an important stage in the development of the self-consciousness of our species.

CONCLUSIONS

In many ways this discussion of the symbolic aspects of the Neolithic revolution is intended more to suggest a direction for future research than to discuss results from previous studies. While it can be argued by some researchers that there are insufficient archaeological data sets to engage in the kinds of debates presented here, it should also be pointed out that archaeologists, like all scholars, tend to find only what they are seeking. Nothing is more limiting and blinding in the accumulation of information than the *a priori* acceptance of select data sets, philosophical approaches, and research avenues. Contextual archaeology has been critiqued on these grounds: an accusation that is, despite the intrinsic interest of its ideas, favored by the very theoretical and ideological perspective in which it is situated. We should note, however, that the "materialism" of the New Archaeology is itself a philosophical *a priori,* one that very much resembles what psychoanalysts call a "blockage." Needless to say, such a blockage is also encouraged in prehistorians as a result of the quasi-"residual" nature of our data, for we lack the texts exploited by historians and the living words gathered by ethnologists. In the case of late prehistory and, specifically, the Near East Neolithic, the existence of art and the development of rich ritual and funerary practices provide a valuable, and in some ways largely unexplored, source of information. To this end, "symbolic archaeology" has contributed, by its new interpretations of material facts themselves, toward the process of freeing of archaeological research from the spell of its positivist reflexes that has been going on for nearly twenty years in the Near East (Cauvin 1978). It is still important to verify in the field the value of these ideas and especially to see whether they clarify facts that are otherwise inexplicable. Although still in its infancy, the models and interpretations presented here aid us in explaining the order of stratigraphically observed phenomena, the results of which radically refute strictly materialist interpretations.

On the other hand, it is an exaggeration to see the hermeneutic approach as only an encouragement of subjective arbitrary affirmations (Binford and Binford 1968). Certainly this danger exists, but we are clearly in a time when in order to face the anarchic invasion of irrational forms of thought, science has undertaken a revision of the foundations of scientific rationality itself. As French historians of the École des Annales have demonstrated, in developing the importance of symbolic structures in the history of societies (for example, Dumezil, Le Goff and Vernant), such an interest in "mentalities" is only outrageous for scholars who support an outdated ideology. Although the difficulties of extending such an approach to the Neolithic are real, we risk missing an essential dimension of human history by ignoring this path.

REFERENCES

Anderson-Gerfaud, P., Deraprahamian, A. and G. Willcox, 1991, Les Premières Cultures de Céréales Sauvages et Domestiques Primitives au Proche-Orient Néolithique: Résultats Préliminaires d'Expériences à Jalès (Ardèche). *Cahiers de l'Euphrate,* 5–6: 191–232.

Bar-Yosef, O., 1980, A Human Figurine from a Khiamian Site in the Lower Jordan Valley. *Paléorient,* 6: 193–299.

Bar-Yosef, O., 1985, *A Cave in the Desert:Nahal Hemar.* Jerusalem, The Israel Museum.

Bar-Yosef, O., Gopher, A., and Goring-Morris, N., 1980, Netiv Hagdud: A "Sultanian" Mound in the Lower Jordanian Valley. *Paléorient,* 6: 201–206.

Bienert, H. D., 1991, Skull Cult in the Prehistoric Near East. *Journal of Prehistoric Religion,* 5: 9–23.

Binford, S. R., and Binford, L. R. (eds.), 1968, *New Perspectives in Archaeology.* Aldine, Chicago, Illinois.

Bintcliff, J., 1991, Post Modernism, Rhetoric and Acholasticism at TAG: The Current State of British Archaeological Theory. *Antiquity,* 65: 274–278.

Caquot, A., Szycer, M., and Herdner, A., 1974, *Textes Ougaritiques. I—Mythes et Légendes,* Editions du Cerf, Paris.

Cauvin, J., 1977, Les Fouilles de Mureybet (1971–1974) et Leur Signification pour les Origines de la Sédentarisation au Proche-Orient. *Annual of the American School of Oriental Research,* 44: 19–47.

Cauvin, J., 1978, *Les Premiers Villages de Syrie-Palestine du IXe au VIIe millénaire av. J. C.* CNRS, Paris.

Cauvin, J., 1980, Mureybet et Cheikh Hassan. In *Le Moyen Euphrate, Zone de Contacts et d'échanges,* edited by J. Margueron, pp. 21–34. Universitè des Sciences Humaines, Strasbourg.

Cauvin, J., 1987, L'apparition des Premières Divinités. *La Recherche,* 194:1472–1480.

Cauvin, J., 1988, La Néolithisation de la Turquie du Sud Est dans Son Contexte Proche-Oriental. *Anatolica,* 15:70–80.

Cauvin, J., 1989, La Stratigraphie de Cafer Höyük-Est (Turquie) et les Origines du PPNB du Taurus. *Paléorient,* 15(1):75–86.

Cauvin, J., 1994, *Naissance des Divinités. Naissance de l'Agriculture. La Révolution des Symboles au Néolithique.* CNRS, Paris.

Chauvet, J. M., Brunel Deschamps, E., and Hillaire, C., 1995, *La Grotte Chauvet, à Vallon Pont d'Arc,* Ed. du Seuil, Paris.

de Contenson, H., 1992 , Les Coutumes Funéraires dans le Néolithique Syrien. *Bulletin de la Société Préhistorique Française,* 89(6): 184–191.

Delporte, H., 1979, L'*Image de la Hemme dans l'Art Préhistorique*. Picard, Paris.

Digard, D., 1990, *L'Homme et les Animaux Domestiques. Anthropologie d'une Passion*. Fayard, Paris.

Ducos, P., 1972, The Oriental Excavations at Mureybet, Syria, Preliminary Report on the 1965 Campaign. Part V: Les Restes de Bovidés. *The Journal of Near Eastern Studies,* 31(4) 295^–304.

Feyerabend, P. K., 1979, *Contre la Méthode* (French translation). Ed. du Seuil, Paris.

Flannery, K. V., 1972, The Origins of the Village as a Settlement Type in Mesoamerica and the Near East: A Comparaison Study. In *Man Settlement and Urbanism,* edited by P. J. Ucko, R. Tringham and, G. W. Dimbley, pp. 23–54. Duckworth, London.

Gardin, J. C., 1992, Semiotic trends in Archaeology. In *Representations in Archaeology,* edited by J. C. Gardin, and C. S. Peebles, pp. 87–106. Indiana University Press, Bloomington, Indiana.

Hauptmann, H., 1988, Nevali Çori Architektur, *Anatolica,* 15: 99–110.

Helmer, D., 1992, *La Domestication des Animaux par les Hommes Préhistoriques*. Masson, Paris.

Hodder, I. , 1989, *The Meanings of Things*. Unwin Hyman, London.

Hodder, I., 1990, *The Domestication of Europe*. Blackwell, Oxford..

Isaac, E., 1962, On the Domestication of cCttle. *Science,* 137: 195–204.

Kenyon, K., 1957, *Digging up Jericho*. Ernest Benn, London.

Kirkbride, D., 1966, Fire Seasons at the Pre-Pottery Neolithic Village of Beidha in Jordan. A Summary. *Palestine Exploration Quarterly,* 98(1) 8–61.

Kislev, M. E., 1992, Agriculture in the Near East in the 8th millenium b.c. In *Préhistoire de l'Agriculture: Nouvelles Approches Expérimentales et Ethnographiques,* edited by P. Anderson-Gerfaud, pp. 87–93. CNRS, Paris.

Kuhn, T. S., 1977, *The Essential Tension. Selected Studies in Scientific Tradition and Change*. The University of Chicago Press, Chicago, Illinois.

Kuijt, I., 1995, *New Perspectives on Old Territories: Ritual Practies and Social and The Emergence of Social Complexity in the Levantin Neolithic*. Ph.D. dissertation, Harvard University. University Microfilms, Ann Arbor, Michigan.

Leroi-Gourhan, A., 1965, *Préhistoire de l'Art Occidental*. Mazenod, Paris.

Leroi-Gourhan, Arlette, 1974, Etude palynologique des Derniers 11,000 ans en Syrie semi-Désertique. *Paléorient,* 2(2): 443–451.

Mellaart, J., 1967, *Çatal Hüyük. A Neolithic Town in Anatolia*. Thames and Hudson, London.

Morin, E., 1991, *La Méthode, t. 4: Les Idées, Leur Habitat, Leur Vie, Leurs Moeurs, Leur Organisation*. Ed. du Seuil, Paris.

Moore, A. M. T., and Hillman, G. C., 1992, The Pleistocene to Holocene Transition and Human Economy in Southwest Asia: The Impact of the Younger Dryas. *American Antiquity,* 97(3): 482–494.

Özbek, M., 1988, Culte des Crânes Humains à Çayönü. *Anatolica,* 15: 127–138.

Özdogan, M., and Özdogan, A., 1990, Çayönü, a Conspectus of Recent Works. In *Préhistoire du Levant: Processus des Changements Culturels,* edited by O. Aurenche, M.-C. Cauvin, and P. Sanlaville, pp. 68–77. CNRS, Paris.

Peebles, C. S., 1992, Rooting out Latent Behaviorism in Prehistory. In *Representations in Archaeology,* edited by J. C. Gardin. and C. S. Peebles, pp. 357–384. Indiana University Press, Blomington, Indiana.

Perrot, J., 1966a, Le Gisement Natoufien de Mallaha (Eyran), Israël. *L'Anthropologie,* 70(5–6): 437–483.

Perrot, J., 1966b, La Troisième Campagne de Fouilles à Munhata. *Syria* 43(1–2): 49–63.

Popper, K. R., 1959, *The Logic of Scientific Discovery*. Hutchison, London.

Renfrew, C., 1994a, Towards a cognitive archaeology. In *The Ancient Mind. Elements of Cognitive Archaeology,* edited by C. Renfrew and E. B. W. Zubrow, pp. 3–12. Cambridge University Press, Cambridge.

Renfrew, C., 1994b, The Archaelogy of Religion. In *The Ancient Mind. Elements of Cognitive Archaeology,* edited by C. Renfrew and E. B. W. Zubrow, pp. 47–54. Cambridge University Press, Cambridge.

Renfrew, C., and P. Bahn, 1991, *Archaeology, Theories, Methods and Practices.* Thames and Hudson, London.

Rollefson, G. A., 1983, Ritual and Ceremony at Neolithic 'Ain Ghazal (Jordan). Paléorient, 9(2): 29–38.

Rossignol-Strick, M., 1993, Late Quaternary Climate in the Eastern Mediterranean Region. *Paléorient,* 19(1): 135–152.

Stordeur, D., 1993, Sédentaires et Nomades du PPNB Final dans le Désert de Palmyre. *Paléorient,* 19(1): 187–204.

Van Zeist, W., and Bakker-Heeres, J. H., 1984, Archaeobotanical Studies in the Levant: 1, Neolithic Sites in the Damascus Basin: Aswad, Ghoraife, Ramad, *Palaeohistoria,* 24: 165–256.

Willcox, G., 1995, Wild and Domestic Cereal Exploitation: New Evidence from Early Neolithic Sites in the Northern Levant and South Eastern Anatolia. *Arx,* 1(1): 9–16.

Zohary, D., 1992, Domestication of the Neolithic Near East Crop Assemblage. In *Préhistoire de l'Agriculture: Nouvelles Approches Expérimentales et Ethnographiques,* edited by P. Anderson-Gerfaud, pp. 81–86. CNRS, Paris.

Chapter **11**

Çatal Höyük in Context

Ritual at Early Neolithic Sites in Central and Eastern Turkey

Mary M. Voigt

INTRODUCTION

Any study of religious beliefs and practices of early agricultural communities in the Middle East must consider the rich assemblage of sculpture and mural art from Çatal Höyük. In popular accounts as well as introductory textbooks on archaeology, this well-preserved settlement in central Anatolia is used to support a belief in an Early Neolithic goddess (for example, see the Çatal Höyük website: www/catal.arch.cam.ac.uk/catal/goddess; Fagan 1998). The goddess from Anatolia was initially described by James Mellaart, who directed excavations at Çatal from 1961 to 1965 (Mellaart et al. 1962, 1963, 1964, 1966, 1967, 1989). Using a direct historical approach to interpret both sculpture and mural art, Mellaart inferred a mythological and ritual system centered on the "Great Goddess," who "as the only source of life . . . became associated with the processes of agriculture, with the taming and nourishing of domesticated animals, with the ideas of increase,

MARY M. VOIGT • Department of Anthropology, College of William and Mary, Williamsburg, Virginia 23187-8795.

Life in Neolithic Farming Communities: Social Organization, Identity, and Differentiation, edited by Ian Kuijt. Kluwer Academic/Plenum Publishers, New York, 2000.

abundance, and fertility" (Mellaart 1967:202). More recently he has described this deity as the "source and mistress of all life, the Creatress, the Great Mother, the symbol of life itself" (Mellaart et al., 1989:23).

More complex constructions of the ideology represented by the material remains from Çatal rely on ethnographic comparisons drawn from regions beyond the ancient Mediterranean. Jean-Daniel Forest (1993, 1994) restricts his discussion to the mural art from Çatal, including both figural and geometric designs. Relying on a structuralist approach and drawing analogies with (unspecified) African societies that he considers similar to Çatal (Forest 1994:118), Forest views the arrangement of wall elements as a metaphor for kinship, with females serving as links between male lineages. A female principle is ubiquitous and ambiguous, beneficent and maleficent, associated with life/birth and death (Forest 1994:123). He states that "the whole set of designs and patterns, either figurative or not, either moulded or painted, may be classified into two categories, respectively referring to a generative principle, conceived as female, and to its product, conceived as male" (Forest 1994:127).

An interpretation of the Çatal material that is more firmly grounded in archaeological data is that presented by Ian Hodder (1987, 1990), who began a new program of archaeological research at the site in 1993. Hodder performed a "contextual" analysis, which he defines as an attempt "to 'read' or interpret the evidence primarily in terms of its internal relations rather than in terms of outside knowledge" (1990:20). He also placed Çatal within a broader cultural context, that of the European Neolithic. Ethnographic analogy is again a key element in interpretation, in this case a specific comparison of Çatal imagery with that of the Nuba of Ethiopia and other well-documented groups (Hodder 1990:5–8). He concludes that there are "sufficient contextual links within the Çatal Höyük data to justify the reconstruction of a symbolism concerning a power and danger associated with female representation" (Hodder 1990:8). Adopting a structuralist approach, Hodder sets up a series of oppositions or "sets of rules" expressed within architecture, mural art, and figurines: male/female, inside/outside (houses), death/life, and wild/domestic (1987, 1990:10). He concludes that:

> The varied dimensions of meaning identified at Çatal Höyük can be seen to be involved most closely in the negotiations between men and women within the domestic context. Women were represented, misrepresented or made invisible in different domains. The ambiguous and contradictory meanings of symbols were involved both in establishing women as life-givers and life-takers. The dependence of society on women is incorporated, transformed, and denied (Hodder 1987:52).

Building upon these previous works, I draw upon ethnographic literature and previous considerations of the Hajji Firuz and Gritille Höyük figurines to discuss how patterned distributional data and consideration of figuring form and breakage aid us in interpreting Near Eastern Neolithic figurine industry in general and Çatal Höyük in particular. Moreover, the goals of this chapter are to specify some of the internal relationships that are central to Hodder's analysis and to any alternative interpretation of the Çatal Höyük figurine industry. My initial definition of "context," however, differs from Hodder's. Starting with dirt—the *archaeological context* in which early Near Eastern figures and figurines occur—I will move to a consideration of the human behavior that might account for the location and condition of figures and figurines at any site—that is, their *immediate cultural context*. Finally I consider the *culture historical context*, looking east rather than west and to contemporary and earlier sites rather than the later sites in Europe. Just as Hodder's analysis stems directly from his research among the Nuba and his interest in the spread of agriculture into Europe, my own analysis grew out of my fieldwork at two Neolithic sites located in northwestern Iran and in southeastern Turkey. This chapter begins with a summary of results from Hajji Firuz Tepe and Gritille, each of which contributed to my picture of ritual and religion at early sedentary communities in the Near East.

DEFINING ARCHAEOLOGICAL CONTEXT: HAJJI FIRUZ TEPE

Excavation carried out in 1968 at Hajji Firuz Tepe (northwestern Iran) was designed to recover information on economic and social organization of a sixth millennium farming community. Recovery methods emphasized artifact distributions in an effort to define the economic activities performed by households, providing control over the location of finds relative to architecture, trash deposits, and other features such as hearths and burials. This kind of recording is normal for sites in the Near East excavated over the past couple of decades, but represented an innovation at the time (see, for example, the discussion of provenience problems at Çatal Höyük in Hamilton 1996). The stratigraphy at Hajji Firuz allowed the definition of relatively small chronological units, estimated as around fifteen years based on an architectural cycle: the construction, use, and decay of mud brick buildings (Voigt 1983:18–30). In order to move from architecture and artifact to action, I relied heavily on ethnographic analogy, where possible supplemented by wear patterns. This approach was used in the analysis of all artifacts recovered, including clay figurines. What follows is a brief summary of the evidence and argument published in 1983.

NATURE OF THE HAJJI FIRUZ SAMPLE

All of the small clay objects from the site came from the latest part of the occupation, phases D through A. To a degree this chronological distribution reflects excavation methods rather than any change in behavior on the part of the sixth millennium villagers. For example, phases A through D were sampled in 1968 using small picks and (sometimes) sieves in an effort to recover all types of material remains. Phases before D were excavated in 1961 with large picks and no sieves in an effort to test the nature and depth of the deposit. The portion of the settlement sampled for each of these phases includes a series of small freestanding houses with intramural burials, two small anomalous structures (VI and VII), and exterior areas with cooking and industrial features and trash deposits (Voigt 1983:18–94, 295–321). During artifact analysis, I initially divided the small, lightly fired clay bits into three categories based on form: (1) figurines of humans and animals, (2) geometrics or tokens, and (3) sealings or pieces that been impressed or modeled (Figure 1; see also Voigt 1983:175–185, Figures 101–102, Plate 27, 28a–f). I then looked at the distribution of these artifacts within each of four chronological phases and found that all three classes showed some patterning; there is an overlap in the distribution of figures and sealings, but a very different distribution for geometrics (Figures 2 and 3). Ignoring tokens and sealings, individual figurine fragments are scattered within and between houses, but there is a consistent clustering in and around a series of ashy pits that I have interpreted as the remnants of bonfire kilns (Voigt 1983:69).

One might be tempted to explain this distribution as the result of sampling error: unbaked clay figures could have been widely distributed within the site, but the only examples recovered were those that happened to fall into a fire and were therefore preserved. While this bias in the sample is possible, it does not explain the entire set of actions documented by these artifacts. The figures were always broken, and in only one case were two halves found; moreover, the breaks were not at points of structural weakness, suggesting deliberate breakage and separation of the two halves (Table 1). We therefore have evidence for manufacture and destruction of figurines, perhaps in the same areas as they had been made or at least fired. Finally, most of the lightly baked figures have a delicate surface, but little or no damage to that surface. The sample is small but consistent.

Figurine Form: Application of the Ucko Typology

In order to abstract some meaning from the patterns observed at Hajji Firuz, I relied heavily on the most systematic discussion of figurines available in

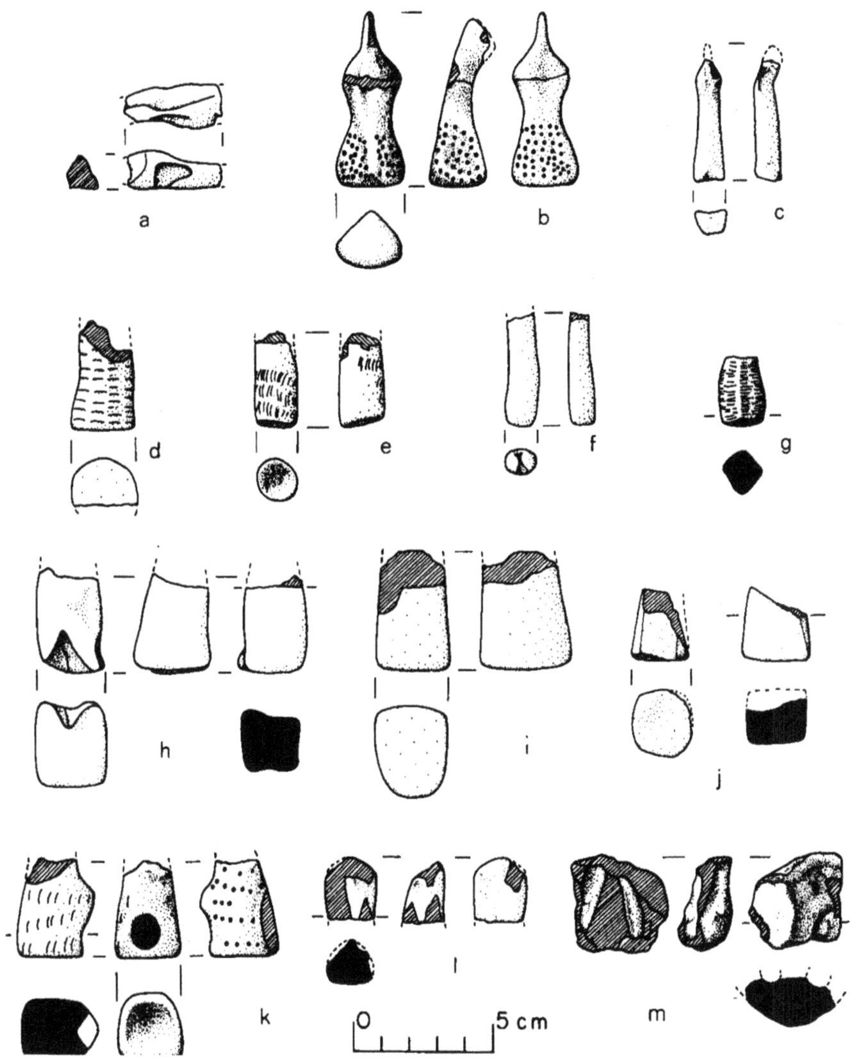

Figure 1. Figurines from Hajji Firuz Tepe (based on Voigt, 1983:Figure 101).

the early 1970s, a monograph by Peter Ucko (1968) that focused on figures found in prehistoric Egyptian sites and tombs, but also discussed Mesopotamian examples.[1] Ucko grouped his material into four functional classes that he defined on the basis of ethnographic accounts:

[1]Figurines have been used in social and economic transactions as symbols of a contract, but this usage is relatively rare and is considered unlikely within Neolithic societies (see Nickerson 1979 and Talalay 1993 for discussion of figurines as contract).

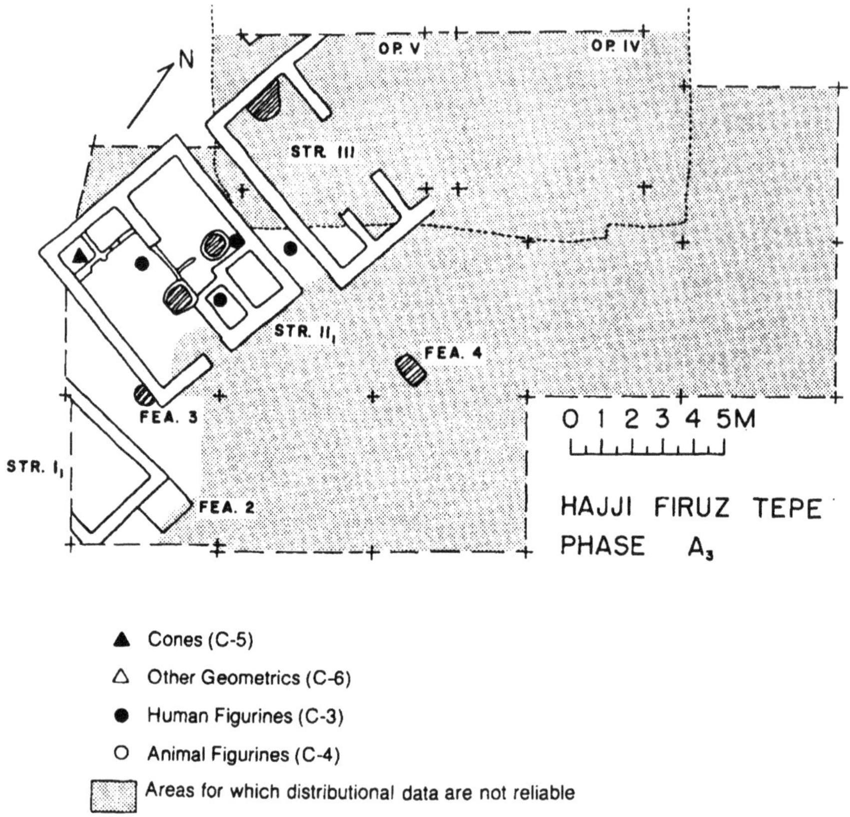

Figure 2. Distribution of small baked clay objects, Hajji Firuz Phase A (based on Voigt, 1983:Figure 103).

1. *Cult figures*, or representations of supernatural beings used primarily as symbols or objects of worship—formal, usually community rituals.
2. *Vehicles of magic*, or figurines used in rituals intended to produce, prevent or reverse a specific situation or state (increased fertility, healthy children, protection of property or crops, harm to one's enemies).
3. *Teaching figures*, including those used in initiation rites to teach adolescent children the proper kinds of behavior.
4. *Toys*, or figures used in entertainment or children's play; the adult equivalent would be ornaments for decoration or aesthetic effect.

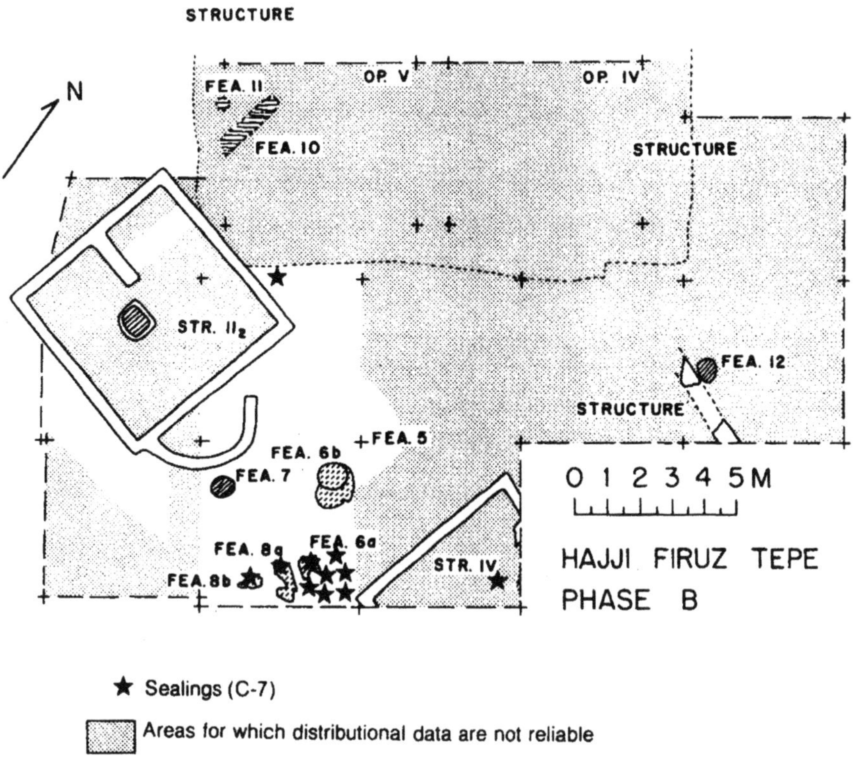

Figure 3. Distribution of small baked clay objects, Hajjir Firuz Phase B (based on Voigt, 1983:Figure 104a).

Ucko then listed the morphological attributes that characterized each group of ethnographic figures, as summarized in Table 2.

Using form as a means of understanding the Hajji Firuz figurines was a discouraging experience. They were simple and similar, with blocky bases and pinched heads. Applying Ucko's attribute sets suggested that we had only one or two functional classes. We could eliminate use of the Hajji Firuz figurines in initiation ceremonies, but form alone was of little use in deciding whether the Hajji Firuz figures were associated with ritual or entertainment. To resolve this question I returned to the distributional data.

Table 1. Wear, Damage, and Depositional Attributes of the Hajji Firuz Clay Figurines (Types 3 and 4)

Field number	Surface decor.[1]	Surface wear[2]	Major damage	Context type, location	Associated material[3]			
					F	S	A	B
Human (Type 3)								
68-26	U	None	Broken at torso, inc.	Lensed trash deposit adjacent to Str. III	O	X	X	O
68-53	U	None?	Brokent at torso, inc.	Unstratified (K10)	M	M	M	M
68-53	F	None?	Broken at torso, inc.	Soft trashy deposit above floor of oven in Str. II_1	O	X	X	O
68-62	U	Chipped base?	Broken at torso, inc.	Wall collapse in courtyard of Str. V	O	X	X	X
68-96	O	Very worn surface chipped	Broken at neck (head fragment)	Trash deposit on floor of Str. II_1, Rm. 1	O	X	X	X
68-100	P	None	Broken at chest, compl.	Unstratified (T2/3)	M	M	M	M
68-112	O	Base very worn, sides worn	Broken at torso, inc.	Trash deposit above floor of Str. II_1, Rm. 2, near Hearth 2	O	X	X	X
68-127	F	None	Broken at torso, inc.	Lensed clay, probably wall collapse; exterior	O	X	X	X
68-145	F	None	Broken at neck, inc.	Lensed ash at bottom of burnt pit (Fea. 6b)	O	X	X	X
68-149	F, O	None?	Broken at torso, inc.	Trash deposits adjacent to burnt pit (Fea. 9)	X	X	X	X
68-186	F, P, O	Chipped?	Broken at torso, inc.	Trash deposit on surface adjacent to Str. VI_1	O	X	O	X
68-218	U	Base worn	Broken at torso, inc.	Unstratified (T2/3)	M	M	M	M
68-222	U	None	Broken at torso, inc.	Trash deposit adjacent to burnt pit (Fea. 9)	X	X	X	X
68-225	U	None	Broken at torso, leg, inc.	Trash deposits adjacent burnt pit (Fea. 9)	X	X	X	X
Animal (Type 4)								
68-220	U	None	Broken at both ends, inc.	Trash deposits adjacent to burnt pit (Fea.9)	X	X	X	X
68-221	F, P	None	Broken at ends, side, inc.	Trash deposits adjacent to burnt pit (Fea 9)	X	X	X	X

[1]Abbreviations used to describe suface: U = undecorated; F = fingernail impressed; P = punctuated; O = ochre wash.
[2]Artifacts stored in the Iran Bastan Museum were not specifically examined for surface wear. Assessments of these pieces are based on drawings and descirptions, but should be considered tentative.
[3]Abbreviations for associated material: F = other figurines; S = sherds; A = other artifact types or chipping debris (see Voight 1983:Figs. 125–139); B = animal bone.
Source: Voigt (1983:Table 31).

Table 2. Figures and Figurines: Attributes of Functional Classes Based on Ethnographic and Ethnohistorical Sources[1]

Class		Attrubutes
Cult figure	Material	May be made of presious materials, or of common materials such as clay or wood
	Morphology	May be technologically superior to other types of figures
		Size is highly variable, ranging from large, stationary figures to small portable ones
		Usually anthropomorphic in form
		May be accompanied by iconic elements such as plants, animals or objects (headdresses, objects held in the hands)
	Use	May be used singly or in groups
		Generally used over an extended period of time
		May be handled (dressed, carried about), but with care
		May be stored/used in a special purpose (ritual) context or in a domestic context
	Disposal	Little data, but treatment probably different from that of ordinary or nonritual artifacts
Vehicle of magic	Material	Made of ordinary materials, including clay, wax, other organic substances; rarely, made of precious materials
	Morphology	Are small, portable
		May take the form of humans or animals; may be male, female, or "sexless"
	Use	Used singly or (rarely?) in groups
		May be used over an extended period (for example, when worn as an amulet) but usually made and disposed of as part of a single behavioral sequence
	Disposal	Frequently destroyed by breaking, burning
		Whole figures or fragments may be deposited within the fabric of domestic structures (within walls or floors, beneath floors, especially at thresholds), in pits in open areas or in bodies of water (streams, pools, wells)
		Fragments (and whole figures:) disposed of in habitation debris
Initiation figure	Material	Made of rare or costly materials as well as of clay or common organic substances
	Morphology	Vary widely in style, technical competence
		Size is variable, ranging from large stationary figures to small portable ones
		May take the form of humans or animals
		Form highly variable, including "strikingly nonconformist" figures
	Use	Used in groups, with each figure having a distinctive form and meaning
		Most used only for a short period, the duration of an initiation ritual, but some stored and used in several successive rites
		Handled during use, but with care
		Stored during use in a special structure (initiation hut); those used in several rites stored in secret place
	Disposal	Often destroyed by burning
		Thrown into bodies of water, habitation debris, rarely in houses

(continued)

Table 2. *Continued*

Class		Attrubutes
Toy	Material	Made of common materials, including clay, wood, or other organic materials
	Morphology	May be crudely or well made
		Portable, may be very small
		Include animals, hu mans, imaginary beings; may be sexless or show sexual characteristics in elaborate detail
		Often have arm stubs rather than arms, or are simple cylinders
	Use	May be used singly or in groups, with a tendency for larger figures to be used alone
		Less durable figures (of unbaked clay or organic materials) used for a relatively brief period; more durable figures may be used for years
		Careless or rough handling not uncommon.
		Used in domestic contexts, both inside houses and in open areas
	Disposal	Treated in similar fashion to any other kind of domestic trash; found in habitation debris, but never in rutal contexts

[1]The sources used to compile this table are those cited within Voigt 1983:186–193. ALthough the structure of this table was derived from Ucko (1962:47–48), its content represents emendations as well as additions to his attribute lists.
Source: Voigt (1983:Table 28).

Figurine Use and Disposal Patterns

Ucko had briefly considered disposal patterns for his functional classes, but could make little headway. For the Neolithic Near East he had only one site with a large corpus of figures, Jarmo, and the Jarmo figurines were said to be distributed throughout the deposit, with a concentration in ashy trash deposits that was "explained by the favorable conditions there for the recovery of figurines, and by the great extent of the ashy area" (Ucko 1968:364 citing Broman 1958:3, 47; see also [Broman] Morales 1983). But the Hajji Firuz figurines did show a patterned distribution and this pattern could be directly related to artifact disposal; they also exhibited consistent damage, which reflected use or disposal or both. I therefore expanded Ucko's approach by searching the ethnographic literature for information on disposal. I also looked for evidence of the way that figures were handled when in use, predicting the kind of wear that might result from specific actions. For example, if the behavior associated with a particular functional class involved a great deal of manipulation or handling, worn areas on the surface of the artifact may be visible (Voigt 1983:Tables 28–29). Suggested disposal and wear patterns for each of Ucko's functional categories are summarized in Table 3. Using a more recent (and less rigid) approach, these patterns can be recast, setting out a series of axes along which Ucko's functional groups can be arranged, showing increasing or decreasing probability of a specific

Table 3. Figures and Figurines: Predicted Patterns of Wear, Ruinous Damage, and Disposition Associated with Functional Classes

Class		Attrubutes
Cult figure	Wear and damage	Intact surface, or minor damage incurred during relatively careful handling
		Localized areas of abrasion or polish may occur on surface from ritual touching (for example, at head or feet)
		Figure may exhibit burning, ruinous fresh breaks due to "killing" at the time of disposal
	Disposition	May be deposited in special purpose (ritual) structure
		May be deposited in inaccessible places (for example, caves, bodies of water)
		Groups of figures which are similar or which differ in morphological characteristics may be associated because of the repeated deposition of figures used sequentially, or to the deposition of a number of cult figures at a single time
		Unlikely to be associated with ordinary refuse
Vehicle of magic	Wear and damage	Either no wear, or abrasion/polish of a type resulting from contact with a person wearing figures as an amulet
		Frequently exhibits burning, ruinous damage which occurred as part of the deposition process
		Characterized by fresh breaks in a consistent location (for ex ample, at neck, waist)
	Disposition	May be placed in the walls or beneath the floors of houses
		May be deposited in burnt features, pits in open areas, bodies of water
		Parts of broken figures are frequently separated at time of deposition, do not occur in same part of settlement
		Groups of figures may be associated by repetition of a rutual in a single locale, or by the use of several figures in a single ritual
		May be associated with ordinary domestic refuse
Initiation figure	Wear and damage	Surface may exhibit minor wear from handling, especially at base
		May be burnt, or intact and unburnt
	Disposition	Frequently disposed of in inaccessible areas such as caves, bodies of water
		Rarely associated with domestic strucutures, houses
		Groups of morphologically different figures usually associated because of deposition of entire teaching group as a unit
		Occasionally associated with ordinary domestic refuse
Toy	Wear and damage	Surface chipped and abraded, especially at base of standing figures
		Appendages frequently broken away
		Broken areas are worn, abraded due to continued use
		No systematic pattern of ruinous damage, except at points of structural weakness
	Disposition	Deposited in ordinary domestic contexts, both interior and exterior
		Figures randomly distributed in debris, not clustered
		Assocated with ordinary refuse, including bones, sherds, other kinds of broken artifacts

Source: Voigt (1983:Table 29).

use given patterns of wear and disposal (Table 4). Returning to the Hajji Firuz figurines, I concluded that the majority were probably vehicles of magic and therefore documented ritual behavior, but the data were more suggestive than convincing (Voigt 1983:193–195). Far more rewarding was my next analysis of Neolithic figurines.

MAGIC AND MEANING: GRITILLE HÖYÜK

From 1981 to 1984 I worked as a member of an archaeological team investigating a small mound on the left bank of the Euphrates near a minor river

Table 4. Probability of Figurine Function Based on Attributes

Material
Cheap (e.g., wood, mud, wax) — Expensive (Material, labor)

<————————Cult Images————————>
<—Vehicles of magic—>
<————Initiation figures————>
<——————Toys——————>

Morphology
Simple — Complex

<——————Cult Images——————>
<———Vehicles of Magic————>
<————————Initiation Figures————————>
<————————Toys————————>

Curation/Duration of Use
Short — Long

<——————Cult Images——————>
<-Vehicles of Magic————>
<———Initition Figures———>
<———Toys————>

Curation/Care in Use
Careless or deliberately destructive handling — Careful Handling

<—Cult Images—>
<——Vehicles of Magic———>
<———Initiation Figures———>
<—Toys——>

Disposal
Casual — Careful

<—Cult Images—>
<——————Vehicles of Magic——————>
<———Initiation Figures———>
<—Toys—>

crossing. The sequence at this site extends from medieval/crusader times back to the Aceramic Neolithic (Ellis and Voigt 1982; Voigt 1985, 1988; Voigt and Ellis 1981). Today both site and surrounding villages lie under water, destroyed as the river rose behind the world's largest earth dam. The Neolithic occupation was sampled along the eroding southern edge of the mound, exposing a narrow strip of the settlement. Four phases of occupation were defined, the earliest (phase D) resting on sterile soil and dated by radiocarbon to the tenth millennium bp (Voigt 1988). The largest sample comes from the third phase (B), which provides information on house form and settlement pattern for the Late Taurus Neolithic during the mid ninth millennium bp. While the sites in southeastern Turkey have distinctive aspects, their chipped stone industries, architecture, and use of lime containers ("white ware") link them to PPNB sites in the Levant, so the sites discussed here are usually referred to as part of the "Taurus PPNB" (see M. Özdogan 1995, 1997a for general discussions of the Neolithic in Turkey).

The Gritille Figurine Sample

The Gritille figurine industry is far more diverse than that from Hajji Firuz. Most of the figures were small and made of clay, but there were also rare stone figurines of humans (Figures 4 and 5; see also Voigt 1985:Figure 10–14). The clay figures included humans and animals, and within each category there was significant variation in form. Animal figurines could be recognized as representations of different species, the most common, being cattle. Human figurines ranged from the highly schematic to relatively realistic.

All but one of the nearly fifty clay figurine fragments recovered was found in a ashy deposits, including roasting pits, other ash-filled pits, and ashy layers of trash. The largest groups of figures and fragments (often accompanied by sealings and tokens) were found inside roasting pits, sealed beneath a cobble floor that was the roasting surface (Voigt 1985:Figure 8a–c). Associated within a single pit were intact figurines, figurine pieces, and shapeless lumps of lightly baked clay. The fragile but intact surfaces of the clay figures suggest that they were carefully handled, and in the case of figures from roasting pits it is clear that figurine manufacture and disposal were closely linked. Based on condition and location I suggest that a large number of figures were made, placed inside the pit, subjected to fire (thus explaining the ash matrix in which they were found), and then capped with cobbles. This scenario would explain the numerous fired fragments and lumps found in roasting pits: unbaked figures sometimes baked and preserved, but sometimes they simply exploded in an uncontrolled firing. The sole exception to this distribution pattern is a battered clay quadruped found on an exterior surface.

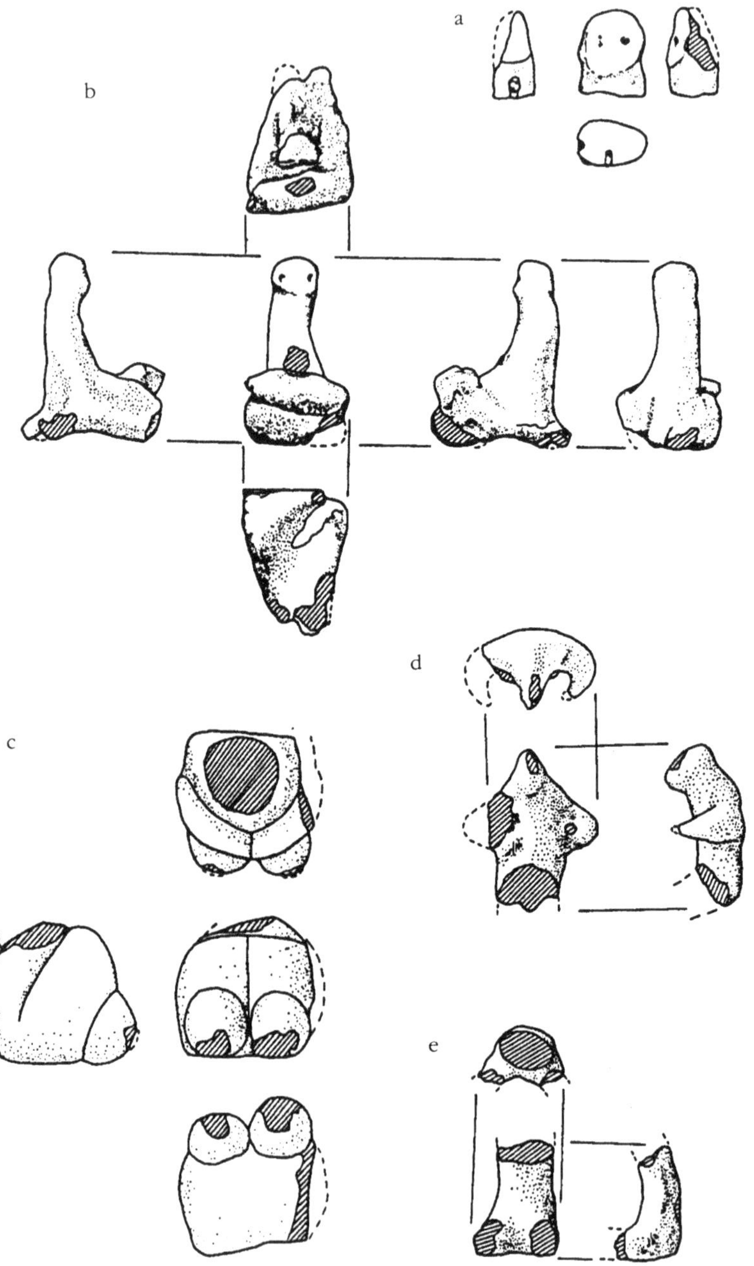

Figure 4. Human figurines from Gritille: (a,c) stone; all others clay; scale: 1/2.

Interpretation of the Gritille Figurines

Viewed collectively, the form, wear, and disposition of the Gritille clay figures conform to the pattern predicted for figures used as vehicles of magic, artifacts made by individuals as part of rituals used to ensure the well-being of the maker(s). This is hardly an earthshaking conclusion, but it is based not on assumption and simple assertion but on matches between archaeological data and a set of predicted attributes based on patterns in *observed* human behavior. Both wear pattern and context suggest that the single heavily worn animal from a living surface was used as a toy. A dual function for figurines—magical vehicle recycled as toy—is not uncommon in the ethnographic literature, but it is also possible that the animal was made and used purely for entertainment.

With the function of clay figurines firmly established, is it possible to use form to suggest ideas or values that might have been associated with their use? Most of the animal figures could be identified as cattle with long, curving horns (Figure 5b; Voigt 1985:Figures 10e, 11). Technical archaeological reports as well as more popular sources often state that small figures of animals represent hunting or herding magic, a desire to maintain the well-being of wild or domestic herds. This argument is particularly attractive when the animals have been stabbed with a sharp object or "killed," as is the case with many of the animals from 'Ain Ghazal (see Rollefson Chapter 7, this volume). If we accept this interpretive framework, rituals employing cattle figurines at Gritille should be related to a concern with cattle fertility and food, an explanation that seems unlikely given the fact that domesticated herds of sheep and goats supplied most of the meat eaten in the village (Stein 1986, 1989). The primary function of the figure is to be a symbol—their form related them to some power or spirit; *but*, by definition, symbols are arbitrary. Making and/or destroying a cow figurine may be related to invocation and/or destruction of a supernatural entity associated iconically with cattle. A second aspect of context and behavior associated with these contexts can be used to argue against an equation of cattle figures and food procurement. Animal figures are found in and around features used by women: roasting or cooking surfaces at Gritille (and for pottery firing at Hajji Firuz). If women manufactured and used these artifacts, it seems unlikely that they were used in magic intended to achieve success in hunting large game, an activity rarely practiced by women in simple farming societies. Given that cattle are prominent in the symbol system of other Anatolian sites of this period (e.g., Çatal Höyük), we can assume that the cattle figurines had a symbolic rather than literal meaning, but what this meaning was cannot presently be determined.

Human figures from Gritille can, on the other hand, be plausibly be

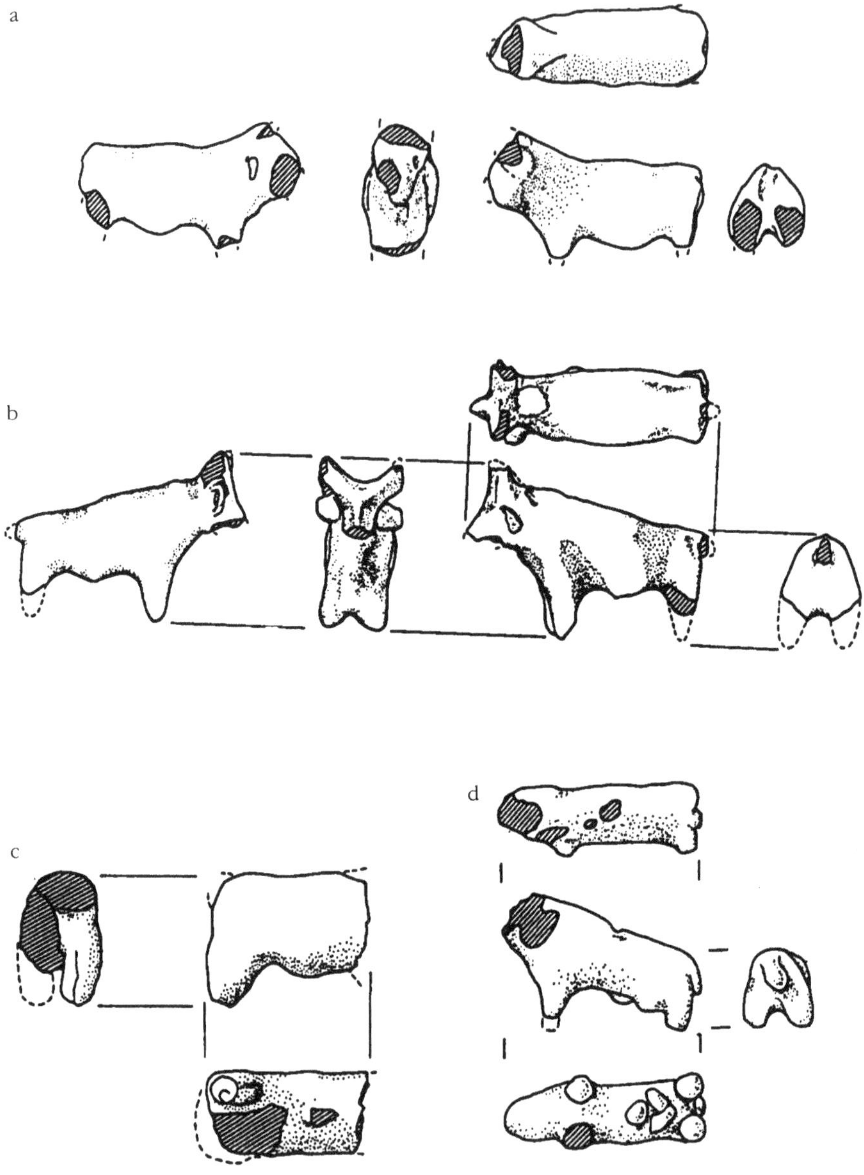

Figure 5. Clay animal figurines from Gritille; scale 1/2.

related to the well-being of females and children. One of the figures shows a skirted person with a protruding stomach and prominent navel but no breasts (Voigt 1985:Figure13a–b). When found, the stomach, which was made from a separate lump of clay, had broken away, showing that the figure was originally modeled with a flat stomach, complete with its own navel! The conditions portrayed by this figure (i.e., before and during pregnancy) seem clear enough, and since the context in which it was found (a trash pit) suggest that it was a magical figure, it may be argued that fertility or healthy children were indeed the object of the ritual act of manufacture and disposal. This is hardly surprising, given that in many premodern societies women were pregnant for most of their adult lives and may have lost half of their offspring in infancy.

Another clay figure shows a seated person with crossed legs (Figure 4b;Voigt 1985:Figure 12a–b). Scars on the body indicate that this figure had something in its lap (perhaps a child?). What symbolic value the snakelike body and flat face of this figure had for its makers is again unknown, but it is one of a group of Neolithic figurines that have a phallic aspect (e.g., Mellaart 1963:Figure 19, discussed below). The distinctive flat face also appears on a tiny limestone head (Figure 4a), which has holes in its flat, finished base, presumably for attachment to a body made of some organic material that has decayed. Use of an unusual material suggests a different function, for example a toy or a cult figure. The head was found in an ash deposit up against a house wall, a context that provides little help in determining function. A second stone figure portrays a very fat seated woman(?) (Figure 4c;Voigt 1985:Figures 10g, 14). The globular feet of this figure have been chipped, as though they had been walked over some surface. Using our interpretive criteria, this figure fits several categories. The fact that it was made of stone and took a greater investment in labor might suggest a cult figure. On the other hand, it was found in household garbage and has wear similar to that expected on a toy. Without more examples of this type, and some kind of distribution and/or wear pattern, there can be no final answer.

To summarize, nearly all of the Gritille figurines are found in contexts and in a condition that strongly indicates use as vehicles of magic. With this established, the precise chronological disposition of the figures may be significant. The figurines are found in roasting pits, but only a few of the roasting pits in the sample contain figurines, and a majority of these date to a narrow stratigraphic zone at the end of phase B, just before a break in the sequence (see Voigt 1988:219–221). Thus there may be a marked increase in rituals designed to alleviate life's problems or ensure the well-being of the community at a specific point in its history. One way to test this inference would be to look for other evidence for environmental or nutritional stress. In the absence of burials, variations in human health cannot be measured directly, but might be evident in animal populations. Detailed analysis of the faunal remains now being conducted by B. Monahan of Northwestern University should provide a means of assessing the meaning of the archaeological distribution of clay figures at Gritille.

FIGURES AND ARCHITECTURE IN THE TAURUS PPNB

Although Gritille supplied little evidence for behavior associated with formal or public ritual within the Neolithic communities of southeastern Anatolia, two nearby and slightly earlier sites within the same cultural tradition do. Nevalı Çori was excavated from 1983 to 1991 by a German team, led by Harald Hauptmann (1987, 1988, 1993, 1997). Beneath the slopes of an eroding gully leading to the right bank of the Euphrates was found a village with substantial stone architecture that is contemporary with part of the Neolithic sequence at Gritille. Most of the Nevalı Çori buildings were long grill- and cell-plan houses, similar to those found at other PPNB sites within the Taurus, especially to domestic structures at Çayönü (Hauptmann, 1988; 1993:Figures 2–3a–c; compare with Özdoğan and Özdoğan 1990:Figures 1–2, Pl. I:2–4). The exceptions were a series of large square structure with elaborate interior fittings that the excavator rightly refers to as "cult buildings" (Hauptmann 1993:Figures 4–13). Göbekli Tepe, 4 km southeast of Sanliurfa, has been excavated by the German group since 1995 (Schmidt 1997). Stone sculpture fragments found on the surface appeared similar in style to finds at Nevalı Çori, and excavation revealed a settlement with another, equally spectacular public building.

The Nevalı Çori Cult Buildings

There were three consecutive nondomestic buildings at Nevalı Çori, constructed in the same location, and associated with each of the first three (of five) Neolithic occupation levels of the site (Hauptmann 1997:131–132, contra Hauptmann 1993:55). The earliest of these structures (Building I) was poorly preserved, consisting of wall fragments. Building II was almost square, nearly 14 m on a side (Hauptmann 1993:42–48, Figures 4–8). A narrow entrance was located in the center of the southwest wall, and a bench 1 m wide and covered with broad stone slabs ran along the northwest, northeast, and niche walls. On the southeast wall, the bench was interrupted by a deep niche, which Hauptmann thinks may originally have held a pedestal (1993:47). Into the bench were set thirteen monolithic slabs or pillars, at least one of which retained its T-shaped capital (Hauptmann 1993:Figure 7). The floor was terrazzo, made up of a 15 cm thick layer of stone chips set in a mortar bed, and the walls were white plastered with traces of red and black suggesting that they may have been painted.

Nevalı Çori Building III was constructed inside Building II, using the same terrazzo floor (Hauptmann 1993:48–55, Figures 9–13; 1997:132). The position of the entrance was retained in the southwest wall, and a new, continuous bench was constructed along the other three walls. The niche in the southeast wall of Building II was filled in, and a new very small niche was built into the northeast wall above the bench. Holes were cut into the floor and in these holes were placed two rectangular slabs or pillars that stood approximately 3 m high. One of the slabs (half of which was still standing *in situ* when the building was excavated) was sculpted in low relief (Hauptmann 1993:Figure 16). It depicts the body of a female, with long rectangular breasts on the narrow face that can be seen from the doorway; below the breasts are stylized hands with five fingers, attached to V-shaped arms on the broad sides of the pillar. Fragments of a T-shaped capital/head were again found (Hauptmann 1993:Figure 11, to left). Like a series of contemporary or slightly later buildings at Çayönü (Hauptmann 1993:Figure 28; Schirmer 1983 and discussion below), this square building can be interpreted as a "community" structure based on plan alone; the presence of monumental sculptures in the Nevalı Çori building strongly suggests use for ritual rather than solely political behavior.

Sculptures and Iconography at Nevalı Çori

Information on stone images used in the Nevalı Çori cult buildings is provided by broken fragments: when the structure was rebuilt, items presumably used within the older building were deposited within the walls and benches of its successor. The most astonishing find was a sculpted column that Hauptmann compares to a Northwest Coast "totem pole" (1997:133; Türe et al. 1999:Plate 21, printed upside down). At the top of the column was a bird with rounded breast, its head broken away (Hauptmann 1993:Figure 24). Below were two similar figures set back to back that represent composite creatures with a human head and a bird's back and tail (Hauptmann 1993:Figure 25 illustrates the better preserved of these images); the lower figure is incomplete, and the total height of the column is unknown. The head of the upper and nearly complete composite creature has narrow eyes that may once have been inlaid, a long narrow nose, and a pursed mouth; an oval form beneath the face appears to be a rounded stomach with incised navel. The bird-like elements are visible only on the sides (and back?), which are difficult to understand from the published photographs. Hauptmann (1997:133) states that the human/bird figures (as well as another monumental head that has not yet been fitted into the column) are female, but this interpretation is hard to justify based on published photos.

A series of sculpture fragments were found in and around the small niche in the northeast wall of Building III, opposite the door. Set into the wall behind the niche was a larger than lifesize human head with a tall rounded skull and ears jutting out to either side (Hauptmann 1993:Figures 12, 19). The face of this sculpture has been destroyed, but the back is well preserved and has a snake with a hemispherical head and a body zigzagging downward, forming what Hauptmann refers to as a "pigtail" (1993:57). The tall, domed head with knob-like ears and snake pigtail certainly suggests representation of a male to a modern viewer, but we cannot be certain of either sex or gender. On the other hand, a human torso found tumbled in front of the niche is in my view unquestionably male: broad shoulders and an absolutely flat chest taper to a narrow waist; the back is beautifully modeled with the spine indicated by a shallow U-shaped channel (Hauptmann 1993:Figure 22a and b). Built into the bench below the niche were more sculpture fragments, and another figure with human head and bird-like body (Hauptmann 1993:Figure 14, 21). This composite figure has the same long nose and pursed mouth as the human/bird from the "totem pole," as well as the same (though less-rounded) bulging area beneath the shoulders: in this case, wings are more clearly depicted on sides and back.

Stone sculptures were also recovered from domestic buildings at Nevalı Çori. Found in House 3 was a human head that had been sculpted in high relief on the upper part of a stele or pillar (Hauptmann 1993:66, Figure 3a, 20; Türe et al. 1999:Plate 22). The face has been (deliberately?) damaged, but narrow, slit-like eyes and a hat or bowl-like hairdo are well preserved. Based on the form of the face and the hair, Hauptmann suggests that this figure is female, and I would certainly agree that the carefully sculpted cheekbones resemble modern images of women. More complex is a high relief carving on a squat pillar with a rounded top that comes from an unspecified context (Hauptmann 1993:67, Figure 26). At the top of the sculpted face is a bulging, oval area, the surface of which has been chipped away; this oval rests on a horizontal edge that can be visualized as the top of an asymmetrical hourglass. The top portion of the hourglass is large, while the bottom is quite small; extending downward from the upper corners of the top of the hourglass are raised V-shaped forms that end at the base of the pillar, spreading away from the small oval bottom of the hourglass. The V-shaped forms clearly resemble human limbs, in part because the terminal/basal area of each has 4 incised lines 5 digits. Hauptmann (personal communication) sees this figure as a human torso and legs, with hips and thighs extending out at a right angle to either side or the damaged human torso and dangling feet; the raised area between the feet is seen as a child, so that the entire image is that of a woman giving birth. This is certainly a reasonable interpretation, and is consistent with the way in which lower bodies of humans are depicted on another sculpture described below (Hauptmann 1993:Figure 27). I would suggest, however, that the figure can also be considered as ambiguous, another example of a male/female image: the constriction between large oval at the top of the piece and the broad triangle below could be viewed either as a waist or as a neck; in the latter case, the upper part of the hourglass depicts a male torso with broad shoulders and narrow waist and hips, bordered by arms bent at the elbows.

A final sculpted piece, a stone bowl fragment found in the foundations of House 3 (Hauptmann 1993:67, Figure 27), is especially important since it provides the kind of internal relationship that is central to Hodder's analysis of Çatal Höyük. Three individuals are shown on the bowl's exterior: a triangular-headed figure with a spherical body and four short straight limbs is flanked by two humans, one larger than the other. The humans have oval heads with excisions to form deep-set eyes and nose, and the smaller figure has an incised linear mouth; they too have spherical bodies, with limbs curving up to form hands with incised fingers, and down, ending in blob-like but firmly grounded feet. To the viewer's left of the larger figure is a raised crescent, truncated by a broken edge, and to the right of the smaller figure is a more complex horned(?) area that again ends in a break. Hauptmann sees all three as a symbols of fertility (1993:67). Alternatively, the bulging torsos might be related to food, either full stomachs or fat deposits; in either case, the issue would be abundance, a concept that is certainly related to fertility. Based on a difference in size, one could argue (as Hauptmann does) that the figure to the left is male, and that to the right is female. If so, identifying the sex of stone sculptures at Nevalı Çori is likely to be unusually difficult, as is suggested by the discussion above. Taking a broader perspective, these figures are particularly interesting because of their raised-arms—calling to mind a series of wall sculptures found at Çatal Höyük that Mellaart and others have identified as images of "the goddess," but that I consider genderless and not necessarily human (see the discussion of Giöbekli Tepe below).

The Nevalı Çori sculptures found in Buildings II and III have attributes associated with cult figures, and their disposition—broken and incorporated into the fabric of the buildings—strengthens this interpretation. This site has, therefore, produced evidence of community ritual in the form of a special purpose structure with associated cult statues. The systematic damage exhibited by other sculptures that were apparently disposed of with some care in domestic buildings suggests that cult images, or at least images considered to have great power, were not restricted to public spaces. Small "male and female figures in sun-dried and fired clay" (Hauptmann 1997:133) suggest that individual ritual acts similar to those inferred for Gritille were also practiced at Nevalı Çori.

Göbekli Tepe: Sculpted Reptiles an the Urfa PLain

Göbekli, located to the southeast of Nevalı Çori, was initially recorded by Peter Benedict in 1963 as part of the joint survey in southeastern Anatolia conducted by Istanbul University and the University of Chicago (Benedict 1980). This mound, located on a rock outcrop, caught the attention of the German team from Heidelberg University because of the stone sculpture and pillar fragments and flint tools that lay scattered on its surface. Excavation quickly established that this is an early PPNB site with a building that contained sculpture fragments, and slab-like pillars with animals carved in low relief (Gates 1997:246,Figure 1; Schmidt 1997; Türe et al. 1999:Plate 26–28). The stone sculptures are in a style similar to that at Nevalı Çori, but depict very different images: reptiles are common, including one with a long snout and prominent teeth (Schmidt 1997:Figure 4). Both this animal and a second reptile with prominent tail (sculpted in high relief on a pillar fragment; Schmidt 1997:Figure 5) share the upright arms and legs and swollen body that are familiar from the Nevalı Çori plaque as well as Çatal wall reliefs. Other finds mentioned by Schmidt are a human head over 20 cm high, a standing male with erect penis, a wolflike animal, and a lion or bear holding a human head between its paws (1997:75–78, Figure 6). Both sculptures and the decoration of the new pillared structure suggest a cult budding, but with images of supernatural figures quite different from those at Nevalı Çori.

Public Buildings and Mortuary Practices at Çayönü Tepesi

The PPNB settlement in southeastern Turkey for which we have the largest sample as well as the longest sequence is Çayönü near modern Diyarbekr. The site was excavated between 1964 and 1991, exposing nearly 5,000 m^2 (*c.* 22%) of the Pre-Pottery Neolithic settlement, with six occupation phases dated between 10,250 and 8,000 bp (A. Özdoğan 1995:81–82; M. Özdoğan 1997b:444). Domestic architecture changes throughout this period, beginning with round houses, succeeded by "channel," grill-plan, and cell-plan buildings similar to those already discussed from Nevalı Çori (Çambel and Braidwood 1980; M. Özdoğan 1997b; see also A. Özdoğan 1995; Özdoğan and Özdoğan 1990; Schirmer 1988). Also present are three large structures that differ in plan from contemporary domestic structures, the Flagstone, Skull and Terrazzo Buildings (Schirmer 1983; plans republished in Hauptmann 1993:Figure 28). The stratigraphy at the site is complex, and the relationship of these structures has not always been clear, but detailed analysis has now shown that the buildings succeed each other chronologically and that one such structure existed during most of the PPN sequence (A. Özdoğan 1995:84–87; M. Özdoğan 1997b:445; Özdoğan and Özdoğan 1990:70–72). The earliest is the Flagstone Building, with stone walls, a floor paved with large flat stones, and stele-like pillars (Çambel and Braidwood 1980:39, Pl. 1–2); the overall plan and pillars link it to Nevalı Çori Building III. A comparison between Nevalı Çori Building III and the latest of the Çayönü communal structures, the Terrazzo Building (Çambel and Braidwood 1980:40, Pl. 1–2, 41, Pl. 1), is preferred by Schmidt, who reconstructs two pillars in the center of a large pit, which destroyed the central part of the Terrazzo Building floor (1997:Pl. 3). The only large stone sculpture from Çayönü that has been published was found in the Terrazzo Building, a stone slab with a human head sculpted in high relief on one side (Çambel and Braidwood 1980:41, Pl. 2). Hemispherical in plan, the head does have the distinctive long straight nose of the human faces from Nevali Çori.

The Skull Building, chronologically intermediate within the Çayönü sequence, was in use for the longest period and was rebuilt six times with significant change in plan (M. Özdoğan 1997b:445, Figure 2). It consists of a large room with a second very narrow room to the rear of the structure. On and under the floors of the Skull Building were the disarticulated remains of nearly 300 people, including a deposit with seventy skulls recovered "during the first exposure of the building" (M. Özdoğan 1997b:445). While there is no equivalent mortuary structure known from other PPNB sites in southeastern Turkey, disarticulated human remains have been found buried beneath house floors at Nevalı Çori, including one deposit with eight skulls (Hauptmann 1993:57, Pl. 17–18). At Gritille, a fragment of a human cranium was found in a pit dated to phase B. The practice of intramural burial of humans whose skeletons had presumably been exposed elsewhere until the bones were relatively clean can thus be added to the list of common ritual practices at Taurus PPNB sites.

ÇATAL HÖYÜK: THE MYTH OF THE GODDESS REVISITED

Çatal Höyük is the most famous Neolithic site in Turkey by virtue of the large scale excavation of well-preserved buildings that were sometimes burnt, leading to excellent preservation of organic and inorganic materials. A large corpus of mural and portable art, including a diverse collection of figures made of clay and stone, provides our best evidence from central Anatolia for a religion centered on anthropomorphic deities. In his preliminary reports, Mellaart illustrated the most interesting and elaborate figures as well as representative examples of simpler types, and gives us some information on the contexts in which they were found (Mellaart 1962, 1963, 1964, 1966).[2] My analysis, originally presented as a paper in 1995, supports Mellaart's identification of the stone and large clay figures as cult statues, but at the same time it moves the discussion of these artifacts from speculation based on figurine form and a direct historical approach relying on European mythologies to an argument based on form, archaeological context, the condition of artifacts at the time of deposition, and generalizations based on ethnographic parallels.

Figurine Form, Context, and Function at Çatal

A consideration of figurine attributes (form, size, and material) results in four distinct groups within the Çatal industry: small clay figures of animals and humans (referred to by Mellaart as *ex voto* figures, and by Hamilton [1996] as "humanoid"); large clay figures of humans; stone figures of humans; and natural or slightly modified rocks that suggest human forms (referred to here as "concretions," "pebble," or "columnar" figures to indicate form, presumably corresponding to Hamilton's "schematic" figures). Mellaart assigned these morphological groups to two functional classes based in part on the archaeological contexts from which they were recovered—small clay figures (invariably referred to as "crude") and "statuettes" (Mellaart 1967:180). Mellaart meticulously documents most of the large figures or statuettes in his publications and provides abundant high quality photographs and drawings (Mellaart 1962, 1963, 1964, 1966). He does not, however, provide a complete catalogue, and in her new study of the Çatal material Hamilton states that she has been able to locate a total of 254 figurines and fragments from the Mellaart excavations (Hamilton 1996:215–217). Most or all of the unpublished pieces appear to be small clay figures of humans and animals (Hamilton 1996:215). In addition to looking at published sources, I have been able to view most of the large figures and a sample of the small clay figurines that have been on display in the Ankara Museum of Anatolian

[2]In early summer 1997, after the first version of this article had been completed and sent to Ian Kuijt, Peter and Wendy Matthews kindly lent me the British Institute's still uncatalogued copy of *On the Surface* (Hodder 1996). In this volume Naomi Hamilton reports on her study of all of the Çatal figurines, both those excavated by Mellaart and those recovered by the new excavations. Hamilton has located and catalogued 254 figures from the Mellaart excavations, assigning new numbers for analytical purposes (1996:215), and only these new numbers are used to identify figures within her discussion of the sample. When discussing individual figures the findspot is given so that some of the distinctive figures can be recognized based on location and description, but it seems foolhardy to *guess* about the match between new numbers and Mellaart's publications, and I have therefore not systematically tied Hamilton's discussion of individual figures to my own.

Civilizations, allowing an assessment of wear and breakage. It is, therefore, possible to reconsider the Çatal figurines within the analytical framework set out for Hajji Firuz and Gritille.

Small Clay Figures of Animals and People

Small lightly baked clay figures were found at Çatal Höyük in levels IX to IV, stuck between the bricks in house walls and clustered in trash pits (Table 5; Mellaart 1962:Pl. VIIa–b, 1963:Pl. XVIIIa, 1967:180, Figure 66; see also Hamilton 1996:217–219, Figure 12.1:544–545,498). Forms include humans with pinched heads and cylindrical bodies with outstretched arms, seated figures with pinched out legs, horned animals, and animals with pronounced snouts. The clay figures are simple, and coincidence is therefore an issue; nevertheless, similarities between the Çatal small clay figures and contemporary figures from Gritille are striking, and may indicate a set of beliefs and practices that extended over large areas of Anatolia in the ninth millennium bp (compare Figure 4d with Mellaart 1963:XVIIIa, center). The fact that fragile clay figures such as a bovid and a human with extended arms remained intact (Hamilton 1996, Table 12.4) indicates disposal without much handling, perhaps soon after manufacture, thus eliminating interpretation as toys. Disposal in inaccessible places and in large groups indicate use as vehicles of magic, that is, in personal or household rituals (Tables 2 and 3). Based on the uniform treatment of the small clay figures from Çatal—their association in the ground—I would argue that representations of humans, wild animals and domesticated animals are linked as part of a single symbolic system and any attempt to assign meaning to the figures must take the entire range of beings into account.

Large Clay Figures

Relatively large and realistically modeled figures come from levels VI through II, and over half (eight of fourteen) come from a single structure, AII.1 (Table 5). The largest clay figure from Çatal Höyük (16.5 cm high without its head, estimated at *c.* 20 cm when complete) was found in a (grain?) bin in a small room of AII.1 (Mellaart 1963:93, 95, Figures 31–32; 1967:Pls. IX, 67–68). The figure depicts a monumentally fat woman seated in a chair; between her feet protrudes a small human head, presumably a child to whom she has just given birth. The unique size and form of this statuette, its careful(?) disposition at the bottom of a bin, and the missing heads suggest use as a cult figure (Tables 2 and 3). Another element supporting the interpretation of this figure as a deity is the presence of symbolic or iconic elements: the arms of the chair are formed by two large standing cats,

TABLE 5. Çatal Höyük Figures by Location

Level	Bldg	Mat	Gend	Type	Year exc.	Mellaart no	Ankra mus no	Publication
II	A II.1	C	F	Standing human	1962	Goddess 13	79-249-65	Mellaart 1963:Pl. XXIIIb; 1967:Pl. 67
		C	F	Seated human	1962	Goddess 27	79-803-65	Mellaart 1963:46, 93, Pl. XXIIIc-d
		C	F	Seated human	1962	Goddess 31	79-244-65	Kulaçoğlu 1992:43, #32
		C	F	Seated human	1962	Goddess 28	79-248-65	Kulaçoğlu 1992:39, #27
		C	F	Seated human	1962	Goddess 32	79-247-65	Kulaçoğlu 1992:43, #33
		C	F	Seated human	1962	Goddess 30	79-245-65	Kulaçoğlu 1992:42, #31
		C	F	Seated human	1962	Goddess 29	79-246-65	Kulaçoğlu 1992:42, #30
		C	F/A	Seated human w/animals	1962	Goddess 11	79-250-65	Mellaart 1963:Pl XXIV; 1967:Pl. IX, 67-68
		S	I	Standing human	1962	Goddess 26	?	Mellaart 1963:46, 93
III	A III.1	C	F/A?	Seated human w/animal?	1961	Goddess 10	79-4-65	Mellaart 1967:Fig. 49
		S	F	Seated human	1961	Goddess 23	79-8-65	Mellaart 1962:Pl. VIIIc
	A III.2	C	F	Standing human	1961	None	79-27-65	Mellaart 1962:Pl. VIIIa
	C III	S	I	Human head	1961	None	?	Mellaart 1962:Pl. IXd
IV	E IV.1	C	F	Standing human	1961	None	79-28-65	Mellaart 1962:Pl. VIIId
	E IV.4	C	F	Seated human	1961	Goddess 12	79-20-65	Mellaart 1962:Pl. VIIIb
		S	F	Standing human	1961	Goddess 25	79-22-65	Mellaart 1962:Pl.. IXa-c
IV-VIII?		C	I	Humans	1961	None	?	Mellart 1962:Pl. VIIa
V	?	C	F	Seated human	?	None	79-656-65	Kulaçoğlu 1992:36, #23
VI	?	C	F?	Seated human?	?	None	79-224-65	Kulaçoğlu 1992:22, #6
	?	C	F	Seated human	?	None	79-272-65	Kulaçoğlu 1992:23, #7
	?	C	A	Animals	1962	None	?	Mellaart 1963:78; 1967:Pl. 66

Continued

TABLE 5. *Continued*

Level	Bldg	Mat	Gend	Type	Year exc.	Mellaart no	Ankra mus no	Publication
	A VI.61	C	F	Seated human	1962	Goddess 14	79-251-65	Mellaart 1963:Pl. XXIIIa; 1967:Pl. 79
	E VI.10	C	I,A	Animals, humans	1962	None	?	Mellaart 1963-78, Pl. XVIIIa
	E VI.28	S	F?	Concretion	1962	None	?	Mellaart 1963:Pl. XIXa
	VI.10	C	F	Seated human	1962	None	79-170-65	Kulaçoğlu 1992:79, #13
	EVIA.10	S	M	Seated human	1962	God 3	79-800-65	Mellaart 1963:Pl. XXI; 1967:Pl. 85
		S	M	Columnar human	1962	Goddess 15	79-804-65	Mellaart 1963:82ff; 1967:Pl. 90
		S	M/A	Standing human w/animal	1962	Goddess 9a	79-162-65	Mellaart 1963: 83-86, Fig.24; 1967:Pl. X, 75-76
		S	M/A	Seated human w/animal	1962	Goddess 9c	79-167-65	Mellaart 1963:83, Fig. 22: 1967:Pl. X
		S	M/A	Seated human w/animal	1962	God 6	79-191-65	Mellaart 1963: 82ff; 1967 Pl. 91
		S	M/A?	Pebble figure	1963	Goddess 2a	?	Mellaart 1964:78, Fig. 30b
		S	M/A?	Pebble figure	1962	Goddess 2b	?	Mellaart 1963:Pl. XIXc; 1967:Pl. 69
		S	M?	Columnar human	1962	Goddess 1	?	Mellaart 1963:Pl. XIXb; 1967:Pl. 65
		S	M?	Columnar human	1962	None		Mellaart 1963:Pl. XIXd
		S	F(2)	Seated human pair	1962	Goddess 5	79-798-65	Mellaart 1963:Pl. XX; 1967:Pl. 70-72
		S	F?	Pebble figure	1962	Goddess 22	79-163-65	Mellaart 1963:Pl. XXX; 1967:Pl. 72
	EVIA.10	S	F/A	Standing human w/animal	1962	Goddess 9b	79-162-65	Mellaart 1963: Fig. 23; 1967:Pl. X
		S	M/F	Columnar human	1962	Goddess 16	79-799-65	Mellaart 1963: 82ff; Kulaçoğlu 1992:29, #15
		S	I	Human torso	1962	Goddess 4	?	Mellaart 1963:Pl. XXb; 1967:Pl. 78

Continued

TABLE 5. Continued

Level	Bldg	Mat	Gend	Type	Year exc.	Mellaart no	Ankra mus no	Publication
VI	VIA.25	S	F	Standing human	1962	Goddess 3a	79-802-65	Mellart 1963:Pl. XXIIc-d; 1967:Pl. 80, 82
		S	M	Seated human	1962	God 2	79-801-65	Mellaart 1963:Pl. XXII; 1967 Pl. 84
		S	A	Animal	1962	Goddess 3b	?	Mellaart 1963:Fig. 26
	VIA.30	S	M,F	Seated humans (4)	1963	Goddess 6	79-1-65	Mellaart 1963:Pl. XXId; 1967 Pl. 83
	VIA.44	S	F	Standing human	1963	Goddess 8a	79-452-69	Mellaart 1964: Fig. 26, Pl. XV
		S	F?	Standing human	1963	Goddess 8b	?	Mellaart 1964:75, Fig. 27
		S	F?	Pebble Figure	1963	Goddess 21	?	Mellaart 1964: Fig. 30a, Pl. XVIb
		S	M	Pebble Figure	1963	God 4	?	Mellaart 1964: Fig. 28b-c, Pl. XVIa
		S	M/A	Seated human w/animal	1963	God 6a	79-457-65	Mellaart 1964: 75, 78; Fig. 29
		S	I	Concretions and stalagtites	1963	None	?	Mellaart 1967:78
		S	I	Human hand	1963	None	?	Mellaart 1964:78, Fig. 31b, Pl. XVIc
	VIB.45	S	I	Pebble figure	1963	Goddess20	79-451-65	Mellaart 1964:78, Fig. 31d, Pl. XVIIa
		S	F	Columnar human	1963	Goddess 19	79-450-65	Mellaart 1964: Fig. 31c, Pl. XVId
VII	E VII.21	S	M/A	Seated human w/animal	1963	God 5	?	Mellaart 1964: 78, Fig. 32, Pl. XVIIc-d; 1967:Pl. 89
	VII.24	S	I	Seated human w/animals	1963	Goddess 18	?	Mellaart 1964:78, Fig. 31a, Pl. XVII
		S	M?	Concretion	1963	God 7	?	Mellaart 1964:78, Fig. 28a
VIII		C	A	Animals	1961	None	?	Mellaart 1962:51, Pl. VIIb
IX		C	I, A	Animals, humans	1961	None		Mellaart 1964:73

whose tails curve up over her shoulders; her hands rest on their heads, showing her dominance over these fierce animals. In the main room of AII.1, clustered around a hearth, were seven seated and one standing clay figures, some quite beautifully modeled (Kulaçoğlu 1992:38–39, 42–43, cat. nos. 26–28, 30–33; Mellaart 1963:46, 93, Figures 29–30, Pl. XXIIIb–d). The seated figures are emphatically female and have swollen bellies that seem to me to indicate pregnancy rather than ample fat deposits.[3] The standing figure has a blocky but not visibly pregnant body and is clothed; small breasts suggest a girl rather than a mature woman, and the spots on her upper garment suggest leopard skin. All but one of the figures in this group is missing its head, and this exception was broken at the neck and mended. Scars on the shoulders suggest the breakage was deliberate; in some cases part of the shoulders were removed with the head, so this is not a simple case of breakage at a weak point.

Taken together, the clay figures from structures AII.1 can be interpreted as cult figures. Related clay figures extend back to level VI at Çatal Höyük but are not common (Table 5), despite the fact that a far larger area of the site was excavated for VI than for later levels. The predominance of carefully modeled female figures continues in later central Anatolian settlements, and the form of the female clay figures from Çatal is clearly similar to the larger and more diverse corpus from the slightly later occupation at Hacilar (Mellaart, 1970).

Stone Figures

The distribution of stone figures is in complementary distribution to that of clay figures. Based on Mellaart's publications, only four stone pieces (three statuettes and a human head) were found in levels V through II (Table 5 with references). The two statues from these levels that have been illustrated are both female, and although neither appears to be pregnant they share iconographic elements with the clay figures (Mellaart 1962:Pls. VIIIc, IX). Most of the stone figures (thirty-one of thirty-five figures documented by Mellaart) are from levels VI and VII, and the majority of these are from two buildings in level VIA (Table 5). Both large groups from level VIA

[3]It is frequently stated that these fat/pregnant figures exhibit a "normal" body type for central Anatolian females (e.g., Hamilton 1996:225). Observation of women in two Turkish villages where I lived for periods of two to three months during each of fourteen years field seasons between 1981 and 1997 has given me the strong impression that most women of childbearing age are relatively slim and almost always pregnant. This does not represent a scientific conclusion, but it is supported by conversations with both women and men in Biriman and Yassi Höyük and by comments made about young American women who resembled the Çatal clay figures and were emphatically "too fat" for local tastes.

include relatively realistic figures and highly schematic figures. The condition of realistic human figures varies, but many had been broken, and not necessarily at the structurally weakest point, the neck. Thus at least some of the stone figures were deliberately broken before disposal (see also Hamilton 1996:219–221, Table 12.4).

Each of the level VI structures that contained stone figures was burned, and Mellaart notes that after the fire the "Leopard Shrine" (VIA.44) was deliberately filled with trash (Mellaart 1964:78). The available evidence suggests that sculpted figures on the walls of both buildings with large figurine clusters were also destroyed before being damaged by fire: in the Leopard Shrine Mellaart states that before the fire the leopards had been covered with a layer of white plaster (above more than forty layers of painted plaster) (1964:42, 45); in "Shrine VIA.10" a relief of a quadruped "goddess" was found collapsed above a series of bulls heads, with no trace of an upper body (Mellaart 1963:70–73, Pl. XIII). The cause of the fires that destroyed part of level VIB and all of VIA cannot be determined without more detailed evidence (from new excavation or old field books), but Mellaart does tell us that below level VI there were no fires, and that fires ended each of the following building levels, V to 0 (Mellaart 1964:115). This consistent pattern suggests design rather than accident. On a practical level, burning would have destroyed vermin infesting walls and roofs, reducing threat (bites, disease) from snakes, insects, and other vermin. On the other hand, there is a clear and undeviating destruction of at least one type of relief when buildings were abandoned (the "modeled goddess" discussed below); it is, therefore, possible that ritual played some role in the level VI destruction, a suggestion also made by Hamilton (1996:219) and Hodder (1996:365).

To summarize, the material used in manufacture, iconic elements, recovery of groups of statues that vary in form, and ritual destruction all point to the use of the stone figures as cult figures (Tables 2 and 3). Whether this cult was domestic or associated with multihousehold groups (as suggested for Nevalı Çori) cannot be determined. Mellaart's division of the Çatal buildings into "shrines" and "houses" has always seemed arbitrary based on his own publications, and micromorphological evidence from the site "serves to further blur the distinction" (Hodder 1996:362; see also Hamilton 1996:226–227) since to date there are no large buildings that do not have evidence of domestic use. This is not to say that large communal buildings used for ritual were not present within the settlement. Large-scale excavations at the earlier but closely related site of Asikli Höyük, to the northeast of Çatal, have revealed a settlement with two distinct quarters divided by a broad street (Esin 1993, 1998). To the north of the street are one- to three-room mud-brick houses built side by side, the "pueblo" plan familiar from Çatal; to the south of the street are two large structures built of stone as well as brick, with traces of paint on the floors of the smaller structure (Esin 1993:Figure 1; 1998:97). Esin considers these to be ritual structures, an inference strengthened by a similarity in plan of the larger building from Asikli (which has a series of narrow compartments to the north of the large main room) and the Skull Building from Çayönü (see above).

Male or Female or Both?

The next step it to look for clues to the nature of the cult represented by the figures from Çatal, and the aspect that has been emphasized in the past is gender. The difficulties inherent in the concept of gender, and especially in sexing figurines, are notorious (see Knapp and Meskell 1997). Mellaart saw most stone figures as female (1967:202–203), an interpretation that has generally been accepted by nonspecialists. If, however, we look at specific attributes of the figures and use these to form classes that differentiate males and females, the results are quite different (Table 6). Viewing the range of variation in the figures, several attributes seem most important or at least most frequently displayed within the Çatal corpus of large clay and stone figures. Two attributes have discrete distributions: beards and breasts. Beards are unambiguously male, but the distinction between breasts and well-developed male pectorals on figurines can be quite difficult. This difficulty does not affect our conclusions about the number of males represented since it results in males being classified as females, artificially inflating the number of female figures and diminishing the number of males. Other consistent differences can be established using the beard/breast dichotomy: males tend to have triangular torsos with thin waists and legs, while females have heavy bellies and legs. When the shaping of an image was sloppy or perfunctory, the presence of these attributes is not clear, and in such cases I have made an assessment based on overall form in comparison to figures with similar posture, followed by a question mark. Within Tables 5 and 6, figures with animals are denoted "M/A" and "F/A," a male and female couple as "F,M," and an ambiguous figure as "M/F." Hamilton (1996:225) discusses the gender of specific figures, and, based on her description of the figures, her identifications do not seem to be very different from my own. It will, however, only be possible to verify this observation with the eventual publication of her catalogue linked to Mellaart's illustrations and/or Ankara museum numbers.

Applying these morphological criteria, we find that all of the Çatal clay figures are female or indeterminate (Table 6; see also discussion of clay figures above). For the stone figures, however, there are more securely identified males than females, about the same number of questionable males as questionable females, and eight indeterminate figures (pebble figures,

TABLE 6. Çatal Höyük Figures by Gender

Gender	Type	Material	Comp?	Dimensions	Level	Publication
A	Animal	White marble	N	Pres H:6 cm	VI	Mellaart 1963: Fig. 26
	Animals	Baked clay	Y	?	VI	Mellaart 1963: 78; 1967:Pl. 66
	Animals	Baked clay	N	?	VIII	Mellaart 1962: 51, Pl. VIIb
A, I	Animals, humans	Clay	?	?	IX	Mellaart 1964: 73
	Animals, humans	Baked clay	Y	?	VI	Mellaart 1963: 78, Pl. XVIIIa
F	Columnar human	Alabaster	Y	H:5.5 cm; W:3 cm	VI	Mellaart 1964: Fig. 31c, Pl. XVId
	Seated human	Baked clay	N	Pres H:6.6 cm; W:6.6 cm	II	Kulaçoğlu 1992: 43, #33
	Seated human	Baked clay	N	Pres H:3.4 cm; W:4 cm	II	Kulaçoğlu 1992: 43, #32
	Seated human	Baked clay	N	Pres H:7.2 cm; W:7.1 cm	II	Kulaçoğlu 1992: 39, #27
	Seated human	Baked clay	N	Pres H:4.3 cm; W:5.7 cm	II	Kulaçoğlu 1992: 42, #31
	Seated human	Baked clay	Y	H:10.2 cm; W:6.0 cm	II	Mellaart 1963: 46, 93, Pl. XXIIIc-d
	Seated human	Baked clay	N	Pres H:4.4 cm; W:5.8 cm	II	Kulaçoğlu 1992: 42 #30
	Seated human	Limestone	N	H:6.6 cm; W:6.5	III	Mellaart 1962: Pl. VIIIc
	Seated human	Baked clay	N	Pres H:5.8 cm; W:8.2 cm	IV	Mellaart 1962: Pl. VIIIb
	Seated human	Baked clay	N	Pres H:4.6 cm; W:4.7 cm	V	Kulaço?lu 1992: 36, #23
	Seated human	Baked clay	Y	H:2.6 cm; W:1.4 cm	VI	Kulaçoğlu 1992: 29, #13
	Seated human	Baked clay	N	Pres H:4.2 cm; W:4.5 cm	VI	Kulaçoğlu 1992: 23, #7
	Seated human	Baked clay	N	Pres H:4.1 cm; W:5.5 cm	VI	Mellaart 1963: Pl. XXIIIa; 1967: Pl. 79
	Seated humans (2)	White marble	Y	H17.2 cm; W:9.7 cm	VI	Mellaart 1963: Pl. XX; 1967:Pl. 70-72
	Standing human	Baked clay	N	Pres H:5.6 cm; W:4.7 cm	II	Mellaart 1963: Pl. XXIIIb; 1967:Pl. 67
F	Standing human	Baked clay	N	Pres H:4.6 cm	III	Mellaart 1962: Pl. VIIIa

Continued

TABLE 6. *Continued*

Gender	Type	Material	Comp?	Dimensions	Level	Publication
	Standing human	Baked clay	N	Pres H:4.9 cm; W:3.7 cm	IV	Mellaart 1962: Pl. VIIId
	Standing human	Alabaster	Y	H:12.4 cm; W:6.0 cm	IV	Mellaart 1962: Pl. IXa-c
	Standing human	White marble	Y	H:17.8 cm; W:7 cm	VI	Mellaart 1963: Pl. XXIIc-d; 1967:Pl. 80, 82
	Standing human	Black stone	Y	H:15.5 cm; W:11.3	VI	Mellaart 1964: Fig. 26, Pl. XV
F?	Concretion	Limestone	Y?	H:9.5 cm	VI	Mellaart 1963: Pl. XIXa
	Pebble figure	White marble	Y	H:5.1 cm; W:2.2 cm wide	VI	Mellaart 1963: Pl. XXa; 1967: Pl. 72
	Pebble figure	Brown limestone	Y	H:4.5 cm	VI	Mellaart 1964: Fig. 30a, Pl. XVIb
	Seated human?	Baked clay	N?	H:6.6 cm; W:6.5 cm	VI	Kulaçoğlu 1992: 22, #6
	Standing human	Basalt	Y	H:12.0 cm	VI	Mellaart 1964: 75, Fig. 27
F/A	Seated human w/animals	Baked clay	N	Pres H:16.5 cm	II	Mellaart 1966: Pl. XXIV; 1967: Pl. IX, 67-68
	Standing human w/animal	Brown limestone	N	H:10.6 cm; H:8.8 cm	VI	Mellaart 1963: Fig. 23; 1967:Pl.X
F/A?	Seated human w/animal?	Baked clay	N	Pres H:6.7 cm; W:6.2 cm	III	Mellaart 1967: Fig. 49
I	Concretions, stalagtites	Stone	?	?	VI	Mellaart 1964: 78
	Human head	Alabaster	N	Pres H:4 cm	III	Mellaart 1962: Pl. IXd
	Human head	Alabaster	Y	H:5.4 cm	VI	Mellaart 1964: 78, Fig. 31b, Pl. XVIc
	Human torso	White marble	N	Pres H:9.5 cm	VI	Mellaart 1963: Pl. XXb; 1967: Pl. 78
	Humans	Baked clay	Y	Various	IV-VIII	Mellaart 1962: Pl. VIIa
	Pebble figure	Chalk	Y	H:7.3 cm; W:5.6 cm	VI	Mellaart 1964: 78, Fig. 31d, Pl. XVIIa
	Standing human	White limestone	?	H:5.5 cm	II	Mellaart 1963: 46, 93
I/A	Seated human w/animals	Calcite	Y	H:4.2 cm	VII	Mellaart 1964: 78, Fig. 31a, Pl. XVII

Continued

TABLE 6. *Continued*

Gender	Type	Material	Comp?	Dimensions	Level	Publication
M	Columnar human	Black limestone	Y	H:8.8 cm; W:4.1 cm	VI	Mellaart 1963: 82ff; 1967:Pl. 90
	Pebble figure	Alabastser	Y	H:9.8 cm	VI	Mellaart 1964: Fig. 28b-c, Pl. XVIa
	Seated human	White marble	Y	H:12.2 cm	VI	Mellaart 1963: Pl. XXI; 1967:Pl. 85
	Seated human	White marble	Y	H:21.8 cm	VI	Mellaart 1963: Pl. XXII; 1967 Pl. 84
	Seated human w/animals	Calcite	Y	H:19.0 cm	VII	Mellaart 1964: 78, Fig. 32, Pl. XVIIc-d; 1967: Pl. 89
M?	Columnar humn	Limestone	Y	H:19.6 CM	VI	Mellaart 1963: Pl. XIXb; 1967: Pl. 65
	Columnar human	Gray Limestone	Y	H:8.5 cm	VI	Mellaart 1963: Pl. XIXd
	Concretion	Limestone	?	H:10 cm	VII	Mellaart 1964: 78, Fig. 28a
M/A	Seated human w/animal	Brown Limestone	N	Pres H:5.9 cm; W:6.3 cm	VI	Mellaart 1963: 83, Fig. 22; 1967:Pl. X
	Seated human w/animal	Alabaster	Y	H:10.5 cm; W:9.2	VI	Mellaart 1964: 75, 78; Fig. 29
	Seated human w/animal	Blue Limestone	Y	H:10.8 cm; W:8.3	VI	Mellaart 1963: 82ff; 1967 Pl. 91
	Standing human w/animal	Blue Limestone	N	Pres H:11 cm	VI	Mellart 1963:83-86, Fig. 24; 1967:Pl. X, 75-76
M/A?	Pebble figure	Blue Limestone	Y	H:11.2 cm; W:11.7	VI	Mellaart 1963: Pl. XIXc; 1967: Pl. 69
	Pebble figure	Limestone	Y	H:16 cm	VI	Mellaart 1964: 78, Fig. 30b
M/F	Columnar human	Limestone	Y	H:8 cm; W:3.5 cm	VI	Mellaart 1963: 82ff; Fig. 19 Kulaçoğlu 1992: 29, #15
M,F	Seated human (4)	Schist	Y	H:11.6 cm; W:11.3 cm	VI	Mellaart 1963: Pl. XXId; 1967 Pl. 83

modified concretions). Both males and females are shown in repetitive postures. Stone females usually stand, while clay females may stand but usually sit; these repetitive postures may suggest different deities or different aspects of a single deity. Stone males are nearly always seated, often on an animal's back; Mellaart identifies many of these animals as bulls, but the absence of horns, and similarity of spotted and unspotted heads, suggests to me that the animals are felines. Among the securely identified figures, both males and females are associated with leopards. More important, these beings control the animals, which are clearly dangerous based on the red painted mouth and paws of the leopards in the building where so many stone figures were found. By extension, gods and goddesses control the wild and the dangerous.

A Religious Revolution

At present, archaeological data that would allow us to link social, economic, and political change to changes in Neolithic ritual and ideology are limited; however, a consideration of the squence at Çatal Höyük provides some preliminary insight into the timing and possible importance of such changes. The chronological distribution of the figurines and mural art suggests that there is a shift in religious practice at Çatal Höyük between levels VI and V. Level VI is dated to approximately 7,500 bp or 5,800 BC by radiocarbon dating (Mellink 1992:Table 4), roughly contemporary with the transition between the Final PPNB and Pottery Neolithic in eastern Anatolia, northern Syria, and somewhat later than this transition in the south-central Levant. Before level V stone figures of males and females were made and used, and these figures have some stylistic links to figures from PPNB site of Nevalı Çori to the east, where context suggests use in community-wide ritual. Personal rituals at Çatal and the Euphrates sites are documented by small clay vehicles of magic. The stone gods of level VI are destroyed, gathered, and effectively entombed. The precise reasons behind this deposition cannot be determined, but a rare ethnoarchaeological study of figurines conducted by Warren DeBoer (1995) indicates that the deposition and destruction of a number of images with ritual significance will *only* occur when the images lose power.

In levels V and above, a new set of images appears, accompanied by murals depicting people in association with animals. It is at this point that fat females, pregnancy, and sexuality become one of several dominant themes visible not only in the figurine industry, but also in wall paintings; for example, in "Shrine" FV.1, a steatopygous figure with a distended stomach but thin waist stands below the great bull on the north wall (Mellaart 1966:Pl. LI,

LIVa–b, LVIb, LVIII), and a less well-preserved figure of a male with an erect phallus standing next to a seated female(?) with legs spread on the south wall (Mellaart 1966:Pl. LXIIb). While the meaning of these images is often glossed as "fertility," I would rather stress "abundance" and an assured supply of food and offspring. This makes a great deal of sense based on a significant shift in the subsistence economy of tenth to eighth millennium sites in Turkey, with early Neolithic settlements relying heavily on wild resources, and a significantly increased reliance on domesticates during the later Neolithic (M. Özdoğan 1997a). Whether we interpret the clay figures from level II as fat or pregnant or both, they clearly have more than enough to eat, and whatever activities they perform they do not expend all of the calories they take in; in other words, these figures may represent spirits (or perhaps mythological humans) who have a relatively high amount of leisure time and are exempt from the kind of heavy labor performed by village women today. Hamilton (1996:225) makes the intriguing suggestion that the clay figures after level VI emphasize "femaleness" (a concept that includes birth and motherhood) and an "increasing concern with women's roles" (Hamilton 1996:226). From this point of view, the figurines could be related to the increased value of female labor in more fully agricultural subsistence systems and to the value of children as potential laborers. But whatever the female figures mean, I would emphasize that the absence of male figures and figurines does *not* mean that there were no male deities or spirits in the religion of the upper levels at Çatal: male figures are predominant in the "hunting" murals found in levels V and above, and these are reasonably interpreted as ritual in origin.[4]

AN EASTERN PERSPECTIVE

Tracing the spread of ideas over large geographical areas is an activity involving risk, but encouraged by Hodder's wish to hear "many voices," I would like to point out some of the links between Neolithic communities of eastern and central Anatolia. Few would question the fact that at least two major cultural traditions are represented, as clearly argued by Mehmet Özdoğan (1995, 1997a). But at the same time, there is some material evidence for contact, especially during the tenth and ninth millennia bp. For example, in her recent study of chipped stone tools from Neolithic sites in eastern and central Anatolia, Nur Balkan-Alti concludes that there are clear

[4]The "new" wall paintings from Çatal (Mellaart 1990; Mellaart et al. 1989) are not fully documented and should not be considered in a scholarly discussion of the evidence from that site (see Collon 1990; Voigt 1991; Eiland 1993 with additional refs.)

affinities between the lithic assemblage at Asikli and PPNB sites to the east (Balkin-Alti1994:143). Copper beads from Asikli and Çayönü are apparently made using copper from different sources (Esin 1995:66), but at both sites the same distinctive technique of manufacture was used (rolling flattened sheets of copper to form tubular beads). Obsidian was apparently widely traded during this period (although the problem of distinguishing sources is far more complex than had been thought; see M. Özdoğan 1996), and when people meet to trade, an exchange of ideas and stories may well be part of the transaction. Among the most obvious nonmaterial links are the depiction of snakes, use of stone birds in ritual contexts, and the practice of secondary burials at Çatal and eastern sites.

To return to the earlier Neolithic religion documented by Nevalı Çori, Göbekli, and the lower excavated levels at Çatal Höyük, does the fact that pregnancy is not (or is rarely) explicitly represented mean that fertility and reproduction were not matters for concern, recognized by myth and alleviated by ritual? Information on this topic can perhaps be derived by another group of sculptures from Çatal: on the walls of buildings assigned to levels VI and below are sculpted figures with arms and legs raised (Mellaart 1963:Pl. IXa, XIII; 1964:Pl. IIIa, c, IV; 1967:Pl. 24–26, VII). The heads of these sculptures are always destroyed, and in many cases hands and feet have also been damaged. Mellaart and others have interpreted these sculptures as goddesses (Mellaart 1967:84–130; Forest 1993, 1994). Mellaart's own highly imaginative reconstructions of these figures show females with small rounded breasts (e.g., 1967:Figures 23, 26), but the photographs all show flat torsos with no indication of sex. Those who see these figures as recording birth rest their argument on the swollen(?) bull's-eye navel of one well-preserved example (Mellaart 1967:Pl. VII) and the position of the spread-legged figures above (or next to) animal skulls in some (but not all) cases. The deliberate and systematic destruction of these sculptures links them to the figurine, and again suggests representation of a powerful spirit. But there is no evidence that this spirit is specifically female, and the form of the figures (scars for heads, position of the limbs) suggests animals rather than humans—an interpretation that Mellaart himself must have shared since most of the restored images have cat's heads (1967:Figures 23, 26–28, 38).

In order to propose another interpretation of these wall sculptures, I would like to return to the issue of ambiguity. At Neolithic sites from central Anatolia to eastern Iran, there are a series of explicitly male/female images. From the sites discussed here, there is the phallic/large thighed figure from Gritille, a male/female relief from Nevalı Çori, and from Çatal Höyük a phallic stone (cylindrical with a line incised around its circumference below the rounded tip) that is also carved to represent a female with pendulous breasts (Kulaçoglu 1992, cat no. 15; Mellaart 1963:Figure 19). The posture and prominent navels of the Çatal wall reliefs with upraised arms link them to the reptile sculptures from Göbekli Tepe, but more significantly to the plaque from Nevalı Çori, which depicts a reptile flanked by two humans, arguably a male and a female. Whether we see the message as a concern with "fertility" or more generally as a concern with abundance and food, males and females as well as animals are involved. During the seventh millennium, the spirit world of sedentary communities in Anatolia was gender balanced. An emphasis on females and pregnancy is a later, but long-lasting, element in the religion of Anatolia and in the rest of the Middle East as well.

ACKNOWLEDGMENTS

Excavations at Hajji Firuz Tepe (1961, 1968) were conducted as part of the Hasanlu Project of the University of Pennsylvania Museum under the overall direction of Robert H. Dyson. Field director in 1961 was T. Cuyler Young, Jr., and in 1968, Mary M. Voigt. Funding for the 1968 season was provided by the Kevorkian Foundation, the Ford Foundation, and generous private donors. Four seasons of excavations at Gritille (1982–1984) were carried out under the direction of Richard Ellis. The project was sponsored by Bryn Mawr College, in cooperation with the University of North Carolina at Chapel Hill and the University of Pennsylvania Museum. Funding was provided by the National Endowment for the Humanities, the Haines Foundation, and generous private donors. This chapter could not have been written without the exemplary preliminary reports written by James Mellaart, who has always been generous, hospitable, and encouraging, from our first meeting at his house on the Bosporus in 1966 to our most recent in London in April 1998. We have disagreed in print, but he has consistently produced some of the most intellectually interesting archaeology in the Middle East. I would like to thank Carol Kramer for her comments on an early draft of this chapter, and for her interest in Hajji Firuz, which began when she was a "guest excavator" in 1968. Credit for the appearance of this chapter goes to Ian Kuijt, who was persistent, patient, supportive, and a good critic. Finally, thanks to my colleagues at William and Mary, who have provided a home for a once wandering teacher–scholar.

REFERENCES

Balkan-Alti, N., 1994, *La Neolithisation de l'Anatolie*. Varia Anatolica Volume 7. Institute Français d'Etudes Anatoliennes d'Istanbul.

Benedict, P., 1980, Survey Work in Southeastern Anatolia. In *Prehistoric Research in South-*

eastern Anatolia I, edited by H. Çambel and R. Braidwood, pp. 151–191. Istanbul University, Faculty of Letters No. 2589, Istanbul.

Bienert, H., D., 1991, Skull Cult in the Prehistoric Near East. *Journal of Prehistoric Religion* 5:9–23.

Çambel, H., and Braidwood, R., 1980, The Joint Istanbul-Chicago Universities' Prehistoric Research Project in Southeastern Anatolia. Comprehensive View: The Work to Date, 1963–1972. In *Prehistoric Research in Southeastern Anatolia I,* edited by H. Çambel and R. Braidwood, pp. 33–64. Istanbul University, Faculty of Letters No. 2589, Istanbul.

Collon, D., 1990, Subjective Reconstruction? The Çatal Hüyük Wall Paintings. *Hali* 53:119–123.

DeBoer, W., 1995, The Use of Anthropomorphic Figurines by Chachi Shamans. Paper delivered at the annual meeting of the Society for American Archaeology, Minneapolis, May 3–7, 1995.

Eiland, M., 1993, The Past Re-made: The Case of Oriental Carpets. *Antiquity* 67:859–863.

Ellis, R., and Voigt, M., 1982, 1981 Excavations at Gritille, Turkey. *American Journal of Archaeology* 86(3):319–332.

Esin, U., 1998, The Aceramic Site of Asikli and Its Ecological Conditions Based on Its Floral and Faunal Remains. *Tüba-Ar* 1:95–104.

Esin, U., 1995, Early Copper Metallurgy at the Pre-Pottery Site of Asikli. *Readings in Prehistory: Studies Presented to Halet Çambel,* pp. 61–77. Graphis, Istanbul.

Esin, U., 1993, Zum Ursprung der Kastenbauweise in Anatolien. *Istanbuler Mitteilungen* 43:123–128.

Fagan, B., 1998, *People of the Earth: An Introduction to World Prehistory.* Longman, New York.

Forest, J. D., 1994, Towards an Interpretation of the Çatal Höyük Reliefs and Wall Paintings. *1993 YiliAnadolu Mediniyetleri Müzesi Konferanslari,* pp. 118–136. Museum of Anatolian Civilizations, Ankara.

Forest, J. D., 1993, Çatal Höyük et son Decor: Pour le Dechiffrement d'un Code Symbolique. *Anatolia Antiqua* 2:1–42.

Garfinkel, Y., 1995, *Human and Animal Figurines of Munhata (Israel).* Les cahiers des Missions Archéologiques Françaises en Israël No. 8.

Gates, M. H., 1997, Archaeology in Turkey. *American Journal of Archaeology* 101:241–305.

Hamilton, N., 1996, Figurines, Clay Balls, Small Finds, and Burials. In *On the Surface: Çatalhöyük 1993–1995,* edited by I. Hodder, pp. 215–263. McDonald Institute of Archaeological Research and the British Institute of Archaeology at Ankara, Cambridge.

Hauptmann, H., 1997, Nevalı Çori. *The Oxford Encyclopedia of Archaeology in the Near East* 4:131–134.

Hauptmann, H., 1993, Ein Kultgebäude in Nevalı Çori. In *Between the Rivers and over the Mountains,* edited by M. Frangipane, H. Hauptman, M. Liverani, P. Matthai, and M. Mellink, pp. 37–69. Universiti di Roma la Sapienza, Rome.

Hauptmann, H., 1988, Nevalı Çori. Arkitektur. *Anatolica* 15:99–110.

Hauptmann, H., 1987, Nevalı Çori. *Anatolian Studies* 37:206–207.

Hodder, I., 1996, Conclusions. In *On the Surface: Çatalhöyük 1993–1995,* edited by I. Hodder, pp. 359-366. McDonald Institute of Archaeological Research and the British Institute of Archaeology at Ankara, Cambridge.

Hodder, I., 1990, *The Domestication of Europe: Structure and Contingency in Neolithic Societies.* Basil Blackwell, Oxford.

Hodder, I., 1987, Contextual Archaeology: An Interpretation of Çatal Hüyük and a Discussion of the Origins of Agriculture. *Bulletin of the Institute of Archaeology* 24:43–56.

Knapp, A. B., and Meskell, L., 1997, Bodies of Evidence on Prehistoric Cyprus. *Cambridge Archaeological Journal* 7(2):183–204.

Kulaçoğlu, B., 1992, *Anadolu Medeniyetleri Müzesi Tanrilar ve Tanriçilar*. T. C. Kültür Bakanligi, Anitlar ve Müzeler Genel Müdürlügü, Ankara.
Mellaart, J., 1990, The Earliest Representations of the Goddess of Anatolia and Her Entourage. In *Anatolische Kelims: (Symposium Basel), die Vorträge*. edited by J. Rageth, pp. 27–46. Galerie Rageth, Basel.
Mellaart, J., 1970, *Excavations at Hacilar*. British Institute of Archaeology at Ankara Occasional Publications 9–10.
Mellaart, J., 1967, *A Neolithic Town in Anatolia*. McGraw-Hill, New York.
Mellaart, J., 1966, Excavations at Çatal Hüyük, 1965: Fourth Preliminary Report. *Anatolian Studies* 16:165–191.
Mellaart, J., 1964, Excavations at Çatal Hüyük, 1963: Third Preliminary Report. *Anatolian Studies* 14:39–119.
Mellaart, J., 1963, Excavations at Çatal Hüyük, 1962: Second Preliminary Report. *Anatolian Studies* 13:43–103.
Mellaart, J., 1962, Excavations at Çatal Hüyük: First Preliminary Report, 1961. *Anatolian Studies* 12:41–65.
Mellaart, J., Hirsh, U., and Balpinar, B., 1989, *The Goddess from Anatolia*. Eskenazi, Milan.
Mellink, M., 1992, Anatolia. In *Chronologies in Old World Archaeology*, edited by R. Erich, pp. 107–220. University of Chicago Press, Chicago, Illinois.
Morales, V. B., 1990, *Figurines and Other Clay Objects from Sarab and Çayönü*. Oriental Institute Communications No. 25. The University of Chicago, Chicago, Illinois.
Morales, V. B., 1983, Jarmo Figurines and Other Clay Objects. In *Prehistoric Archeology along the Zagros Flanks*, edited by L. Braidwood, R. J. Braidwood, B. Howe, C. A. Reed, and P. J. Watson, pp. 369–423. The University of Chicago Oriental Institute Publications, Volume 105. The University of Chicago, Chicago, Illinois.
Nickerson, J., 1979, *The Study, Analysis, and Interpretation of the Human and Animal Figurines from Tal-e Malyan (Anshan), Iran*. B.A. thesis, Ohio State University.
Özdoğan, A., 1995, Life at Çayönü during the Pre-Pottery Neolithic Period (according to the artifactual assemblage). In *Readings in Prehistory: Studies Presented to Halet Çambel*, pp. 79–100. Graphis, Istanbul.
Özdoğan, M., in press, Buildings of Cult and Cult of Buildings. In *Halet Çambel Festschrift*, edited by G. Arsebük and W. Schirmer. Graphis, Istanbul.
Özdoğan, M., 1997a, Anatolia from the Last Glacial Maximum to the Holocene Climatic Optimum: Cultural Formations and the Impact of the Environmental Setting. *Paléorient* 23(2):25–38.
Özdoğan, M., 1997b, Çayönü. *The Oxford Encyclopedia of Archaeology in the Near East* 1:444–446.
Özdoğan, M., 1996, Obsidian in Anatolia: An Archaeological Perspective on the Status of Research. In *Archaeometry 94*, edited by S. Demirci, A. Özer, and G. Summers, pp. 423-431. Proceedings of the 29th International Symposium on Archaeometry, Ankara, May 9–14, 1994. TÜBITAK, Ankara.
Özdoğan, M., 1995, Neolithic in Turkey—The Status of Research. *Readings in Prehistory: Studies Presented to Halet Çambel*. Graphis, Istanbul.
Özdoğan, M. and Özdoğan, A., 1990, Çayönü: A Conspectus of Recent Work. *Paléorient* 15(1):387–396.
Özdoğan, M., Özdoğan, A., Bar-Yosef, D., and van Zeist, W., 1994, Çayönü Kazisi ve Güneydogu Anadolu Karma Projesi: 30 Yillik Genel Bir Degerlendirme, *XV Kazi Sonuçlari Toplantisi* 1:103–122.
Schmidt, K., 1997, "News from the Hilly Flanks": Zum Forschungsstand des obermesopotamischen Frühneolithikums. *Archäologisches Nachrichtenblatt* 2(1):70–79.

Schirmer, W., 1988, Zu den Bauten des Çayönü Tepesi. *Anatolica* 15:139–159.

Schirmer, W., 1983, Drei Bauten des Çayönü Tepesi. In *Beiträge zur Altertumskunde Kleinasiens: Festschrift für Kurt Bittel*, edited by R. Boehmer and H. Hauptman, pp. 463–476. Von Zabern, Mainz am Rhein.

Stein, G., 1989, Strategies of Risk Reduction in Herding and Hunting Systems of Neolithic Southeast Anatolia. In *Early Animal Domestication and its Cultural Context*, edited by P. Crabtree, D. Campana, and K. Ryan, pp. 87–97. MASCA Research Papers in Science and Archaeology, Special Supplement to Vol. 6.

Stein, G., 1986, Herding Strategies at Neolithic Gritille. *Expedition* 28(2):35–42.

Talalay, L., 1993, *Deities, Dolls, and Devices: Neolithic Figurines from Franchthi Cave, Greece.* Excavations at Franchthi Cave, Greece Fascicle 9. Indiana University Press, Bloomington and Indianapolis, Indiana.

Türe, Fatma (Project Coordinator) et al., 1999, *Auf der Suche nach Verschwundenen Zeiten: Die Ausgrabungen des Deutschen Archäologischen Instituts in der Türkei.* Istanbul: Yapl Kredi Kültür Sanat Yayincilik, for the Yapi Kredi Bankasi A.S.

Ucko, P., 1968, *Anthropomorphic Figurines of Predynastic Egypt and Neolithic Crete with Comparative Material from the Prehistoric Near East and Mainland Greece.* Royal Anthropological Institute Occasional Paper No 24. Andrew Szmidla, London.

Ucko, P., 1962, The Interpretation of Prehistoric Anthropomorphic Figurines. *Journal of the Royal Anthropological Institute* 92:38–54.

Voigt, M., 1983, *Hajji Firuz Tepe, Iran: The Neolithic Settlement.* Hasanlu Excavation Reports, Volume I. University Museum Monograph No. 50. The University of Pennsylvania Museum of Archaeology and Anthropology, Philadelphia.

Voigt, M., 1985, Village on the Euphrates: Excavations at Neolithic Gritille, Turkey. *Expedition* 27(1):10–24.

Voigt, M., 1988, Excavations at Neolithic Gritille. *Anatolica* XV:215–232.

Voigt, M., 1990, Reconstructing Neolithic Societies and Economies in the Middle East: An Essay. *Archaeomaterials* 4(1):1–14.

Voigt, M., 1991, The Goddess from Anatolia: An Archaeological Perspective. *Oriental Rug Review* 11(2):32–39.

Voigt, M., and Ellis, R., 1981, Excavations at Gritille, Turkey: 1981. *Paléorient* 7(2):87–100.

Chapter **12**

The Pottery Neolithic Period

Questions about Pottery Decoration, Symbolism, and Meaning

Estelle Orrelle and Avi Gopher

INTRODUCTION

The appearance of cultivated grains and domesticated animals in the southern Levant has traditionally been related in some form or another to the realm of subsistence. An increasing number of archaeologists, however, (e.g., Bender 1985; Hayden 1990) assert that social motivation may have been the impetus for economic intensification rather than the optimization of ecological, demographic, or utilitarian considerations. Keswani (1994), for example, describes how in most societies the acquisition and maintenance of food resources are motivated not merely by concerns for survival but by a variety of social and ideological requirements that determine the specific dimensions of change in subsistence practice. Ethnographic and archaeological studies suggest that it is artificial to separate the social and subsistence aspects of prestate societies (Ferguson 1985; Hodder 1984). Thus, to understand changes associated with food production, we must look for structural aspects of the relationship between social and subsistence behav-

ESTELLE ORRELLE and AVI GOPHER • Institute of Archaeology, University of Tel Aviv, Ramat Aviv, Israel 69978.

Life in Neolithic Farming Communities: Social Organization, Identity, and Differentiation, edited by Ian Kuijt. Kluwer Academic/Plenum Publishers, New York, 2000.

iors. The change from a hunter–gatherer socioeconomic system to that of the early agriculturalists involved more than a change in the method of food procurement; indeed, it involved a transformation that touched not only the mechanics of food procurement but the very fabric of stability and values upon which hunter–gatherer society was based.

The Pottery Neolithic (PN) period of the southern Levant (*ca.* 7,500–6,000 bp uncalibrated C-14 dates) sees the end of this transformation - and consists of largely two cultural configurations: the Yarmukian and the following Lodian (Jericho IX) entity, and the Wadi Raba cultures (Gopher and Gophna 1993). They represent, on the one hand, the end of a process, the move to food production, and the end of hunting as seen in the dwindling quantities of wild bones and arrowheads in the faunal and lithic assemblages, and represent on another level the new transformed socioeconomy of animal and crop elevation and its accompanying symbolic system. During this period, the domesticated cow, the large manufactured horned animal, appears in the archaeological record at the same time that the last collected resource (hunted gazelle or deer) disappeared. We describe briefly the main characteristics of the two Pottery Neolithic cultures, some of their symbolic aspects, and then raise questions about the social implications of these data sets.

THE YARMUKIAN CULTURE

The Yarmukian was recognized as a cultural entity in the early 1950s (Stekelis 1950/51, 1972) and additional layers were excavated during the 1950s and 1960s (Perrot 1968; Kaplan 1958, 1965, 1978), and during the last decade both in Israel (Garfinkel 1993; Gopher 1996; Gopher and Tsuk 1990) and in Jordan (Kafafi 1985, 1986, 1988; Muheisen et al. 1988; Simmons et al. 1989; Rollefson et al. 1992). Recent summaries make it clear that the Yarmukian is a very distinct entity restricted to a specific area of *ca.* 10,000 km^2 of the southern Levant and dated to the last two thirds of the sixth millennium BC (see Garfinkel 1992, 1993; Gopher 1995; Gopher and Gophna 1993; Kafafi 1987, 1993).

Deposits of the Yarmukian are recovered from above Pre-Pottery Neolithic occupations. It has a well known distinct assemblage of pottery with painted and incised designs as well as a distinct flint tool assemblage. Its economy is thought to represent a form of a settled agriculture with animal exploitation (e.g., Rollefson and Köhler-Rollefson 1989; Simmons et al. 1988; and for general summaries see Garfinkel 1993; Gopher 1995; Gopher and Gophna 1993)—ovicaprines, cattle, and raising the possibility of pastoralism. Pigs were reared, and cereals, pulses, and flax cultivated

(Simmons et al. 1988; Köhler-Rollefson et al. 1988). Architecture includes both circular and rectilinear buildings and the few burials exposed in Yarmukian levels include both adults and children. It was recently suggested that assemblages similar to Garstang's (1935, 1936) stratum IX and Kenyon's PNA (Kenyon 1981) deserve a place as an independent cultural entity in the Pottery Neolithic sequence of our region named the "Lodian." This entity with its distinct pottery and lithic assemblages postdates the "normative Yarmukian" and predates the Wadi Raba culture (Blockman 1997; Gopher and Gophna 1993). This entity, however, has very little imagery, and it is unclear how this assembage should be incorporated into regional reconstructions.

THE WADI RABA CULTURE

The Wadi Raba culture is known from some thirty-five archaeological assemblages from Israel, Jordan, and Lebanon and is described as "normative" following the description of the original discovery by Kaplan (1958). Stratigraphically, Wadi Raba occupations are later than the Yarmukian and earlier than the Chalcolithic Ghassulian. The paucity of radiometric dates for Wadi Raba makes it difficult to date precisely, but the 'normative' has been assigned a time span of about 500 C-14 years in the seventh millennium bp, and the larger culture, including variants, occupies much of the seventh millennium bp. The 'normative' Wadi Raba has a limited geographical range of about 10,000 km^2, but together with other contemporaneous variants extends this area considerably occupying the Mediterranean zones of the southern Levant (Gopher and Gophna 1993). Faunal assemblages from settlements in and near the Jezreel Valley show a dominance of domestic sheep and goats, followed by cattle and pigs (Davis, personal communication). Spindle whorls, loom weights, and other spinning and weaving equipment suggest the intensive use of animal hair, possibly goat hair since there is a high frequency of goat bones in one of the faunal assemblages collected and analyzed (Davis in press).

A prototype of a churn from the late Wadi Raba layer at Nahal Zehora I might indicate the preparation of milk products, in addition to the cereal cultivation attested by phytoliths (Rosen personal communication 1991) and recently discovered seeds. Research at late eighth and in seventh millennium bp settlements off the coastal plain has recently demonstrated the intensive use of olives (Galili and Sharvit 1994–5; Galili et al. 1989). Lithic assemblages are almost devoid of arrowheads with an abundant sickle blade and bifacial tool assemblage. Residential structures are rectangular, with some buildings being quite large and showing evidence of internal subdivision. Features include a variety of pits, circular, paved or lined in stone, brick, or both. Circular basins and small irregularly shaped paved areas are sometimes found outside walls. As in the Yarmukian, the few adult burials are intramural. The first appearance of burials of fetuses, babies, and children in jar burials and cist graves is, however, the more conspicuous phenomenon here.

SYMBOLIC ASSEMBLAGES OF THE POTTERY NEOLITHIC

The social manipulation of symbolic material culture is a previously unexamined source of information on the nature and mechanisms of social and economic change in Pottery Neolithic communities. The two main categories of symbolic artifacts to appear in the Pottery Neolithic period are the figurines, which we prefer to refer to as "imagery" à la Conkey (Conkey 1987), and the assemblages of decorated and undecorated pottery vessels that appear in a rich variety of sizes and shapes in both cultures. Although pottery assemblages are usually treated from a functional perspective by archaeologists, we prefer to regard the first pottery assemblages to appear in the southern Levant as fulfilling a social role, as "sociotechnic" (Goren and Gopher 1995), and to treat them as a symbolic assemblage. In the rest of this chapter we briefly describe the main elements of these symbolic assemblages, discuss some of the differences between them, and then proceed to examine some of the possible implications of these artifacts for the socioeconomic milieu from which they came.

Yarmukian Pottery, Imagery, and Interpretations

The Yarmukian imagery assemblage was described originally by Stekelis (1950/1951, 1972) and more recently by Garfinkel (1992, 1995), with alternative interpretations offered by Gopher and Orrelle (1996). The assemblage comprises a number of categories of stone and clay items representing males, females, and androgynous persons, as well as sexual organs. A concentration of genital imagery characterizes both the stone and clay items. We have argued elsewhere (Gopher and Orrelle 1996) that some elements of the Yarmukian imagery encode information associated with age and reproductive status relating to gender categorization and social discourses such as control of reproduction. Other elements engage, we believe, in symbolic contest between male and female blood rituals. The Yarmukian pottery assemblage is distinctive mainly in form and decoration. Types include bowls of various sizes, including short pedestal bowls, amphora, platters, necked jars and hole-mouth jars. A variety of handles include knobs, pseudoledge handles, and loop types.

Analysis of the technology of Yarmukian vessels reveals no evidence of heat resistance for cooking, and it has been suggested that production was aimed mainly at achieving a visual effect and not a pyrotechnological one (Goren and Gopher 1995). Most of the vessels are plain, but there is a highly decorated element on some 12.5% of the assemblage, for example at Munhata layer 2B (Garfinkel 1992:12). The decoration found on the pottery runs the gamut of variations from plain incised, to painted with plain incised, to painted only. It includes combinations of red painted triangles arranged in zigzag formations and interspersed with plain bands incised with nested V motifs (usually referred to as herringbone motifs); this design appears with or without the red paint. There is a plain incised weave design in varying degrees of complexity. Bands appear in a variety of formations, and some vessels bear the red painted design with no incised element. The assemblage is very much dominated by the undulating variations of triangles and V motifs and the contrasting red and white grounds. The Lodian, though identified as an independent cultural unit appearing at the end of the Yarmukian, is still relatively unclear (e.g., Blochman 1997; Gopher and Gophna 1993). Characteristic features of its pottery assemblage are a burnished glossy paint motif in brown and red-brown hues on unburnished cream background and a long narrow converging neck to the jars. The painted motifs of red triangles are absent. The practice of burnishing starts most probably at the end of the Yarmukian, continues, and grows in the Lodian and becomes more frequent in the Wadi Raba culture pottery assemblage.

What we have here, then, is a new category of produced artifact made of a transformed raw material, common clay, that has been mixed with temper and water, formed, individually decorated, and passed through fire to produce an artifact with a high degree of investment. It is found in *domus* contexts as described by Hodder (1990:44–45, 52). Its decoration consists of the triangle and V motif appearing with and without red color. We believe that the triangle, V motif, and zigzag are actually the same motif and derive from age old 'Paleolithic shorthand' vulvar representations that appear in Ice Age symbol systems (e.g., Marshak 1991). Triangles appear in Upper Paleolithic European cave and mobile art sometimes engraved in pubic position on images of women, and more often alone as near iconic images. We interpret these as vulvae, especially in view of fourth millennium BC Sumerian pictorial script where the slit triangle icon means "woman" (Green and Nissen 1987). Often traces of red color are found on these images and, thus, can be interpreted as menstruating vulvae.

We have argued elsewhere (Gopher and Orrelle 1996) that these and similar almond shapes near iconic motifs stand as symbols for a social system governing rules of access to women. They are what Ortner describes as "summarizing symbols," visual meaning-laden symbols, schematized versions of the natural model (thus near iconic) that have entered a sacred category and stand for a complex system of ideals in communities. They form part of what Knight describes as a "transformational template" (Knight 1991)—a time-resistant hunter–gatherer division of labor, the meaning of which these symbols represent. By the Neolithic, these schematized versions of vulvae have become incorporated into patterns and appear on a large range of artifacts in many countries (e.g., Gimbutas 1991). Synchronized vulva triangles bearing red color, traditionally blood, must still suggest menstruating vulvae, female ritual symbols connected with female inviolability during menstruation, a taboo deeply embedded in hunter–gatherer symbolic systems (Knight 1991; Knight et al. 1995). Given that these symbols were related to ancient female hunter–gatherer rituals expressed on an innovative artifact in an agricultural–horticultural–pastoral society, then this probably represents deliberate attempts to defend the old order (Gopher and Orrelle 1996). High investment in decoration of any artifact class suggests "loud ritual" (elaboration is employed to support the claims of a particular social group) (Sperber 1975; Knight 1994; Peltenberg 1994). It can serve to introduce change or to defend against change in the battle of symbols (Harrison 1992). We believe that the recovery of several different kinds of artifact classes in the Yarmukian illustrates that discourse of this kind occurred simultaneously on several different levels.

Wadi Raba Pottery, Imagery, and Interpretations

In contrast to the Yarmukian, anthropomorphous imagery is very sparse in the Wadi Raba culture. The only known example is a carinated hole-mouth jar from Ein el Jarba with an applied image of a human figure with an animal headdress. This and similar items from Tepe Gawra are interpreted by Kaplan (1969:18) as the prototype of a figure related to a fertility cult concerned with the renewal of youth or the bringing of rain. Small clay animal figurines are known from a number of Wadi Raba sites, and we will refer to these later.

The Wadi Raba pottery assemblage is very rich and diverse (Orrelle 1993) and differs from the Yarmukian repertoire in elements of morphology and decoration. It includes a variety of bowls, rounded, straight upright, V-shaped or carinated, with inverted, everted, or cutoff rims. A small thin highly fired carinated bowl of grit-free fabric, usually slipped and burnished in a deep glossy black or red is characteristic of this assemblage. This ware is usually referred to as dark-faced burnished ware after Braidwood and Braidwood (1960) and includes larger bowls and rare platters with inverted or rounded sides. Pedestaled bowls occur and are usually large and heavy. The typical Wadi Raba jar has a characteristic bow-shaped rim that appears in almost all of the assemblages assigned to this entity or its variants. Hole-mouth jars are also present, both large and small, and sometimes slipped and burnished. Pithoi have thickened triangular section rims, sometimes with knobs or lugs. Handles include characteristic large loop handles with splayed attachments, knob handles, and small pierced handles. Vessels stand on flat or disklike bases. In this assemblage, too, the highly decorated element is small, accounting for a very similar proportion to that in the Yarmukian (Orrelle 1993). There are, however, many more categories and kinds of decoration. The broad range of surface treatments includes primary plain smoothed, slipped, slipped and burnished, and various incised, impressed, combed, painted, or applied plastic motifs. A red band around the inside and outside of vessel rims is a common feature (see Gopher and Gophna 1993).

The incised decoration consists of five main groups—the nested versus wavy line, various parallel lines, lunate, and raised incisions. The painted element is far less dominant and consists usually of single or more bands, often positioned on the point of carination (Orrelle 1993). There is a very small element of plastic decoration of raised rope type that will continue in the later Chalcolithic period. Occasional small nipple-like bosses appear as well. A striking expansion of the Yarmukian repertoire is seen in the variety of color in the pottery assemblage. In contrast to the Yarmukian red and white-buff, blacks, browns, reds, and oranges appear on Wadi Raba vessels in a variety of combinations and hues, some matte and some deeply lustrous from applied burnish. Black color which appeared on "rods"—phallic type artifacts in the Yarmukian (see Garfinkel 1995; Gopher and Orrelle 1996)—appears in small quantities on pottery vessels in the Wadi Raba assemblage.

DISCUSSION

Organization of Design Space on Vessels

Although Wadi Raba pottery assemblages are consistently fragmented, observations of the sherds and published reconstructions permit us to examine the meaning(s) of these objects on the basis of ethnographic studies such as (Douglas 1966) and place them into an interpretive context. Decoration appears to occupy horizontal bands around the rim, neck, shoulder, or carinated "waist" of vessels. One heavily decorated area of the vessel is the "handle" (sometimes up to 50%; Eyal personal communication). The most consistently decorated area, however, is the vessel rim (some 60%), and of these the most common decoration is a painted red band around the 'mouth' hinting at the use of red color/blood around an "orifice" of a "pot/

woman." If we extend Barley's (1994) research on African pottery (in "speaking with pots" logic), then this orifice can be interpreted as a menstruating vulva.

On vessels of both the Yarmukian and Wadi Raba assemblages are found appendages, which we regard as small plastic shapes for symbolic information rather than serving some practical carrying function (Orrelle 1993). We suggest that these might mirror images of bodily appendages resembling fleshy flaps surrounding orifices such as ears, noses, and mouths, which incorporate triangular and oval vulva-like shapes. Very often in folklore, magic and superstition, ears, noses, and mouths have great symbolic significance as entrances to the body—orifices of the pot/woman, or the social body as it may represent, and they are likened to vulvae (e.g., Lewis 1980), although see Orrelle (1993) for a consideration of the difficulties of applying folklore or popular 'magical' interpretations in an archaeological context. The high degree of "orifice elaboration" could well relate to issues concerning access to the woman/social body—to different patterns of marriage arrangements (see Orrelle 1993).

A marked change in the decoration from the Yarmukian to the Wadi Raba assemblage is the sharp reduction of synchronous triangles and V patterns and raises a number of questions. It is possible, for example, that the decorative transition from the Yarmukian triangle/vulvae arranged in a unified pattern to the single isolated, Wadi Raba vulva symbols on the woman pot, represent a deliberately altered use of the vulva symbol from one of solidarity to one of isolation. By extension this may imply that the view of menstruation changed from one of empowering to that of polluting. Moreover, we believe that the replacement of a dominant pattern of Yarmukian synchronized red triangles on the "woman" vessel, with a pattern of the division into horizontally demarcated symbolic space on the Wadi Raba vessels also may have been linked to a reorganization of a woman's rights, roles, and position within Pottery Neolithic communities.

In sum, we argue that these vessels of clay, in the Pottery Neolithic cultures, traditionally identified with the female body and by definition with the social body (Barley 1994:136; Welbourn 1984, 1989), acted as a kind of blueprint onto which the norms and beliefs of society were displayed and that were expressions of restrictions and guidelines of social behavior for the Yarmukian and Wadi Raba cultures. We suggest that Yarmukian imagery indicates a number of processes in action: some linked with the reestablishment of a cohesive society after a degree of dispersement, and others associated with the needs of the new economy in which reproduction was of central importance (Gopher and Orrelle 1996).

What, then, might be the role of the pottery assemblage in these communities? In horticultural/pastoral economies, the demands for labor are clearly felt, and, thus, an imagery in which genital symbolism is central might indicate tightening controls of women's rights, labor and offspring. Women themselves would have had an added workload and different scheduling from agricultural duties to child care. Naveh (1996) illustrates such a trend through the changing context of processing activities in Yarmukian and Wadi Raba strata of the Nahal Zehora II village. In the Yarmukian, pounding/grinding activities were conducted outside of the houses in a seemingly communal open space, while in the Wadi Raba it seems to have moved to a more isolated single household context. It is possible that this hints at females becoming increasingly isolated from each other, and the symbolic elaboration of Yarmukian pottery vessels may have been related to this. The ceramic repertoire of Wadi Raba contains a rich decorated element and has an overall greater diversity of shapes and forms and is more plentiful than in the Yarmukian. The symbols of menstrual rituals, the red color, the V motifs, and the lunate shapes still persist but are not now so dominant as they were in the Yarmukian.

Orrelle (1993) has likened the pottery assemblage found in the Wadi Raba culture to the "grid" of society (Miller 1985; Douglas 1966). Douglas suggests that in small pre-industrial societies, personal relations are structured by two independent variables—the relationship of an individual with her/his group, a bounded unit contained in space, known by a common name sharing a common interest in property, and her/his position on a grid, which controls the flow of behavior within the group to the extent that roles within it are allocated according to various principles such as sex age and seniority. The notion of group encompasses a vast range of allegiances from the lowest possible associations to the most tightly knit closed groups. The temporal dimension of people's association and range of permanence, of boundedness is also expressed in this concept. Included in the internal group organization are the forms of structures, such as hierarchy of command and delegation of responsibility from the center, all of which can be envisioned as a social grid. This grid may also express the extent to which an individual may or may not be bound in different ways—to what extent they are constrained by gender, age, or hierarchy. In such societies, social codes and ideology are likely to carry over into the creation of the material world. Using the principle of symbolic replication of a social state, Miller (1985) traced appropriateness between symbolic forms and social forms in Dangawara pottery and shows how even the most trivial codes can be articulated with almost any other aspect of conceptualization.

Assuming that pottery is used in such a way and that Wadi Raba vessels still symbolize women's bodies and the social body, then this provides us a material means of attempting to interprete social relations in these communities. Specifically, we relate the decoration (brown, black, and red slips burnished to a high lustrous glaze with raised incisions) on Neolithic pots to rubbed-in fat and mud mixtures and cicatrices embossed on women and pots and cattle in modern African contexts (e.g., Barley 1994; Welbourn 1989), with these decorative features serving as metaphors relating women to the cattle in which marriages were validated and social roles were established. In this context, the decorations on pots may have served as means of communicating the value of women within communities, and these changes ultimately represent changes in the ritual uses and meanings of symbols associated with menstruation. In this light women, cattle, and pots can be envisioned as being embedded in a grid of a horticultural/agricultural pastoral society, where an increasingly powerful gerontocracy accumulated resources, later expressed in the emergence of social differentiation.

Does the appearance of clay images of domesticated animals in the Pottery Neolithic fit into this scheme? One possible link is that this structural grid of female resource and bovine, and other animal capital, formed a social mesh upon which relations were built, and variations within it led to the development of local archaeological cultures. When one observes the archaeological cultures of the eighth to fifth millennia bp, the Yarmukian, Lodian, Wadi Raba, and the Ghassulian, one cannot but notice the ever increasing variation in the scale and material expression. In social terms, this may reflect the existence of successful exchange systems in which neighbors exchange women and livestock to gain access to vital resources such as land, water, and grazing pastures. Collectively, this raises the possibility that Early Pottery Neolithic communities envisioned social relations from the perspective of control of reproductive resource within society and focused on young women. In such a system young men may have been bound in debt to their older kin, forced to find livestock for bride-price to acquire a wife, bound in debt to elder males accumulating power and authority through feasting and control of grain, cloth, and other goods (Orrelle 1993:133). It is possible, then, that some of the difference between the Yarmukian and the Wadi Raba symbolic assemblages can be viewed as contests over ritual power, access to resources, and control of progeny. If this is the case, then the seemingly sparse imagery but highly homogenous pottery decoration of the Wadi Raba culture may indicate the resolution of these tensions.

Death and Ancestors in the Pottery Neolithic

From the Natufian to the end of the PPNB period, a large part of the ritual paraphernalia was centered on mortuary ritual—from the elaborate dead in the Natufian graveyards to the highly decorated skulls and animal/human burials of the PPNB. Seemingly, the dead and the authority of the ancestors was harnessed to support the power of the living—an authority that seems to transform from a nebulous other-worldly spiritual power to a very material relic. We agree with Kuijt (1996) that these were used in powerful communal acts in the Pre-Pottery Neolithic context, but disagree that these were used as means of social integration. Rather, we believe that the dead and authority of the ancestors served as tools and mechanisms for accumulating power in the hands of those controlling ritual. In each entity, the highly decorated material relic element was " . . . finite in quantity and enduring . . . " (Harrison 1992:231)—as in the few highly ornamented Natufian graves, or the PPNB individual skulls selected for plastering and painting. We believe that these highly elaborated relics may represent an original embodying of ancestral power, which later transformed to "elder power" by the Pottery Neolithic, representing a constancy of form and simultaneous transition in the meanings and players (Boas 1955:128; Bloch 1986; Hodder 1984).

The wresting of authority from ancestors in ancestral time and its conversion into authority of elders in historical time (Criado 1989), or for that matter the transition from "la pensee sauvage" (Levi-Strauss 1966) to domesticated thought (Goody 1977), is one of the marks of the cognitive change between hunter–gatherer and agriculturist perceptions. The finite but enduring highly decorated element of the Natufian and Pre-Pottery Neolithic mortuary ritual had a specific purpose and meanings, and some of these practices reappeared in the Pottery Neolithic with new meanings. In the mortuary context of the Pottery Neolithic, one finds new variations in the mortuary treatment of children, infants, and fetuses. For example, the burials of children in settlements and fetuses in "womb" pot graves may be another variation of Wadi Raba mortuary rituals: practices that hint at inherited wealth and authority under the control of successful elder individuals and the emergence of social differentiation. As noted earlier, once the restraints of hunter–gatherer society's norms begin to lift, the pace of social differentiation accelerated, a process that would increase in the later Chalcolithic period. Studies of this and other topics are in their infancy, but we feel that research is progressing sufficiently to allow us to argue that by the Pottery Neolithic period of the sixth and fifth millennia BC elder members of communities controlled ritual, social and economic resource in a way that was profoundly different from that of egalitarian hunter–gatherer communities, as well as earlier agriculturists of the Pre-Pottery Neolithic periods.

REFERENCES

Barley, N., 1994, *Smashing Pots Feats of Clay from Africa.* British Museum Press, London.

Bender, B., 1985, Emergent Tribal Formations in the American Midcontinent. *American Antiquity* 50(1):52–62.

Bloch, M., 1986, *From Blessing to Violence History and Ideology in the Circumcision Ritual of the Merina of Madagascar.* Cambridge University Press, Cambridge.

Blockman, N., 1997, *The Lodian Culture (Jericho IX) Following the Excavations at Newe Yarak, Lod.* M.A. thesis, Tel Aviv University.

Boas, F., 1955, *Primitive Art.* Dover, New York.

Braidwood, R. J., and Braidwood L. S., 1960, *Excavations in the Plain of Antioch.* Oriental Institute Publications. University of Chicago Press, Chicago, Illinois.

Conkey, M. W., 1987, New Approaches in the Search for Meaning? A Review of Research in "Paleolithic Art." *Journal of Field Archaeology* 14:413–428.

Criado, F., 1989, "We the post-megalithic people . . . " In *The Meaning of Things, Material Culture and Symbolic Expression,* edited by I. Hodder, pp. 78–89. Harper Collins Academic, London.

Davis, S., forthcoming, The Animal Bones from Nahal Zehora I and II. In (ed.) A. Gopher, *The Nahal Zehora Sites: Pottery Neolithic Villages in the Menashe Hills Israel.* Monograph Series of the Institute of Archaeology, Tel Aviv University.

Douglas, M., 1966, *Purity and Danger. An Analysis of the Concepts of Pollution and Taboo.* Routledge, London.

Ferguson, J., 1985, The Bovine Mystique: Power, Property and Livestock in Rural Lesotho. *Man* (New Series) 20(4):647–674.

Galili, E., Weinstein-Evron, M., and Zohary, D., 1989, The Appearance of Olives in Submerged Neolithic Sites along the Carmel Coast. *Mitekufat Haeven* (2):95–97.

Galili, E., and Sharvit, J., 1994–5, Evidence of Olive Oil Production from the Submerged Site at Kfar Samir, Israel. *Journal of the Israel Prehistoric Society* 26:122–133.

Garfinkel, Y., 1992, *The Material Culture in the Central Jordan Valley in the Pottery Neolithic and Early Chalcolithic Periods.* Ph.D. dissertation, The Hebrew University, Jerusalem.

Garfinkel, Y., 1993, The Yarmukian Culture in Israel. *Paléorient* 19:115–134.

Garfinkel, Y., 1995, *Human and Animal Figurines of Munhata (Israel).* Cahiers du Centre de Recherche Francais de Jerusalem, Volume 8. Association Paléorient, Paris.

Garstang, J., 1935, Jericho City and Necropolis. Fifth Report. *Liverpool Annals of Archaeology and Anthropology* 22:143–168.

Garstang, J., 1935, Jericho City and Necropolis. Report for Sixth and Concluding Season. *Liverpool Annals of Archaeology and Anthropology* 22:143–168.

Gimbutas, M., 1991, *The Language of the Goddess.* Harper, San Francisco, California.

Goody, J., 1977, *The Domestication of the Savage Mind.* Cambridge University Press, Cambridge.

Gopher, A., 1995, Early Pottery-Bearing Groups in Israel—The Pottery Neolithic Period. In *Archaeology of Society in the Holy Land.* edited by T. E. Levy, pp. 206–225. Sheffield University Press, Sheffield.

Gopher, A., 1996, *The Nahal Qanah Cave,* Monograph No. 12 of the Institute of Archaeology. Tel Aviv University, Tel Aviv.

Gopher, A., and Gophna, R., 1993, Cultures of the Eighth and Seventh Millennia BP in the Southern Levant: A Review for the 1990s. *Journal of World Prehistory* 7(3):297–353.

Gopher, A., and Orrelle, E., 1996, An Alternative Interpretation for the Material Imagery of the Yarmukian: A Neolithic Culture of the Sixth Millennium B.C. in the Southern Levant. *Cambridge Archaeological Journal* 6(2):255–279.

Gopher, A., and Tsuk, T., 1990, Earliest Gold Artifacts in the Levant. *Current Anthropology.* 31(4):436–443.

Goren, Y., and Gopher, A., 1995, The Beginnings of Pottery Production in the Southern Levant: A Model. In *The Ceramics Cultural Heritage. Proceedings of the Internatioal Symposium "The Ceramic Heritage" of the 8th SIMPEC World Ceramic Conference and*

Forum on New Materials, Florence, Italy, edited by P. V. Vincenzini, pp. 21–28. Faenza, Techna.

Green, M. W., and Nissen H. J., 1987, *Zeichenliste de Archaischen Texte aus Uruk, Ausgraven der Deutschen Forschunggsgemeinschaft in Uruk-Wark,* Volume 12. Gebr. Mann Verlag, Berlin.

Harrison, S., 1992, Ritual as Intellectual Property. *Man* 27(2):225–244.

Hayden, B., 1990, Nimrods, Piscators, Pluckers and Planters: The Emergence of Food Production. *Journal of Anthropological Archaeology* 9(3):225–236.

Hodder, I., 1984, Burials, Houses, Women and Men in the European Neolithic. In *Ideology, Power and Prehistory*, edited by D. Miller and C. Tilley, pp. 51–68. Cambridge University Press, Cambridge.

Hodder, I., 1990, *The Domestication of Europe*, Blackwell, Oxford.

Kafafi, Z., 1985, First Season of Excavations at Jebel Abu Thawwab (Er-Rumman), 1984 Preliminary Report. *Annals of the Department of Antiquities of Jordan* 19:31–41.

Kafafi, Z., 1986, Second Season of Excavations at Jebel Abu Thawwab (Er-Rumman), 1985 Preliminary Report. *Annals of the Department of Antiquities of Jordan* 20:57–67.

Kafafi, Z., 1987, The Pottery Neolithic in Jordan in Connection with other Near Eastern Regions. In *Studies in the History and Archaeology of Jordan, Volume III*, edited by A. Hadidi, pp. 33–39. Routledge and Kegan Paul, London.

Kafafi, Z., 1988, Jebel Abu Thawwab: A Pottery Neolithic Village in North Jordan. In *The Prehistory of Jordan*, edited by A. N. Garrard and H. G. Gebel, pp. 451–471. BAR International Series 396. BAR, Oxford.

Kafafi, Z., 1993, The Yarmoukians in Jordan. *Paléorient* 19:101–113.

Kaplan, J., 1958, *The Neolithic and Chalcolithic Settlement in Tel Aviv and Neighbourhood.* Ph.D. dissertation, Hebrew University, Jerusalem.

Kaplan, J., 1965, Hamadiya. *Revue Biblique* 72:543–544.

Kaplan, J., 1969, 'Ein el Jarba, *Bulletin of the American Schools of Oriental Research* 194:2–39.

Kaplan, J., 1978, Habashan Street. Encyclopaedia of Archaeological Excavations in the Holy Land. *The Israel Exploration Society* 4:1159–1161.

Kenyon, K. M., 1981, *Excavations at Jericho. Volume III: The Architecture and Stratigraphy of the Tell.* British School of Archaeology in Jerusalem, London.

Keswani, P. S., 1994 The Social Context of Animal Husbandry in Early Agricultural Societies: Ethnographic Insights and an Archaeological Example from Cyprus. *Journal of Anthropological Archaeology* 13:255–277.

Knight, C., 1991, *Blood Relations. Menstruation and the Origins of Culture.* University Press, New Haven and London.

Knight, C., 1994, *Ritual and the Origins of Language.* Department of Sociology, University of East London.

Knight, C., Power, C., and Watts, I., 1995, The Human Symbolic Revolution: A Darwinian Account. *Cambridge Archaeological Journal* 5(1):75–114.

Köhler-Rollefson, I., Gillespie, W., and Metzger M., 1988 The Fauna from Neolithic 'Ain Ghazal. In *The Prehistory of Jordan*, edited by A. N. Garrard and H. G. Gebel, pp. 423–430. British Archaeological Reports International Series 396. BAR, Oxford.

Kuijt, I., 1996, Negotiating Equality through Ritual: A Consideration of Late Natufian and Pre-Pottery Neolithic A Period Mortuary Practices. *Journal of Anthropological Archaeology* 15(4):313–336.

Levi-Strauss, C., 1966, *The Savage Mind.* Weidenfeld and Nicolson, London.

Lewis, G., 1980, *Day of Shining Red. An Essay on Understanding Ritual.* Cambridge University Press, Cambridge.

Marshak, A., 1991, *The Roots of Civilization.* Moyer Bell, New York.

Miller, D., 1985, *Artifacts as Categories*. Cambridge University Press, Cambridge.
Muheisen, M., Gebel, H. G., Hannss, C., and Neef, R., 1988, 'Ain Rahub, a New Final Natufian and Yarmoukian Site near Irbid. In *The Prehistory of Jordan*. edited by A. N. Garrard and H. G. Gebel, pp. 472-502. BAR International Series No. 396. BAR, Oxford.
Neveh, D., 1996, Patterns of Activity in the Pottery Neolithic Sites of Nahal Zehora II Based on Groundstone Tools and Grain Distribution. Proceedings—Forli. 8–14th September, 1996. Volume 3:469–474.
Orrelle, E., 1993, *Nahal Zehora I and II: Fifth millennium B.C. Villages of the Wadi Raba Culture. Quantitative Studies on the Pottery Assemblages*. M.A. thesis, Tel Aviv University.
Peltenburg, E., 1994, Constructing Authority: The Vounous Enclosure Model, *Op Ath* 20:157–162.
Perrot, J., 1968, "La Prehistoire Palestinienne" *Supplement au Dictionnaire de la Bible* 8:286–446. Letouzey et Ane, Paris.
Rollefson, G., and Köhler-Rollefson, I., 1989, The Collapse of Early Neolithic Settlements in the Southern Levant. In *People and Culture Change: Proceedings of the Second Symposium on Upper Paleolithic, Mesolithic and Neolithic Populations of Europe and the Mediterranean Basin*, edited by I. Hershkovitz, pp. 59–72. British Archaeological Reports, International Series 508, Oxford.
Rollefson, G., Simmons, A., and Kafafi, Z., 1992, Neolithic Cultures at 'Ain Ghazal, Jordan. *Journal of Field Archaeology* 19(4):443–470.
Simmons, A., Rollefson, G., Kafafi, Z., and Moyer, K., 1989, Test Excavations at Wadi Shu'eib, a Major Neolithic Settlement in Central Jordan. *Annual of the Department of Antiquities of Jordan* 33:27–42.
Simmons, A., Rollefson, G., Köhler-Rollefson, I., Mandel, R., and Kafafi, Z., 1988, 'Ain Ghazal: A Major Neolithic Settlement in Central Jordan. *Science* 240(4848):35–39.
Sperber, D., 1975, *Rethinking Symbolism*. Cambridge University Press, Cambridge.
Stekelis, M., 1950–51, A New Neolithic Industry: The Yarmukian of Palestine. *Israel Exploration Journal* 1:1–19.
Stekelis, M., 1972, *The Yarmukian Culture of the Neolithic Period*. Magnes Press, Jerusalem.
Welbourn, A., 1984, Endo Ceramics and Power Strategies. In *Ideology, Power and Prehistory*, edited by D. Miller and C. Tilley, pp.17–25. Cambridge University Press, Cambridge.
Welbourn, A., 1989, "Endo." In *Kenyan Pots and Potters*, edited by J. Barbour and S. Wandibba, pp. 59–64. Oxford University Press, Nairobi.

Part V

Conclusion

Chapter **13**

Near Eastern Neolithic Research

Directions and Trends

IAN KUIJT

SOCIAL PROCESS, SCALE, AND THE NEOLITHIC

Although researchers have long acknowledged that the foraging and farming transition of the Near Eastern Neolithic was an important economic event, only recently have studies begun to explore the nature of changes in social organization over this period and the nature of social organization at different scales. Building on our understanding of the broader evolutionary trajectory of this transition, archaeologists are now directing new attention to the social context of Neolithic life at the household (Byrd 1994, Chapter 4, this volume; Flannery 1972; Voigt Chapter 11, this volume), community (Hodder 1987, 1990; Goring-Morris Chapter 5 this volume; Kuijt 1996; Rollefson 1997; Rollefson et al. 1992), and regional scales (Bar-Yosef and Belfer-Cohen 1991; Cauvin 1994; Moore 1985). Collectively, these studies offer an alternative perspective on the Neolithic transition by shifting the point of debate from the questions of how and when did plant and animal

IAN KUIJT • Department of Anthropology, University of Notre Dame, Notre Dame, Indiana 46556.

Life in Neolithic Farming Communities: Social Organization, Identity, and Differentiation, edited by Ian Kuijt. Kluwer Academic/Plenum Publishers, New York, 2000.

domestication occur to what was the nature of Neolithic social organization throughout this period, and how might these social frameworks be linked with new systems of food production. This redirection of the discussion focuses our attention on different research areas, exploring how changes in the scale of communities and the nature of civic leadership and social complexity relect how Neolithic peoples created new ways of living in the economic context of food production. From this context, therefore, Neolithic social arrangements cease to be a passive by-product of food production and become conceptualized as the intellectual cornerstone upon which food production exists.

Beyond placing a greater emphasis on the complexities of social change at different scales in the Neolithic, this approach encourages researchers to move beyond consideration of a single system of Neolithic social organization to explore the nature of, and variations within, Neolithic social relations through time and space. For example, in the case of the south-central Levant, researchers are investigating the archeological contexts of three very different, interrelated social processes in the broader Pre-Pottery Neolithic: (1) the initial founding of early agricultural villages at around 10,000 bp with the emergence of some form of civil and ritual leadership (Byrd 1994, Chapter 4, this volume; Cauvin 1994, Chapter 10, this volume; Rosenberg and Redding Chapter 3, this volume); (2) the subsequent consolidation of villages into large aggregate communities with expanded needs for leadership, probably expressed in ritual practices, and increased primary evidence for social differentiation (Gebel and Bienert 1997; Nissen et al. 1987; Kuijt 1995; Rollefson 1987; Rollefson et al. 1992; Simmons Chapter 9, this volume); and (3) the abandonment of the large aggregate communities at around 8,000 bp (Köhler-Rollefson and Rollefson 1990; Rollefson 1996; Rollefson and Köhler-Rollefson 1989). The conceptualization of these social events as distinct, yet clearly interrelated, social processes encourages us to address the different social, economic, and political foundations as the context of these events and moves archaeologists toward the recognition that these events or processes are likely to have material manifestations. This consideration of the complex social processes of the Neolithic, as well as how material culture and the built environment may reflect these transitions, offers an alternative and very productive approach for archaeologists interested in issues of social complexity and the origins of agriculture.

Coexistence of Hierarchical and Egalitarian Elements

Recent Neolithic studies suggest that social relations, as well as their material manifestations, can be envisioned as amalgamations of social practices that alternatively serve to highlight elements of social differentiation and egalitarianism in communities. This perspective relies on ethnographic and anthropological research that illustrates three aspects to social relations in present and past middle-range communities: (1) social inequality is ubiquitous and found in all societies (Blanton 1995; Feinman 1995; Hayden 1995a, b; Kelly 1993; Paynter 1989); (2) "egalitarian" social systems require highly complex codes of social behavior, codes that are as complex as those seen within cultural contexts where systems of hereditary power exist (Flanagan 1989; Gerlach and Gerlach 1988; Rayner 1988); and (3) hierarchy and egalitarianism are fundamentally interrelated and coexist in many, if not most, social systems (Berreman 1981; Kan 1989; McKinnon 1991; Myers 1986; Plog 1995). A number of recent ethnographic studies (Flanagan 1989; McKinnon 1991; Schiller 1997) have clearly demonstrated that most forms of governance in small-scale agricultural or horticultural communities combine hierarchical and egalitarian dimensions, and several archaeologists have applied this framework to explore social relations in different archaeological contexts (e.g., Blanton 1995; Feinman 1995; Plog 1995; Renfrew 1974). For example, Plog (1995:190) notes that ethnographic accounts from historical periods illustrate egalitarian dimensions in Pueblo society as well as hierarchical ones, leading him to conclude that "rather than trying to characterize Pueblo social relations using a single label, it seems more accurate to concluded that there are both egalitarian and hierarchical aspects of Pueblo societies, a point that has tended to be under-emphasized, if not over looked, during much of the previous discussion in the archaeological literature."

As with Plog's recognition of the coexistence of hierarchical and egalitarian dimensions in Pueblo societies, the authors in this volume introduce important implications for how researchers can understand Neolithic social systems. On one level, adoption of this approach moves discussions beyond consideration of the initial appearance of social differentiation, or, for that matter, how archaeologists should label Neolithic specific social systems, to explorations of the nature of social relationships. Second, this perspective encourages researchers to understand variations in Neolithic social organizations through time and space and at different scales. In the models we have developed to describe Neolithic social systems, we too often fail to acknowledge variations in time and space, and as a result often produce highly simplistic and broad formulations of social systems. Such a trend risks the intellectual homogenization of the Neolithic and the multiple pathways to power and authority, ultimately producing monolithic reconstructions that overlook the subtle, yet significant, differences between different kinds of Neolithic communities in different places and times. Finally, researchers recognize that internal social relations in Neolithic communities were more dynamic and complex than most of our models allow. Despite the rich archaeological record of the Neolithic of the Near East, remarkably little literature has focused on modeling social relations at any scale. Acknowledging that egalitarian and hierarchical relations are likely to have coexisted in select periods releases researchers from the focus on labeling societies and facilitates the development of realistic and comprehensive models of cultural dynamics, including the possible pathways to power and authority in Neolithic communities.

Neolithic Social Organization: Heterarchy, Hierarchy, or Both?

There are many possible ways to conceptualize how power and authority might have been controlled and/or shared in Neolithic communities. While often unrecognized, many discussions of Neolithic social systems are also situated within the much broader discussion of if, or how, social relations in agricultural communities are organized along hierarchical and heterarchical lines (Crumley 1987). In the case of the Neolithic, these positions differ in whether the pathways of power existed as either a single hierarchical system or one in which there were numerous coexisting hierarchical power structures. While there is no consensus on the matter, I suspect that most archaeologists working on the Near Eastern Neolithic would agree that there is no convincing evidence for some form of organized central social hierarchy, exemplified by the existence of hereditary elites, and ethnographically exemplified by chiefdom-level organizations. To the same extent, however, many researchers note evidence for some form of social differentiation among individuals, households, or communities, especially in the later periods of the Neolithic. While almost no archaeological research has directly addressed the topic, I suggest that it is likely that social differentiation in the Neolithic was derived from the authority of ritual practitioners, civic leaders, or perhaps community/household elders.

If there are insufficient grounds to argue for a hierarchically organized system of leadership focused on an individual or group of individuals with differential access to resources and authority, then it may be more profitable to alternatively envision Neolithic social relations as an organized series of interrelated, coexisting hierarchical units. From some perspectives, archaeological evidence from the Near Eastern Neolithic reflects several forms of hierarchical ritual and civic administration. For example, it appears that some dimensions of ritual practice found expression on the community level and would have undoubtedly involved ritual practitioners who controlled the timing, nature, and context of some, but not necessarily all, community rituals (see Chapter 5, this volume; Kuijt 1996; Mellaart 1967; Rollefson 1986; Voigt 1983). Researchers have also reflected on the importance of civic leadership for other tasks, such as the construction and maintenance of community buildings (Byrd 1994, Chapter 4, this volume; Kafafi and Rollefson 1995; Özdögan and Özdögan 1989; Schirmer 1990; Voigt Chapter 11, this volume). In light of the number of people living in some of these Neolithic communities, it is also possible that some form of civic, community-oriented leadership would have been necessary for organizing the planting and harvesting of crops. Based on spatial patterning of lithic waste materials from 'Ain Ghazal, Quintero and Wilke (1995) note that there is evidence for stone tool workshops in the Middle Pre-Pottery Neolithic and that the high degree of standardization may well reflect some from of craft specialization.

Given the absence of differentation in residential architecture and mortuary practices, however, I would suggest that collectively we construct a picture of Neolithic communities in which we see a balance of economic centralization and autonomy and of coexisting dimensions of egalitarianism and hierarchy. Moreover, to understand the shifting nature of these relationships through time, it is helpful to conceptualize Neolithic community relations as focused on a series of interrelated coexisting social units that might have included, but were not limited to, kin groups, the household, ritual sodalities, and the community. Ultimately the point here is not to argue that Neolithic community relations should be viewed only as heterarchically organized, for this argument is admittedly more observed than demonstrated. Instead the papers in this volume encourage researchers to explore the diversity of social relations through material culture and the built environment, and to abandon the predetermined position that these material manifestations are linked to a single hierarchical thread. I believe that the studies in this volume provide an initial foundation upon which to understand some dimensions of different, yet interrelated, pathways of power and authority existed in the Neolithic of the Near East.

FUTURE DIRECTIONS

Over the last 30 years archaeological research by a wide range of scholars has transported us beyond the general recognition that the Neolithic was a pivotal economic and social event to the point where researchers are able to explore the more detailed nature of social relations at many different scales. Essentially this journey demonstrates the relevance of our reconstruction of the Neolithic revolution to critical anthropological issues of social differentation, demography, and ritual and civic organization. These studies have been instrumental in expanding our understanding of the Neolithic in the Near East while reaffirming our exploration of new and old topics from the standpoint of new data or methodological developments. Although hardly an exhaustive list, I want to briefly draw attention to several topics and issues that are likely to become central in our future investigations of Neolithic social organization—most importantly the nature of civic governance and social differentiation in these communities, and the role of households and lineages in creating the framework for life in Neolithic settlements.

Frameworks of Governance and Social Differentiation in Neolithic Society

In light of the relative wealth of information we have on the nature of Neolithic subsistence economies, I think many researchers (e.g., Baird 1997; Rollefson 1998) would agree that we know comparatively little about the critical topic of leadership and governance in Neolithic households and communities. While there are exceptions, few studies have directly addressed the nature of leadership and governance at different scales, such as the individual, household, community, and regional levels, or how these frameworks change through different periods of the Pre-Pottery and Pottery Neolithic. Although a number of works have illustrated how many, if not most, Neolithic communities shared material practices at the regional level (e.g., Bar-Yosef and Belfer-Cohen 1989; Bar-Yosef and Meadows 1995; Cauvin 1994), in many ways our understanding of governance remains highly theoretical, abstract, largely removed from the specifics of archaeological data sets from individual sites (e.g., Hayden 1995a), and rarely moves beyond consideration of community, if not regional, ideology. A portion of this gap is unquestionably linked to the relatively limited amount of excavation at Neolithic sites and the importance of addressing broader anthropological issues such as the emergence of social inequality, but it is also the result of archaeologists, including this author, struggling with the much broader problem of how to connect various bodies of anthropological theory on human behavior with Neolithic material culture in a meaningful way.

We see a growing consensus among researchers, I believe, that social practices existed at certain points in the past to differentially distinguish individuals within the overall Neolithic community. On a material level, many discussions of Neolithic social organization focus on the issues of how, or if, select Neolithic material culture reflects the interests, behavior, and social role(s) of individuals versus a collective group of individuals (e.g., Bienert 1991; Garfinkel 1994). In the broadest of senses, this question challenges us to understand some of the ways in which social practices highlight aspects of cohesiveness and integration or, conversely, the hierarchical organization of power and authority. In cases where we have some idea of how the selection of individuals from the broader community occurred, we have only a limited understanding of why it occurred, the basis for this selection, and how this helps us understand the nature of Neolithic social organization. Recently archaeologists have devoted considerable attention to mortuary and ritual practices as a way to reconstruct broader Neolithic social organization and the existence of some degree of social differentiation. As illustrated by many of the chapters in this volume, Neolithic ritual practices provide important insights about social arrangements at the household and/or community level. In contrast to many other world areas, relatively few studies have explored the possible material correlates for ritual or community leaders in Neolithic communities, nor have they reached some form of consensus of the existence of social differentiation. It is, therefore, not surprising that there are no clear answers to this question, and it is likely that considerable future debate will revolve around this issue in the future.

Arguably some of the most dramatic advancements in our understanding of Neolithic social systems have centered on identifying the spatial location of community or household rituals. While often based on field work conducted many years ago, recent reflections (e.g., Banning and Byrd 1987; Byrd 1994; Byrd and Banning 1988; Özdögan and Özdögan 1989; Schmidt 1997; Rollefson 1997; Rollefson et al. 1992) of observed patterning of residential and nonresidential architecture have enhanced our understanding of Neolithic social organization through the exploration of how space was constructed and used by households and families and by the investigation of the relationships between residential and nonresidential spaces in these communities.

Detailed consideration of Neolithic architecture, mortuary practices and ritual actions collectively brings researchers to the point where we can start to reflect upon how ritual and civic leadership might have been organized in different Neolithic communities. New data, as well as synthesis studies of earlier publications, elicit a number of important questions related to the issue of governance in these communities. For example, were the elaborate mortuary practices of the south-central Levantine MPPNB, such as at 'Ain Ghazal or Jericho, organized at the household, kin-group, or community level? In the case of Çatal Höyük, Çayönü, and Nevali Çori, were some forms of ritual practices organized by household members and others organized by, and oriented toward, the broader community? If so, how might archaeologists distinguish these in an archaeological context? These questions focus on the nature of the relationships among housholds, ritual, and civic structures. Moreover, if Neolithic social practices did differentiate some members from the community to perform ritual or civic tasks, why do we not see more evidence for differential power and authority? Critical examination of these questions in the future, as well as expanded discussion of issues related to the nature of Neolithic governance and leadership, will be central to expanding out understanding of Neolithic soical complexity and the origins of agriculture.

Neolithic Households, "Houses," and Links to Economy

One of the key ways to explore issues of governance in Neolithic communities is to explore how households served as arenas for everyday life, such as with ritual practices and collective labor. Specifically, I think that future research will benefit from renewed attention to how Neolithic social practices might have been focused on either individual households, or the broader House as a social and economic unit (Lévi-Strauss 1983). For the most part discussions of Neolithic social organization remain focused on the scale of the community, rarely addressing the existence of the household or House as a social and economic unit. While there are exceptions, most research has focused on how to differentiate between nuclear or extended family households as a form of classification. One alternative to this perspective is to explore the possibility that Neolithic social frameworks were organized along lines similar to a House Society (Lévi-Strauss 1983). Lévi-Strauss (1983:174) defines the House as "a corporate body holding an estate made up of both material and immaterial wealth, which perpetuates itself through the transmission of its name, its goods and its titles down a real or imaginary line, considered legitimate as long as this continuity can express itself in the language of kinship or of affinity and, most often, of both." In this context, then, the House emphasizes elements of temporal continuity, the hereditary transfer of valued property and authority, and the strategic exploitation of the language of kinship and affinity, existing simultaneously as a social, ritual, and economic unit. Moreover, this social and economic unit can be composed of multiple residential units dwelling in separate structures and can serve as a physical and symbolic place of origin for fictive and real ancestors. In considering how communities and living units were structured, this perspective challanges archaeologists to define and explain the building blocks of social and economic relations within Neolithic communities.

Furthering our understanding of the links between economic and social change at different points in the Neolithic is likely to become another important avenue for research in the future. In the broadest of scales, we can note that the initial development and later entrenchment of food production must have radically altered the nature of ownership, labor, and civic organization in Neolithic communities. In the case of large agricultural communities, especially of the later periods, organizational changes might have occurred in the nature of agricultural labor in certain periods of the year, such as harvesting crops in the fall. This raises the question of whether this labor was organized along community or household lines. Increases in the scale of some communities, especially those that might have been the focus of regional economic or ritual activities, might have required multiple households, or even communities, to combine forces to meet the challenge of some project, such as the construction of the PPNA tower at Jericho or harvesting of temporally limited food resources. Who, for example, organized people to undertake farming, planting, herding, and harvesting? There is no question that these issues are central to our understanding of how economic developments, such as the appearance of domesticated plants and animals, and how the control of these resources would have been linked to social changes.

Despite all that archaeologists know about Neolithic subsistence practices, we have a very poor understanding of the social aspects of Neolithic economic practices, such as the production and distribution of shell and stone beads or other nonlocal objects. With the exception of the sourcing of obsidian in the Near East, researchers have only a limited understanding of the sourcing of nonlocal materials and have yet to examine how trade and exchange might have been organized. Future scholars will, for example, have the opportunity to explore how to envision Neolithic communities from select periods as being socially and economically independent, or if physically separate communitis were highly interlinked through household marriage, economic practices, and ideological beliefs. It will be important, moreover, for us to understand how leaders or households might have controlled trade. The resolution of such topics is pivotal to understanding the nature of social systems in the Neolithic and the broader trajectory of human development in the transition from foragers and cultivators to agriculturalists. Collectively, these studies highlight the importance of future studies exploring Neolithic social relations through a variety of artifact classes, integrating our understanding of economic practices with that of social organization, and the need to develop more sophisticated models to explain the dynamic nature of civic and ritual leadership.

SUMMARY

As seen in this volume, archaeologists are making important strides forward in the task of reconstructing Neolithic social organization. This volume certainly provides us with a more detailed understanding of some of the complexities of the archaeological record of the Neolithic of the Near East and challenge us to view the Neolithic as an economic event and to explore the interrelationships between the development, entrenchment, and expansion of systems of food production with the nature of social organization at the household, kin-group, and community levels. An improved awareness of the nature of Neolithic social organization will require broad consideration of interrelated lines of archaeological evidence at different scales, including regional settlement patterns, systems of regional trade and exchange, contact within and between different regional communities, the nature of shared systems of belief and ritual at the regional scale, consideration of regional and community architectural systems, and the nature of mortuary practices. Synthesis and interpretation of these diverse archaeological patterns will also require consideration of complementary and conflicting aspects of behavior within these communities. These rich lines of archaeological evidence contribute to our broader anthropological understanding of issues of social agency, household compositions, and ritual practices, and aid us in exploring the complex, rewarding, and challenging interface of archaeological data and anthropological theory and the links between different forms of human behavior and material culture in the past.

REFERENCES

Baird, D., 1997, Goals in Jordanian Neolithic Research. In *The Prehistory of Jordan II. Perspectives from 1997*, edited by H. G. K. Gebel, Z. Kafafi, and G. O. Rollefson, pp. 371–381. Studies in Early Near Eastern Production, Subsistence, and Environment 4. Ex Oriente, Berlin.

Banning, E. B., and Byrd, B. F., 1987, Houses and Changing Residential Units: Domestic Architecture at PPNB 'Ain Ghazal. *Proceedings of the Prehistoric Society* 53:309–325.

Bar-Yosef, O., and Belfer-Cohen, A., 1989, The Levantine "PPNB" Interaction Sphere. In *People and Culture in Change: Proceedings of the Second Symposium of Upper Paleolithic, Mesolithic, and Neolithic Populations of Europe and the Mediterranean Basin*, edited by I. Hershkovitz, pp. 59–72. British Archaeology Review, International Series 508. BAR, Oxford.

Bar-Yosef, O., and Belfer-Cohen, A., 1991, From Sedentary Hunter-Gatherers to Territorial Farmers in the Levant. In *Between Bands and States*, edited by S. A. Gregg, pp. 181–202. Center for Archaeological Investigations, Carbondale, Illinois.

Bar-Yosef, O., and Meadows, R., 1995, The Origins of Agriculture in the Near East. In *Last Hunters–First Farmers: New Perspectives on the Prehistoric Transition to Agriculture*, edited by T. D. Price and A. G. Gebauer, pp. 39–94. School for American Research, Santa Fe, New Mexico.

Berreman, G. D., 1981, Social Inequality: A Cross-Cultural Analysis. In *Social Inequality: Comparative and Developmental Approaches*, edited by G. D. Berreman, pp. 3–40. Academic Press, New York.

Bienert, H. D., 1991, Skull Cult in the Prehistoric Near East. *Journal of Prehistoric Religion* 5:9–23.

Blanton, R. E., 1995, The Cultural Foundations of Inequality in Households. In *Foundations of Social Inequality*, edited by T. D. Price and G. M. Feinman, pp. 105–127. Plenum, New York.

Byrd, B. F., 1994, Public and Private, Domestic and Corporate: The Emergence of the Southwest Asian Village. *American Antiquity* 59(4):639–666.

Byrd, B. F., and Banning E. B., 1988, Southern Levantine Pier Houses: Intersite Architectural Patterning during the Pre-Pottery Neolithic B. *Paléorient* 14:65–72.

Cauvin, J., 1994, *Naissance des divinités, Naissance de l'agriculture*. CNRS, Paris.

Crumley, C., 1987, A Dialectical Critique of Hierarchy. In *Power Relations and State Formation*, edited by T. Patterson and C. W. Gailey, pp. 155–169. American Anthropological Association, Washington, D.C.

Feinman, G. M., 1995, The Emergence of Inequality: A Focus on Strategies and Process. In *Foundations of Social Inequality*, edited by T. D. Price and G. M. Feinman, pp. 255–279. Plenum, New York.

Flanagan, J. C., 1989, Hierarchy in Simple "Egalitarian" Societies. *Annual Review of Anthropology* 18:245–266.

Flannery, K. V., 1972, The Origins of the Villages as a Settlement Type in Mesoamerica and the Near East: A Comparative Study. In *Man, Settlement and Urbanism*, edited by P. J. Ucko, R. Tringham, and G. W. Dimbleby, pp. 23–53. Duckworth, London.

Garfinkel, Y., 1994, Ritual Burial of Cultic Objects: The Earliest Evidence. *Cambridge Archaeology Journal* 4(2):159–188.

Gebel, H. G. K., and Bienert, H. D., 1997, Ba'ja Hidden in the Petra Mountains. Preliminary Report on the 1997 Excavation. In *The Prehistory of Jordan II. Perspectives from 1997*, edited by H. G. K. Gebel, Z. Kafafi, and G. O. Rollefson, pp. 221–262. Studies in Early Near Eastern Production, Subsistence, and Environment 4. Ex Oriente, Berlin.

Gerlach, L. P., and Gerlach, U. M., 1988, Egalitarianism, Collectivism, and Individualism: The Digo of Kenya. In *Rules, Decisions, and Inequality in Egalitarian Societies*, edited by J. G. Flanagan and S. Rayner, pp. 113–144. Gower, London.

Hayden, B., 1995a, Pathways to Power: Principles for Creating Socioeconomic Inequalities. In *Foundations of Social Inequality*, edited by T. Douglas Price and G. M. Feinman, pp. 15–86. Plenum, New York.

Hayden, B., 1995b, The Origins of Agriculture in the Near East. In *Last Hunters-First Farmers: New Perspectives on the Prehistoric Transition to Agriculture*, edited by T. D. Price and A. G. Gebauer, pp. 273–299. School for American Research, Santa Fe, New Mexico.

Hodder, I., 1987, Contextual Archaeology: An Interpretation of Çatal Hüyük and a Discussion of the Origins of Agriculture. *Bulletin of the Institute of Archaeology* 24:43–56.

Hodder, I., 1990, *The Domestication of Europe: Structure and Contingency in Neolithic Societies*. Basil Blackwell, Oxford.

Kafafi, Z., and Rollefson, G., 1995, The 1994 excavations at 'Ayn Ghazal: Preliminary Report. *Annual of the Department of Antiquities of Jordan* 39:13–29.

Kan, S., 1989, *Symbolic Immortality: The Tlingit Potlatch of the Nineteenth Century*. Smithsonian Institution Press, Washington, D.C.

Kelly, R. C., 1993, *Constructing Inequality: The Fabrication of a Hierarchy of Virtue among the Etoro*, Unversity of Michigan Press, Ann Arbor, Michigan.

Köhler-Rollefson, I., and Rollefson, G., 1990, The Impact of Neolithic Man on the Environment: The Case of 'Ain Ghazal, Jordan. In *Man's Role in Shaping of the Eastern Mediterranean Landscape*, edited by S. Bottema, G. Entjes-Nieborg, and W. Van Zeist, pp. 3–14. Rotterdam, A Balkema.

Kuijt, I., 1995, *New Perspectives on Old Territories: Ritual Practices and the Emergence of Social Complexity in the Levantine Neolithic*. Ph.D. dissertation, Harvard University. University Microfilms, Ann Arbor, Michigan.

Kuijt, I., 1996, Negotiating Equality through Ritual: A Consideration of Late Natufian and Pre-Pottery Neolithic A Period Mortuary Practices. *Journal of Anthropological Archaeology* 15(4):313–336.

Levi-Strauss, C., 1983, *The Way of the Masks*. Translated by S. Modelski. Jonathan Cape, London.

McKinnon, S., 1991, *From a Shattered Sun: Hierarchy, Gender, and Alliance in the Tanimbar Islands*, University of Wisconsin Press, Madison, Wisconsin.

Mellaart, J., 1967. *Çatal Höyük, a Neolithic Town in Anatolia*. McGraw-Hill, New York.

Moore, A., 1985, The Development in Neolithic Societies in the Near East. In *Advances in World Archaeology*, edited by F. Wendorf and A. E. Close, pp. 1–69. Academic Press, New York.

Myers, F. R., 1986, Reflections on a Meeting: Structure, Language, and the Polity in a Small Scale Society. *American Ethnology* 13:430–447.

Nissen, H. J., Muheisen, M., and Gebel, H. G. K., 1987, Report on the First Two Seasons of Excavation at Basta. *Annual of the Department of Antiquities of Jordan* 31:79–119.

Özdögan, M., and Özdögan, A., 1989, Çayönü. *Paléorient* 15:65–74.

Paynter, R., 1989, The Archaeology of Equality and Inequality. *Annual Review of Anthropology* 18:369–399.

Plog, S., 1995, Equality and Hierarchy: Holistic Approaches to Understanding Social Dynamics in the Pueblo Southwest. In *Foundations of Social Inequality*, edited by T. D. Price and G. M. Feinman, pp. 189–206. Plenum, New York.

Quintero, L., A. and Wilke, P. J., 1995, Evolution and Economic Significance of Naviform Core-and-Blade Technology in the Southern Levant. *Paléorient* 21(1):17–33.

Rayner, S., 1988, The Rules That Keep Us Equal: Complexity and Costs of Egalitarian Organization. In *Rules, Decisions, and Inequality in Egalitarian Societies*, edited by J. G. Flannagan and S. Rayner, pp. 20–42. Gower, London.

Renfrew, C., 1974, Beyond a Subsistence Economy, the Evolution of Social Organization in Prehistoric Europe. In *Reconstructing Complex Societies*, edited by C. B. Moore, pp. 69–95. Bulletin of the American School of Oriental Research.

Rollefson, G. O., 1986, Neolithic 'Ain Ghazal (Jordan): Ritual and Ceremony II. *Paléorient* 12:45–52.

Rollefson, G. O., 1987, Local and Regional Relations in the Levantine PPN Period: 'Ain Ghazal as a Regional Center. In *Studies in the History and Archaeology of Jordan*, edited by A. Hadidi, pp. 2–32. Amman, Department of Antiquities of Jordan.

Rollefson, G. O., 1996, The Neolithic Devolution: Ecological Impact and Cultural Compensation at 'Ain Ghazal, Jordan. In *Retrieving the Past: Essays on Archaeological Research and Methodology in Honor of Gus W. Van Beek.* edited by J. S. Seger, pp. 219–229. Eisenbrauns: Winona Lake, Indiana.

Rollefson, G. O., 1997, Changes in Architecture and Social Organization at 'Ain Ghazal. In *The Prehistory of Jordan II, Perspectives from 1997*, edited by H. G. K. Gebel, Z. Kafafi, and G. O. Rollefson, pp. 287–307. Studies in Early Near Eastern Production, Subsistence, and Environment 4. Ex Oriente, Berlin.

Rollefson, G. O., 1998, The Aceramic Neolithic. In *The Prehistoric Archaeology of Jordan*, edited by D. O. Henry, pp. 102–126. Archaeopress, Oxford.

Rollefson, G., and Köhler-Rollefson, I., 1989, The Collapse of Early Neolithic Settlements in the Southern Levant. In *People and Culture Change: Proceedings of the Second Symposium on Upper Paleolithic, Mesolithic and Neolithic Populations of Europe and the Mediterranean Basin*, edited by I. Hershkovitz, pp. 59–72. British Archaeological Reports, International Series 508. BAR, Oxford.

Rollefson, G. O., Simmons, A. H., and Kafafi Z., 1992, Neolithic Cultures at 'Ain Ghazal, Jordan. *Journal of Field Archaeology* 19:443–470.

Schiller, A., 1997, *Small Sacrifices: Religious Charge and Cultural Identity among the Ngaju of Indonesia*, Oxford University Press, New York.

Schmidt, K., 1997, "News from the Hilly Flanks": Zum Forschungsstand des obermesopotamischen Frühneolithikums. *Archäologisches Nachrichtenblatt* 2(1):70–79.

Voigt, M. M., 1983, *Hajji Firuz Tepe, Iran: The Neolithic Settlement.* The University Museum, University of Pennsylvania, Philadelphia, Pennsylvania.

Index

GPSR Compliance
The European Union's (EU) General Product Safety Regulation (GPSR) is a set of rules that requires consumer products to be safe and our obligations to ensure this.

If you have any concerns about our products, you can contact us on

ProductSafety@springernature.com

In case Publisher is established outside the EU, the EU authorized representative is:

Springer Nature Customer Service Center GmbH
Europaplatz 3
69115 Heidelberg, Germany

www.ingramcontent.com/pod-product-compliance
Ingram Content Group UK Ltd.
Pitfield, Milton Keynes, MK11 3LW, UK
UKHW020135160926
475607UK00003B/49

9780306461224